Progress in Optical Science and Photonics

Volume 9

Series Editors

Javid Atai, Sydney, NSW, Australia

Rongguang Liang, College of Optical Sciences, University of Arizona, Tucson, AZ, USA

U. S. Dinish, Singapore Bioimaging Consortium (SBIC), Biomedical Sciences Institutes, A*STAR, Singapore, Singapore

The purpose of the series Progress in Optical Science and Photonics is to provide a forum to disseminate the latest research findings in various areas of Optics and its applications. The intended audience are physicists, electrical and electronic engineers, applied mathematicians, biomedical engineers, and advanced graduate students.

More information about this series at http://www.springer.com/series/10091

Lei Wei
Editor

Advanced Fiber Sensing Technologies

 Springer

Editor
Lei Wei
Nanyang Technological University
Singapore, Singapore

ISSN 2363-5096   ISSN 2363-510X   (electronic)
Progress in Optical Science and Photonics
ISBN 978-981-15-5506-0   ISBN 978-981-15-5507-7   (eBook)
https://doi.org/10.1007/978-981-15-5507-7

This Springer imprint is published by the registered company Springer Nature Singapore Pte Ltd.
The registered company address is: 152 Beach Road, #21-01/04 Gateway East, Singapore 189721, Singapore

Preface

Fiber sensing technologies have enabled both fundamental studies and a wide spectrum of applications in every aspect of life. This book highlights the recent advancements in fiber sensing technologies based on newly developed sensing mechanisms, advanced fiber structures, and functional materials. The book is organized into fifteen chapters. A brief description of each of the chapters is as follows:

- Chapter "Plasmonic Photonic Crystal Fibers" provides a comprehensive review of the recent progress of plasmonic photonic crystal fiber structures in terms of design and applications including refractive index sensing, biosensing, temperature sensing, and polarization and birefringent devices.
- Chapter "Hybrid Fiber-Optic Sensors" reviews the recent fiber-optic plasmonic sensors coupled with the new advances in functional nanomaterials such as 2D materials and gold nanoparticles capped with macrocyclic supramolecular as well as new fiber structure designs and fabrication.
- Chapter "Microstructured Fibers for Sensing" presents the newly developed microstructured optical fibers with different lightwave guiding mechanisms and their applications in sensing, including fiber gratings, interferometers, and surface-enhanced Raman scatterings.
- Chapter "Optical Microfiber Sensors" reviews the recent progress in optical microfiber-based sensors, including the fundamental optical properties, the fabrication techniques, the well-established microfiber-based sensing schemes, and the new effects and strategies for sensing enhancement.
- Chapter "Fiber-Based Infrasound Sensing" discusses a composite diaphragm-type fiber-based external Fabry–Perot interferometer infrasound sensor, aiming at low-frequency or infrasound detection for special applications.
- Chapter "Specialty Fiber Grating-Based Acoustic Sensing" presents a thin core ultra-long period fiber grating sensor with high sensitivities of curvature, high acoustic pressure sensitivity, easy fabrication, simple structure, and low cost.

- Chapter "Electrospinning Nanofibers" covers aspects related to electrospinning nanofibers, including the nanofiber materials, processing mechanism and parameters of electrospinning techniques, special electrospinning techniques, and their applications.
- Chapter "Nanofibers for Gas Sensing" reports a comprehensive review of the nanofiber gas sensors for fast and minimal monitoring of the target gas concentration for public safety, food processing, and environmental monitoring.
- Chapter "Sapphire-Derived Fibers and Optical Fiber Sensing" discusses the recent development of sapphire-derived fibers to achieve new breakthroughs in high-temperature sensing with good mechanical properties.
- Chapter "Thermoelectric Fibers" summarizes a new type of flexible thermoelectric fibers made by a wide range of thermoelectric materials and their applications for fiber-based thermal sensing and energy harvesting.
- Chapter "In-Fiber Breakup" presents the technology of using fluidic instabilities induced in-fiber breakup phenomena to modify the traditional axially invariant in-fiber structure and to achieve in-fiber material engineering for sensing applications.
- Chapter "Nano- and Micro-structuring of Materials Using Polymer Cold Drawing Process" discusses the use of cold drawing in structuring functional materials including crystals, glasses, and polymers with the structured fiber- and film-based geometries.
- Chapter "Fiber-Based Triboelectric Nanogenerators" presents the recent development of fiber-based triboelectric nanogenerators as convenient energy harvesters to power wearable electronics and as self-powered real-time wearable sensors to monitor personal health care and bio-motions.
- Chapter "Fiber-Shaped Energy-Storage Devices" reviews the design principles and device performance of fiber-shaped energy storage devices including supercapacitors, non-aqueous, and aqueous batteries.
- Chapter "Brillouin Fiber Laser Sensors" presents single longitudinal mode Brillouin fiber laser-enabled sensors with the enhanced sensitivity achieved in typical Brillouin distributed sensing systems.

I would like to thank all the chapter authors for their excellent contributions. Also, I would like to thank Nanyang Technological University for the constant support. Finally, I would like to express my sincere gratitude to my family and friends for their understanding, encouragement, and support.

Singapore   Lei Wei
March 2020

Contents

Plasmonic Photonic Crystal Fibers

Dora Juan Juan Hu and Aaron Ho-Pui Ho

Abstract Surface plasmon waves are coupled electron–photon modes at the metal–dielectric interface. They can significantly enhance light–matter interactions that are favorable in many applications including nanophotonics, data storage, microscopy, solar cells, and sensing. Compared with the prism-based coupling configuration, optical fiber-based plasmonic devices offer more compact and robust configuration for exciting the plasmon modes. Photonic crystal fibers (PCFs) are a special class of optical fibers in which the presence of holey structures or periodic microstructures of refractive index modulations can provide a wide scope of flexibility in the control and engineering of the optical properties, thus opening up the potential for many new applications and scientific explorations. Notably, PCFs are a desirable platform to incorporate plasmonic structures for the excitation of surface plasmon resonance (SPR) and localized surface plasmon resonance (LSPR). Three main types of plasmonic PCF structures have been developed and reported in the literature, including metal nano-/microwire-filled plasmonic PCF, metal-coated plasmonic PCF, and nanoparticle-deposited/filled plasmonic PCF. This chapter provides a comprehensive review on the recent progress of these reported plasmonic PCF structures in terms of design and applications. Firstly, the operating principles based on surface plasmon polaritons and localized surface plasmon polaritons are presented. Secondly, the experimental studies of plasmonic PCF structures for various application areas are reviewed, including refractive index sensing, biosensing, temperature sensing, polarization, and birefringent devices. Lastly, design considerations and challenges are discussed.

D. J. J. Hu (✉)
Institute for Infocomm Research, Singapore, Singapore
e-mail: jjhu@i2r.a-star.edu.g

A. H.-P. Ho
Department of Biomedical Engineering,
The Chinese University of Hong Kong, Hong Kong SAR, China
e-mail: aaron.ho@cuhk.edu.hk

© Springer Nature Singapore Pte Ltd. 2020
L. Wei (ed.), *Advanced Fiber Sensing Technologies*,
Progress in Optical Science and Photonics 9,
https://doi.org/10.1007/978-981-15-5507-7_1

Keywords Photonic crystal fiber · PCF · Plasmonics · Surface plasmon
resonance · SPR · Localized surface plasmon resonance · LSPR · Sensor · Filter ·
Polarizer

1 Introduction

Plasmonics is an exciting bridging technology for electronics and photonics. The
development of plasmonic technologies has led to remarkable capabilities in a wide
range of areas such as nanophotonics, magneto-optic data storage, microscopy, solar
cells, and sensing applications such as biological and chemical detection (Homola
2008; Barnes et al. 2003). Surface plasmon (SP) waves are coupled electron–photon
modes at the boundaries between a metal and a dielectric. Generally, plasmonic
sensing devices are categorized into two types: propagating surface plasmon resonance (SPR) sensors and localized surface plasmon resonance (LSPR) sensors.
Because of the momentum mismatch between SPs and photons propagating in
vacuum, special configurations are needed to excite the plasmon modes.

Plasmonics in optical fibers uses optical field to excite the SP waves, offering
better compactness and robustness in comparison with traditional plasmonic devices
based on prism coupling configurations. Optical fiber-based plasmonic devices have
received wide interest for development and deployment for remote and in situ monitoring applications. The invention of PCF was an important milestone in the history
of the optical fiber technology. Compared with conventional fibers, PCFs have the
great flexibility to control in the waveguiding properties by modifying the microstructured geometry. For example, the air filling fraction can be engineered to control
the refractive index contrast between the core and the cladding to obtain endless
single-mode operation. Hollow core guidance can be realized by new light-guiding
mechanisms of photonic bandgap effect in the microstructured cladding. PCFs can
be characterized by various guiding mechanisms, including index-guiding mechanism, photonic bandgap (PBG)-guiding mechanism, inhibited coupling-guiding
mechanism, antiresonance-guiding mechanism, and twist-induced guiding mechanism (Markos et al. 2017). PCFs with hybrid guiding mechanisms of index guiding
and PBG guiding and their unique features as well as applications have been reviewed
(Hu et al. 2019). The peculiar properties of PCFs have attracted enormous research
interest to unlock their great potential in various applications. PCFs have undergone
tremendous development in the past two decades for sensing, communication, and
medical applications (Knight 2003; Russell 2003). In particular, PCF-based sensor
devices have demonstrated superior performance in terms of sensitivity, versatility,
sensing range, etc (Pinto and Lopez-Amo 2012; Villatoro and Zubia 2016; Hu et al.
2018). For example, the photonic bandgap effect in PCFs has enabled hollow core
light guidance for gas sensing application (Debord et al. 2019). Moreover, the air hole
channels in PCFs allow efficient integration of other functional materials to greatly

enhance the tunability of optical properties and significantly expand the functionality of the fibers, such as sensing applications using PCFs as an optofluidic platform (Ertman et al. 2017). PCFs have brought new developments to fiber lasers and nonlinear optics (Knight 2007; Knight and Skryabin 2007; Travers et al. 2011).

Integrating plasmonic structures in PCF platforms has created many opportunities by leveraging on the flexibilities of the PCF platform to achieve better performance or to develop new devices for sensing applications. The works on plasmonic PCFs and their applications were reviewed in Yang et al. (2011), Zhao et al. (2014), Hu and Ho (2017). The reported plasmonic PCF structures can be broadly categorized in three types, i.e., metal nano-/microwire-filled plasmonic PCF structures, metal-coated plasmonic PCF structures, and nanoparticle-deposited/filled plasmonic PCF structures. In this chapter, we start with a discussion of the operating principles of the plasmonic PCF structures, mainly based on surface plasmon polaritons (SPPs) and localized surface plasmon polaritons (LSPPs). The excitation of the SPPs/LSPPs in the three types of plasmonic PCF structures is discussed. Subsequently, a review of applications based on plasmonic PCF structures is presented. The review in this chapter is focused on the experimental investigations. The fabrication techniques of plasmonic PCF structures and simulation studies of plasmonic PCF structures for various applications are referred to an earlier review article (Hu and Ho 2017).

2 Overview of PCFs

PCF technology is a revolutionary breakthrough in optical fiber technology. It has opened up new possibilities of light guidance mechanism and brought tremendous potential in a broad range of applications with performance enhancements. Based on different light-guiding mechanisms, the PCFs can be broadly categorized into two main groups. The first group of PCFs guide light by modified total internal reflection (TIR) or index-guiding mechanism. The effective refractive index of the core region is higher than that of the cladding region. The index contrast can be flexibly controlled by altering the air filling fraction of the holey cladding, resulting in enhanced tunabilities in optical properties or enhanced optical performance. Index-guiding PCFs are finding applications in highly nonlinear optical devices, high power delivery, polarization-maintaining fibers, dispersion-tailored fibers, etc. The second group of PCFs guide light by photonic bandgap (PBG) effect. The effective refractive index of the core region is lower than that of the cladding region; thus, TIR or index guiding is impossible for light confinement in such fiber structures. The light confinement in lower refractive index core region is realized by photonic crystal effect in the microstructured cladding, prohibiting light transmission in the cladding. PBG-guided PCFs such as hollow core PCFs and Bragg fibers are finding applications in high power delivery, gas-based nonlinear devices, guidance in broadened spectral regions, etc. PCFs operating with other guiding mechanisms such as inhibited coupling and antiresonance and twist-induced guidance (Markos et al. 2017) and hybrid PCFs with coexistence of both index-guiding and PBG-guiding mechanisms (Hu et al.

2019) have been reported previously. Notably, the presence of holey structure in PCF offers a desirable platform for incorporating functional materials to achieve new or enhanced functionalities. Examples include liquid crystals, liquids, magnetic fluids, gases, chalcogenide glasses, and metals. Metal-filled or metal-deposited PCF structures are two main types of plasmonic PCF structures that use PCFs as substrates for incorporating plasmonic structures and manipulating plasmonic properties.

3   Operating Principles of Plasmonic PCF Structures

The electromagnetic waves that propagate along the interface between a metal and a dielectric are surface plasmon polaritons (SPPs). They can be optically excited by photons when the parallel optical wave vector matches the propagation constant of the corresponding SPP modes. This wave vector matching requirement is very sensitive to the ambient parameters, which is the fundamental principle of the surface plasmon resonance (SPR) sensors. Figure 1 shows the Kretschmann configuration and the optical fiber-based configuration of the SPR sensors. The Kretschmann configuration has been widely adopted for SPR sensor. The matching of wave vectors takes place at the metal–glass interface in an inverted metal-coated prism. Optical fiber-based SPR sensors enable efficient coupling between the evanescent optical field and the

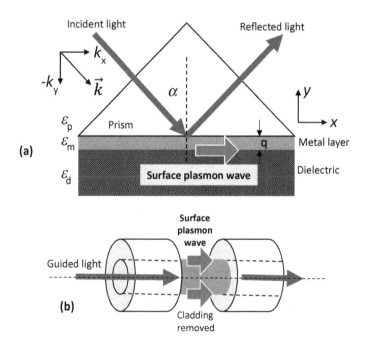

Fig. 1 Plasmonic PCF based on the SPP

SPP modes by exposing the core using side-polishing (Chiu et al. 2007) or selective etching in hydrofluoric acid (Coelho et al. 2015). The optical excitation of SPP modes in metal thin-film-coated PCF and metal nanowire-filled PCF structures can be understood from Fig. 1, relying on the coupling between the fiber modes and the SPP modes.

Localized surface plasmon polaritons (LSPPs) is another operating principle of plasmonic PCF structures that have plasmonic nanoparticles (NPs) or plasmonic nanostructures. Plasmonic nanoparticles (NPs) such as gold or silver nanoparticles have spectral properties in the UV-to-near-IR range. The LSPR is a highly localized electromagnetic field around NPs or the nanostructures, and it can be directly excited as the illumination frequency matches the eigenfrequency of the LSPR (Sauvan et al. 2013). The LSPR-based plasmonic PCF structures can be realized by depositing plasmonic NPs on the inner wall throughout the PCF length (Csaki et al. 2010; Schröder et al. 2012).

4 Review of Applications

- Refractive index sensing and biosensing

Refractive index sensing and biosensing is one of the most exploited applications by plasmonic PCFs. The SPR and LSPR configuration in the PCF structures can be realized at the external/outer surface of fiber, or at the inner walls of the fiber.

Deposition of high-quality metal thin films outside the PCFs could be achieved by the electroless plating method or sputtering technique, using exposed core PCF (Klantsataya et al. 2015), side-hole PCF (Wang et al. 2009), side-polished PCF (Wu et al. 2017), or PCF with collapsed cladding (Wong et al. 2013). These PCF structures provide greater evanescent field compared unperturbed PCF structures, thus enabling more efficient excitation of SPR which occurs at the boundary of the silica–metal thin film. For example, Wu et al. reported a gold-coated side-polished PCF SPR sensor for refractive index sensing (Wu et al. 2017). The sensor was immersed in liquid with varying refractive index. The transmission spectra and the resonance wavelengths were monitored as shown in Fig. 2. The experimental data of the resonance wavelength shift as a function of the refractive index agreed reasonably well with the simulation results, despite the nonlinearity observed in the measurement toward higher refractive index values.

In fact, the process of depositing a metal thin layer in the internal surfaces of PCFs itself is a challenging task. High-pressure microfluidic chemical deposition of semiconductor and metal within the internal surfaces of PCFs was reported by Sazio et al. The work has provided a significant step further toward the development of optoelectronic fiber devices (Sazio et al. 2006). Boehm et al. reported a silver-deposited PCF using an optimized chemical coating method. The method was based on the Tollens reaction, which was optimized to coat 60-nm-thick silver layer over a 1-m-long suspended core fiber with three large holes around the solid core. Slides with

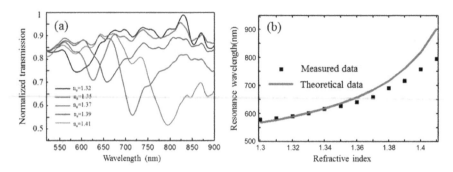

Fig. 2 **a** Experimental normalized transmission spectra of the D-shaped PCF-SPR sensor testing in different RI. **b** Experimental and theoretical values for the wavelength sensitivity. Reprinted with permission from (Wu et al. 2017) © The Optical Society

chemically deposited silver using the optimum recipe as well as sputtering technique were tested for SPR signal measurements. The sensitivity was 160°/RIU. Although the silver-coated PCF was not tested for SPR measurement, the reported technique provided a viable approach to produce plasmonic PCF based on SPR signals (Boehm et al. 2011).

The combination of plasmonic NPs in the PCF platform provides promising potential for LSPR-based sensing applications. Using the self-assembly monolayer technique, the inner walls of the PCF structures could be chemically modified for the deposition of metal NPs. Csaki et al. and Schröder et al. reported the gold/silver NP-deposited PCFs for refractive index sensing with sensitivities up to 78 and 80 nm/RIU, respectively (Csaki et al. 2010; Schröder et al. 2012).

Through the excitation of LSPR for local enhancement of electric field in various PCF structures, the Raman scattering response has been significantly enhanced to sufficient levels for revealing molecular features (Yang et al. 2010). A side-channel PCF structure with strong evanescent field was used as the platform for highly sensitive surface-enhanced Raman scattering (SERS) sensing (Zhang et al. 2016). The fiber design was optimized to ensure light propagation with fundamental mode at 632.8 nm. When the side channel and the air holes in the cladding were filled with liquids, the power of the evanescent field in the side channel increased linearly with the increasing refractive index of the liquid. To test the SERS sensing capability, the gold NPs were mixed in the rhodamine (R6G) solutions, which were filled into the side channel via capillary effect. The SERS spectra of different concentrations of R6G solution were measured, and the results are shown in Fig. 3. A low detection limit of 50 fM R6G solution was achieved.

Hollow core PCFs with the inner wall coated with NPs were used as a SERS sensing platform for biosensing applications, which include detection of EGFR extracted from human epithelial carcinoma cells (Dinish et al. 2012), serological liver cancer biomarkers (Dinish et al. 2014), and leukemia cells (Khetani et al. 2015).

Another technique for depositing metal NPs into the inner walls of PCF was reported by Amezcua-Correa et al., using an organic solvent under high pressure to

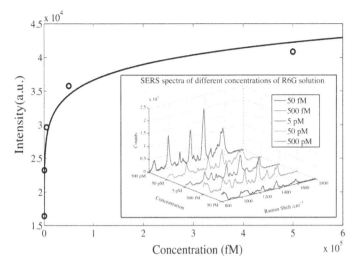

Fig. 3 SERS intensity as a function of concentration of R6G solution. The black solid curve is fitted with the Langmuir isotherm. (Inset) SERS spectra of various concentrations of R6G solutions (Zhang et al. 2016). Reproduced with permission. All Rights Reserved

deliver a silver precursor complex into the fiber holes, followed by a simple thermal reduction of the precursor to form an annular deposition of silver nanoparticles inside the holes (Amezcua-Correa et al. 2007; Peacock et al. 2008). The NP-deposited PCFs demonstrated enhanced SERS sensing performance due to several factors including high numerical aperture for efficient collection and detection of Raman response, and low-loss core-guided modes with large optical component propagating in the voids for large excitation area and long interaction length.

- Temperature sensing

The plasmonic resonance is highly tunable with temperature, which enables developments of functional plasmonic PCF devices for temperature sensing (Yang et al. 2016). In this work, silver nanowire solutions in ethanol and chloroform were filled into the air holes of PCF structure for several centimeters by capillary effect. As shown in the experiment, the silver nanowire colloid was a stable translucent colloidal suspension of silver nanowires in ethanol carrier. The liquid was regarded as a physical mixture of ethanol and chloroform. The diameter of the nanowires was about 90 nm, and the average length was about 30 μm. The mixing volume ratio of the silver nanowire solution and the chloroform was changed to tune the plasmonic resonance and the transmission loss. As shown in Fig. 4, the sensor with a mixing ratio of 1:2 yielded narrower and deeper loss curves at all temperatures compared to that with a mixing ratio of 1:1. The loss curves shifted to shorter wavelengths with increasing temperature from 25 to 60 °C. The temperature sensitivities were −1.8 and −2.08 nm/°C for mixing ratios 1:1 and 1:2, respectively. The temperature sensitivities were lower compared with the results of PCF sensor (−5.5 nm/°C) selectively

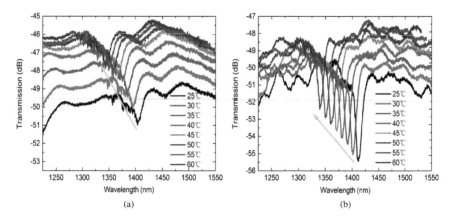

Fig. 4 Transmission spectra of the designed sensor when temperature varies from 25 to 60 °C with a step of 5 °C, filled with different volume ratios of ethanol and chloroform. **a** 1:1. **b** 1:2 (Yang et al. 2016). Reproduced with permission. All Rights Reserved

filled with gold nanoparticles (Peng et al. 2013), and PCF sensor selectively filled with liquid crystal based on coupler configuration (−3.9 nm/°C) (Hu et al. 2012) or hybrid guiding mechanism (4.91 nm/°C) (Xu et al. 2018).

- Polarization and birefringent devices

Plasmonic nanowires or microwires can be embedded in fiber structure in periodic arrangement to form metamaterial. The fabrication is challenging in volume. A feasible drawing technique to fabricate the metallic microwires in a PCF was introduced by Tuniz et al., demonstrating codrawing of polymethyl methacrylate and indium and producing several meters of metamaterial. The wire diameter was down to ~10 μm, and the lattice constant was ~100 μm. The metamaterial fiber transmittance was measured via THz time domain spectroscopy. Experimental results demonstrated that the metamaterial fiber acted as THz high-pass filter and polarizer. Notably, the technique can be used to further reduce the wire diameters down to nanoscales, which means much better prospect for practical plasmonic devices (Tuniz et al. 2010).

Lee et al. reported strong polarization-dependent spectral splitting in the double-nanowire PCF. The attenuation spectra of the fiber were measured and are shown in Fig. 5. The double-peak feature of the attenuation was due to optical mode coupling to SPP modes at phase-matching wavelengths. The polarization dependence in the fiber can be used for polarization-dependent polarizers (Lee et al. 2012). Polarization-dependent attenuation was also observed in selectively metal-coated PCF structure, showing promise for in-fiber absorptive polarizers (Zhang et al. 2007).

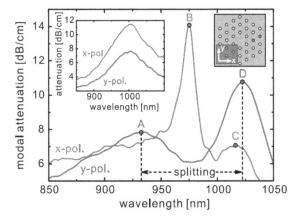

Fig. 5 Measured attenuation spectrum of the x- and y-polarized glass core mode in the double-nanowire fiber. The polarization directions are defined in the right-hand inset. Top left-hand inset: measured attenuation spectrum for the single-wire fiber. **a** 932.2 nm; **b** 974.8 nm; **c** 1015.9 nm; **d** 1022.4 nm. Reprinted with permission from (Lee et al. 2012) © The Optical Society

- New developments and opportunities

The combination of graphene and plasmonic structures has led to new opportunities as significant enhancement of the evanescent field is observed associated with the presence of the graphene. Graphene-based plasmonic sensors have been explored for sensing applications. Zeng et al. reported significant sensitivity enhancements of graphene–gold metasurface architectures (Zeng et al. 2015). The graphene–gold architecture was fabricated in a PCF-based SPR sensor, showing sensitivity improvement by 390 nm/RIU after the introduction of graphene (Li et al. 2019). The fiber sensor structure is shown in Fig. 6. The exposed core fiber was coated by gold film and graphene at the notch, and the measured liquid was covering the graphene.

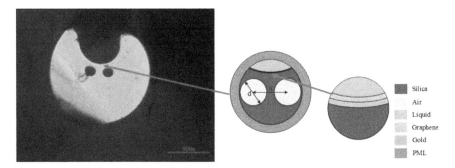

Fig. 6 Structure of the sensor (Li et al. 2019). Reproduced with permission. All Rights Reserved

The average sensitivity performance by the gold film-coated PCF SPR sensor was measured to be 1900 nm/RIU. Subsequently, the graphene was coated on the gold film and the average sensitivity performance of the PCF SPR sensor was measured to be 2290 nm/RIU.

The fabrication challenge of depositing graphene in the inner wall of the PCF structure was addressed by Chen et al. in a recent report. Up to half-meter-long PCF deposited with graphene was demonstrated using the chemical vapor deposition method (Chen et al. 2019). Such capability could enable realization of the graphene–gold architecture in the inner wall of the PCF for developments of new sensors with performance enhancements.

5   Conclusion

The integration of plasmonics in the PCF platform is a promising field of research. By leveraging on the flexibilities offered by PCF, SPP modes and LSPR can be efficiently excited for enhanced sensor performance. The operating principles are based on coupling between fiber modes and SPP modes in metal thin-film-coated PCF and metal nanowire-filled PCF structures, and direct excitation of the LSPR in nanoparticle-deposited/filled plasmonic PCF structures. The plasmonic PCF devices are particularly useful for sensing applications through the detection of refractive index, temperature, and biological events, and as conventional photonic components such as filters and polarizers. New developments in terms of fabrication techniques, which will expand the scope of coupling between the plasmonic effects and the unique photon confinement characteristics of PCFs, should further open up more exciting more opportunities, which were previously not possible, for plasmonic PCF devices.

Acknowledgements The authors wish to thank the Hong Kong Research Grants Council for supporting their plasmonic sensing work through project number AoE/P-02/12.

References

A. Amezcua-Correa, J. Yang, C.E. Finlayson, A.C. Peacor, J. Hayes, P.J. Sazio, S.M. Howdle, Surface-enhanced Raman scattering using microstructured optical fiber substrates. Adv. Func. Mater. **17**, 2024–2030 (2007)

W.L. Barnes, A. Dereux, T.W. Ebbesen, Surface plasmon subwavelength optics. Nature **424**, 824–830 (2003). https://doi.org/10.1038/nature01937

J. Boehm, A. Francois, H. Ebendorff-Heidepriem, T.M. Monro, Chemical deposition of silver for the fabrication of surface plasmon microstructured optical fibre sensors. Plasmonics **6**, 133–136 (2011)

K. Chen, X. Zhou, X. Cheng, R. Qiao, Y. Cheng, C. Liu et al., Graphene photonic crystal fibre with strong and tunable light–matter interaction. Nat. Photonics **13**(11), 754–759 (2019)

M.-H. Chiu, C.-H. Shih, M.-H. Chi, Optimum sensitivity of single-mode D-type optical fiber sensor in the intensity measurement. Sens. Actuators B **123**(2), 1120 (2007)

L. Coelho, J.M. de Almeida, J.L. Santos, R.A. Ferreira, P.S. André, D. Viegas, Sensing structure based on surface plasmon resonance in chemically etched single mode optical fibres. Plasmonics **10**(2), 319 (2015)

A. Csaki, F. Jahn, I. Latka, T. Henkel, D. Malsch, T. Schneider, W. Fritzsche, Nanoparticle layer deposition for plasmonic tuning of microstructured optical fibers. Small **6**(22), 2584–2589 (2010)

B. Debord, F. Amrani, L. Vincetti, F. Gérôme, F. Benabid, Hollow-core fiber technology: the rising of "gas photonics". Fibers **7**(2), 16 (2019)

U.S. Dinish, C.Y. Fu, K.S. Soh, R. Bhuvaneswari, A. Kumar, M. Olivo, Highly sensitive SERS detection of cancer proteins in low sample volume using hollow core photonic crystal fiber. Biosens. Bioelectron. **33**, 293–298 (2012)

U.S. Dinish, G. Balasundara, Y.T. Chang, M. Olivo, Sensitive multiplex detection of serological liver cancer biomarkers using SERS-active photonic crystal fiber probe. J. Biophotonics **7**(11–12), 956–965 (2014)

S. Ertman, P. Lesiak, T.R. Woliński, Optofluidic photonic crystal fiber-based sensors. J. Lightwave Technol. **35**(16), 3399–3405 (2017)

J. Homola, Surface plasmon resonance sensors for detection of chemical and biological species. Chem. Rev. **108**, 462–493 (2008)

D.J. Hu, H.P. Ho, Recent advances in plasmonic photonic crystal fibers: design, fabrication and applications. Adv. Opt. Photonics **9**(2), 257–314 (2017)

D.J. Hu, J.L. Lim, Y. Cui, K. Milenko, Y. Wang, P.P. Shum, T. Wolinski, Fabrication and characterization of a highly temperature sensitive device based on nematic liquid crystal-filled photonic crystal fiber. IEEE Photonics J. **4**(5), 1248–1255 (2012)

D.J. Hu, R.Y.-N. Wong, P.P. Shum, Photonic crystal fiber–based interferometric sensors, in *Selected Topics on Optical Fiber Technologies and Applications* (Intech, London, 2018), pp. 21–41

D.J. Hu, Z. Xu, P.P. Shum, Review on photonic crystal fibers with hybrid guiding mechanisms. IEEE Access **7**, 67469–67482 (2019)

A. Khetani, A. Momenpour, E.I. Alarcon, H. Anis, Hollow core photonic crystal fiber for monitoring leukemia cells using surface enhanced Raman scattering (SERS). Opt. Express **23**(22), 4599–4609 (2015)

E. Klantsataya, A. Francois, H. Ebendorff-Heidepriem, P. Hoffmann, T.M. Monro, Surface plasmon scattering in exposed core optical fiber for enhanced resolution refractive index sensing. Sensors **15**, 25090–25102 (2015)

J.C. Knight, Photonic crystal fibres. Nature 847–851 (2003)

J.C. Knight, Photonic crystal fibers and fiber lasers. J. Opt. Soc. Am. B **24**(8), 1661–1668 (2007)

J.C. Knight, D.V. Skryabin, Nonlinear waveguide optics and photonic crystal fibers. Opt. Express **15**(23), 15365–15376 (2007)

H.W. Lee, M.A. Schmidt, P.S. Russell, Excitation of a nanowire "molecule" in gold-filled photonic crystal fiber. Opt. Lett. **37**(14), 2946–2948 (2012)

B. Li, T. Cheng, J. Chen, X. Yan, Graphene-enhanced surface plasmon resonance liquid refractive index sensor based on photonic crystal fiber. Sensors **19**, 3666 (2019)

C. Markos, J.C. Travers, A. Abdolvand, B.J. Eggleton, O. Bang, Hybrid photonic crystal-fiber. Rev. Mod. Phys. **89**, 045003 (2017)

A.C. Peacock, A. Amezcua-Correa, J. Yang, P.J. Sazio, S.M. Howdle, Highly efficient surface enhanced Raman scattering using microstructured optical fibers with enhanced plasmonic interactions. Appl. Phys. Lett. **92**, 114113 (2008)

Y. Peng, J. Hou, Y. Zhang, Z. Huang, R. Xiao, Q. Lu, Temperature sensing using the bandgap-like effect in a selectively liquid-filled photonic crystal fiber. Opt. Lett. **38**(3), 263–265 (2013)

A.M. Pinto, M. Lopez-Amo, Photonic crystal fibers for sensing applications. J. Sens. **2012**, 598178 (2012)

P. Russell, Photonic crystal fibers. Science 358–362 (2003)

C. Sauvan, J.P. Hugonin, I.S. Maksymov, P. Lalanne, Theory of the spontaneous optical emission of nanosize photonic and plasmonic resonators. Phys. Rev. Lett. **110**, 237401 (2013)

P.J. Sazio, A. Amezcua-Correa, C.E. Finlayson, J.R. Hayes, T.J. Scheidemantel, N.F. Baril, J.V. Badding, Microstructured optical fibers as high-pressure microfluidic reactors. Science **311**, 1583–1586 (2006)

K. Schröder, A. Csáki, A. Schwuchow, F. Jahn, K. Strelau, I. Latka, W. Fritzsche, Functionalization of microstructured optical fibers by internal nanoparticle mono-layers for plasmonic biosensor applications. IEEE Sens. J. **12**(1), 218–224 (2012)

J.C. Travers, W. Chang, J. Nold, N.Y. Joly, P.S. Russell, Ultrafast nonlinear optics in gas-filled hollow-core photonic crystal fibers. J. Opt. Soc. Am. B **28**(12), A21–A26 (2011)

A. Tuniz, B.T. Kuhlmey, R. Lwin, A. Wang, J. Anthony, R. Leonhardt, S.C. Fleming, Drawn metamaterials with plasmonic response at terahertz frequencies. Appl. Phys. Lett. **96**, 191101 (2010)

J. Villatoro, J. Zubia, New perspectives in photonic crystal fibre sensors. Opt. Laser Technol. 67–75 (2016)

A. Wang, A. Docherty, B.T. Kuhlmey, F.M. Cox, M.C. Large, Side-hole fiber sensor based on surface plasmon resonance. Opt. Lett. **34**(24), 3890–3892 (2009)

W.C. Wong, C.C. Chan, J.L. Boo, Z.Y. Teo, Z.Q. Tou, H.B. Yang, K.C. Leong, Photonic crystal fiber surface plasmon resonance biosensor based on protein G immobilization. IEEE J. Sel. Top. Quantum Electron. **19**(3), 460217 (2013)

T. Wu, Y. Shao, S.C. Ying Wang, W. Cao, F. Zhang, C. Liao, Y. Wang, Surface plasmon resonance biosensor based on gold-coated side-polished hexagonal structure photonic crystal fiber. Opt. Express **25**(17), 20313–20322 (2017)

Z. Xu, B. Li, D.J. Hu, Z. Wu, S. Ertman, T. Wolinski, P.P. Shum, Hybrid photonic crystal fiber for highly sensitive temperature measurement. J. Opt. **20**(7), 075801 (2018)

X. Yang, C. Shi, R. Newhouse, J.Z. Zhang, C. Gu, Hollow-core photonic crystal fibers for surface-enhanced raman scattering probes. Int. J. Opt. **2011**, 751610 (2010)

X. Yang, C. Shi, R. Newhouse, J.Z. Zhang, C. Gu, Hollow-core photonic crystal fibers for surface-enhanced raman scattering probes. Int. J. Opt. **2011**, 754610 (2011)

X.C. Yang, Y. Lu, B.L. Liu, J.Q. Yao, Temperature sensor based on photonic crystal fiber filled with liquid and silver nanowires. IEEE Photonics J. **8**(3), 6803309 (2016)

S. Zeng, K.V. Sreekanth, J. Shang, T. Yu, C.-K. Chen, F. Yin, K.-T. Yong, Graphene-gold metasurface architectures for ultrasensitive plasmonic biosensing. Adv. Mater. **27**(40), 6163–6169 (2015)

X. Zhang, R. Wang, F.M. Cox, B.T. Kuhlmey, M.C. Large, Selective coating of holes in microstructured optical fiber and its application to in-fiber absorptive polarizers. Opt. Express **15**(24), 16270–16278 (2007)

N. Zhang, G. Humbert, T. Gong, P.P. Shum, K. Li, J.-L. Auguste, L. Wei, Side-channel photonic crystal fiber for surface enhanced Raman scattering sensing. Sens. Actuators B Chem. **223**, 195–201 (2016)

Y. Zhao, Z.-Q. Deng, J. Li, Photonic crystal fiber based surface plasmon resonance chemical sensors. Sens. Actuators B Chem. **202**, 557–567 (2014)

Hybrid Fiber-Optic Sensors

Nancy Meng Ying Zhang, Kaiwei Li, Miao Qi, and Zhifang Wu

Abstract With the increasing demand of achieving comprehensive perception in every aspect of life, optical fibers have shown great potential in various applications due to their highly sensitive, highly integrated, flexible and real-time sensing capabilities. Among various sensing mechanisms, plasmonics-based fiber-optic sensors provide remarkable sensitivity benefited from their outstanding plasmon–matter interaction. Therefore, surface plasmon resonance (SPR) and localized SPR (LSPR)-based fiber-optic sensors have captured intensive research efforts. Conventionally, SPR or LSPR-based fiber-optic sensors rely on the resonant electron oscillations of thin metallic films or metallic nanoparticles functionalized on fiber surface. Coupled with the new advances in functional nanomaterials as well as fiber structure design and fabrication in recent years, new solutions continue to emerge to further improve the fiber-optic plasmonic sensors performance in terms of sensitivity, specificity and biocompatibility. For instances, 2D materials like graphene can enhance the surface plasmon intensity at metallic film surface so as the plasmon–matter interaction. 2D morphology of transition metal oxides can be doped with abundant free electrons to facilitate intrinsic plasmonics in visible or near-infrared frequencies, realizing exceptional field confinement and highly sensitivity detection of analyte molecules. Gold nanoparticles capped with macrocyclic supramolecules show excellent selectivity to target biomolecules and ultralow limit of detection. Moreover, specially

N. M. Y. Zhang (✉) · M. Qi
School of Electrical and Electronic Engineering, Nanyang Technological University,
50 Nanyang Avenue, Singapore 639798, Singapore
e-mail: mzhang018@e.ntu.edu.sg

M. Qi
e-mail: miao001@e.ntu.edu.sg

K. Li
Institute of Photonics Technology, Jinan University, Guangzhou 510632, China
e-mail: likaiwei11@163.com

Z. Wu
Fujian Key Laboratory of Light Propagation and Transformation,
College of Information Science and Engineering, Huaqiao University, Xiamen 361021, China
e-mail: wzh.fang@gmail.com

© Springer Nature Singapore Pte Ltd. 2020
L. Wei (ed.), *Advanced Fiber Sensing Technologies*,
Progress in Optical Science and Photonics 9,
https://doi.org/10.1007/978-981-15-5507-7_2

designed microstructured optical fibers are able to achieve high birefringence that can suppress the output inaccuracy induced by polarization crosstalk meanwhile deliver promising sensitivity. This chapter aims to reveal and explore the frontiers of such hybrid plasmonic fiber-optic platforms in various sensing applications.

Keywords Optical fibers · Surface plasmon resonance · Localized surface plasmon resonance · Microstructured optical fibers · 2D materials · Graphene · Transition metal oxides · Macrocyclic supramolecules · MoO3 · Cyclodextrin

1 Introduction

Plasmonic fiber-optic sensors have captured intensive research attention in recent years due to their high degree of integration, high sensitivity, flexibility and remote sensing capability. Fiber-optic plasmonic sensors can be generally classified into two categories, surface plasmon resonance (SPR)-based sensors and localized surface plasmon resonance (LSPR)-based sensors. Conventionally, SPR and LSPR-based fiber-optic sensors are realized by depositing thin metal films and metallic nanoparticles on various fiber structures (e.g., fiber gratings, side-polished fiber, microfiber), respectively, to contribute strong plasmon–matter interaction. To further improve the measurement accuracy, sensitivity and selectivity to analyte molecules, specially designed fiber structures or functional materials are normally applied to strengthen the intensity of surface plasmon or the adsorption to target molecules. Specialty optical fibers like microstructured optical fibers (MOFs) can achieve high-level integration that the very small dimension waveguide and the microfluidic channels are able to be integrated within a single fiber with only micrometer-scale diameter, leading to effective plasmon–matter interaction. The recent breakthroughs in 2D materials such as graphene, transition metal dichalcogenides (MX_2), transition metal oxides (TMOs) reveal new opportunities in plasmon–matter enhancement by constructing 2D material/metal hybrid plasmonic structures or heavily doping free carriers in 2D TMOs to realize intrinsic strong plasmonics in frequently used visible or near-infrared (NIR) optical window. In addition, macrocyclic supramolecules have been recently proven to be excellent surface functionalization candidates for metallic nanoparticles, contributing to simple functionalization process, selective target molecules recognition and improved biocompatibility. In this chapter, the background and the state-of-the-art of SPR/LSPR fiber-optic sensors will be reviewed. More importantly, the abovementioned emerging hybrid fiber-optic plasmonic sensing solutions will be illustrated in details. For instances, the exploration of how the highly birefringent MOF-based SPR sensor can suppress polarization crosstalk and improve sensitivity in the meantime, how to integrate graphene-on-gold hybrid structure on fiber-optic platform to strengthen the surface plasmon intensity and to effectively adsorb biomolecules, how to dope abundant free electrons in 2D MoO_3 and achieve highly integrated microfiber-based plasmonic sensing in

NIR optical frequencies, how to synthesize cyclodextrin capped gold nanoparticles in a one-step process and realize microfiber-based highly selective detection of cholesterol in human serum, etc. will be demonstrated.

2   Main Body Text

- Fiber-Optic Surface Plasmon Resonance Sensors

Surface plasmon polariton, in short SPP, is an electromagnetic wave that parallelly propagates at the interface between metal film and dielectric medium. The SPP is TM-polarized. As illustrated in Fig. 1a, the polarization direction of SPP is perpendicular to the metal–dielectric interface (Jana et al. 2016). The SPP has the evanescent nature, which is strongest at the surface of metal film and exponentially decays into the dielectric material. Conventionally, SPR is realized on a Kretschmann–Raether silica prism of which the base is coated with a nanometer-scale thin metal film (Fig. 1b). The ambient medium of thin metal film is dielectric and considered as semi-infinite. SPR can be excited by the TM-polarized total reflected light at silica–metal interface when the phase matching condition is satisfied, that the propagation constant of reflected light equals to the propagation constant of SPP:

$$\frac{\omega}{c}\sqrt{\varepsilon_{prism}}\sin(\theta_{res}) = \frac{\omega}{c}\sqrt{\frac{\varepsilon_{metal}\varepsilon_{dielectric}}{\varepsilon_{metal} + \varepsilon_{dielectric}}} \tag{1}$$

Along with the increasing demand of compact, highly integrated, flexible and even in situ sensing devices, optical fiber-based SPR sensors receive more and more attention. Various fiber-optic SPR configurations have been investigated to achieve highly sensitive SPR sensors. The key point in the design of fiber-optic SPR sensor is to realize the phase matching between the guided mode in fiber and the SPP at metal–dielectric interface. Hence, it is essential to coat the thin metal film at the surface of

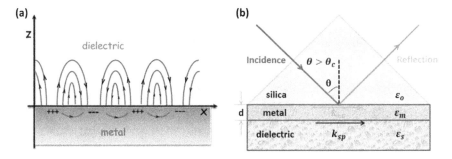

Fig. 1   a The propagating SPP at the metal–dielectric interface (Adapted with permission from Jana et al. 2016). **b** The schematic illustration of conventional Kretschmann–Raether prism configuration (Adapted with permission from Zhao et al. 2014)

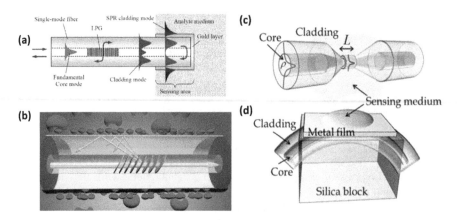

Fig. 2 Fiber-optic SPR sensor based on **a** LPG (Adapted with permission from Schuster et al. 2011). **b** TFBG (Adapted with permission from Shevchenko et al. 2011). **c** Tapered fiber (Adapted with permission from Roh et al. 2011). **d** Side-polished fiber (Adapted with permission from Roh et al. 2011)

fiber structure where strong evanescent field of guide mode can be exposed, leading to the strong SPP at metal surface for effective light–matter interaction. Fiber gratings like long-period fiber grating (LPG) and tilted fiber Bragg grating (TFBG), tapered fiber, side-polished fiber have been demonstrated to be feasible for SPR sensing (Fig. 2).

In recent years, MOFs are favored due to higher degree of integration, longer interaction distance and improved robustness, that the cladding air holes can function as microfluidic channels for liquid or gas analyte infiltration (Yu et al. 2009). With the special design of core dimension and cladding air holes arrangement, the thin metal films coated on the inner surface of air holes can effectively interact with the evanescent field of core mode, which grants access of infiltrated analyte to the strong SPP. Numerous MOF structures have been proposed, including hexagonal MOFs, semicircular-channel MOFs, exposed-core MOFs (Fig. 3). In most MOF-based SPR

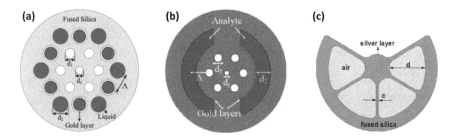

Fig. 3 MOF based SPR sensors with specially designed structures of **a** hexagonal air holes (Adapted with permission from Peng et al. 2012). **b** Semicircular air channels (Adapted with permission from Hassani and Skorobogatiy 2007). **c** Exposed core (Adapted with permission from Yang et al. 2015)

designs, the prime consideration is to facilitate easy analyte infiltration and large interaction area. Hence birefringence commonly exists in MOF-based SPRs. Based on Eq. (1), birefringence leads to the offset between SPR wavelengths corresponding to two orthogonal polarizations of core mode. When external perturbations such as fiber bending, twisting and pressure are applied on fiber, the coupling from the desired mode polarization to undesired mode polarization will occur. Therefore, the overall SPR peak which is the superposition of SPR of two orthogonal polarizations will be unstable, leading to inaccurate sensing results.

To address the issue of birefringence induced measurement unstability, polarization-maintaining MOF-based SPR sensor with high birefringence could be a promising solution. A large birefringence can be realized in a near-panda MOF with the two central air holes of the photonic-crystal arranged cladding holes enlarged (Fig. 4). The enlarged two central holes can facilitate easier thin noble metal film deposition and analyte infiltration. Strong surface plasmons can be excited by the x-polarized fundamental core mode with the thin gold film deposited on the inner walls of central holes. As discussed earlier, SPP can only excited by the TM-polarized incident light (i.e., the polarization perpendicular to the metal film surface), y-polarized core mode corresponds to a much weaker SPP compared with that of x-polarized mode (Fig. 4b, c). This indicates the SPR sensing output is predominated by the plasmonic behaviors of x-polarized code mode. For a low-birefringent MOF, of which the diameter of central holes ($d2$) is comparable to that of other cladding holes ($d1$) (e.g., $d1/d2 = 0.95$), both x- and y-polarized mode can excite relatively strong SPP. As a result, the existence of unwanted polarization could induce an offset of overall resonant wavelength as high as 0.67 nm from that of desired polarization, which means the SPR sensing accuracy is considerably compromised (Fig. 5a). On the contrary,

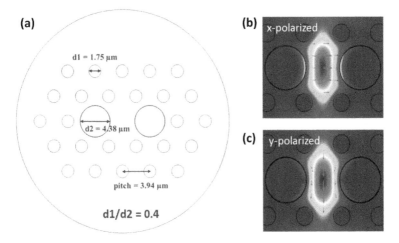

Fig. 4 **a** The configuration of proposed highly birefringent MOF. **b** x-polarized core mode. **c** y-polarized core mode (Adapted with permission from Zhang et al. 2016)

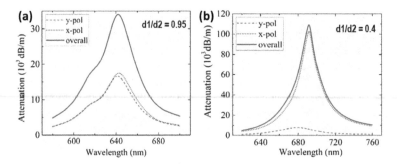

Fig. 5 Attenuation spectra of highly birefringent MOF when **a** $d1/d2 = 0.95$ and **b** $d1/d2 = 0.4$ (Adapted with permission from Zhang et al. 2016)

even though a highly birefringent MOF consists of two modal polarizations corresponding to even larger resonant wavelength difference, the immensely suppressed SPP of unwanted polarization has bare influence on the overall resonant wavelength so as the sensing accuracy. For instance, the wavelength offset of the proposed highly birefringent near-panda MOF with $d1/d2 = 0.4$ is as small as 0.06 nm (Fig. 5b).

Based on the FEM simulation of photonic-crystal arranged MOFs with different $d1/d2$ ratios, the relation between phase birefringence and sensing inaccuracy can be deduced. As shown in Fig. 6a, the resonant wavelength offset could increase to be as large as 18.89 nm when the phase birefringence increases from ~4×10^{-5} to ~1×10^{-4}. When the phase birefringence exceeds beyond a threshold (~1×10^{-4}), the wavelength offset effectively reduces and even tends toward 0 after 4×10^{-4} phase birefringence. The investigation indicates that small birefringence that commonly exists in MOF-based SPR sensors could induce nonnegligible undesired resonant wavelength offset, which affects sensing accuracy. The proposed highly birefringent

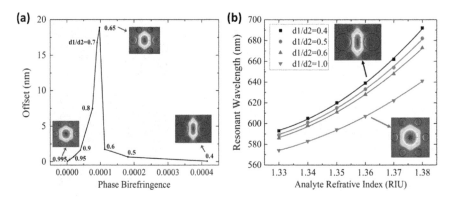

Fig. 6 **a** The variation of wavelength offset along with phase birefringence. **b** The SPR sensitivities when $d1/d2 = 1.0, 0.6, 0.5$ and 0.4 respectively (Adapted with permission from Zhang et al. 2016)

MOF with intentionally introduced large phase birefringence ~4.2×10^{-4} can effectively suppress such impact of polarization crosstalk to be extremely small. In addition, more expanded central holes enhance the plasmon–matter interaction thereby provide a higher sensitivity. Figure 6b compares the sensitivities when $d1/d2 = 0.4$, 0.5, 0.6 and 1.0. It is clear that the sensitivity is improved when the central holes expand. At high analyte refractive index range of 1.37–1.38, the proposed highly birefringent MOF SPR sensor can achieve a sensitivity as high as 3000 nm/RIU.

Besides optimizing the design of fiber structure, integrating functional nanomaterials with fiber-optic platform can also effectively promote the light–matter interaction. In the past decade, 2D materials have drawn extensive attention in various research field including the highly integrated sensors. The extremely large surface-to-volume ratio, in situ plasmonic properties tunability and near-field confinement are the great advantages of 2D materials in sensing applications (Luo et al. 2013; Rodrigo et al. 2015). The plasmonics of most common 2D materials such as graphene and MX_2 fall in MIR or terahertz regions, which are not compatible with the well-developed optical communication window even though they can achieve superior plasmonic sensing performance (Ju et al. 2011; Koppens et al. 2011). Therefore, numerous research efforts focus on enhancing the plasmon–matter interaction by applying 2D material/metal film hybrid structures to SPR configurations. For instances, the thin gold film in conventional Kretschmann configuration has been upgraded to graphene/gold, graphene oxide/gold, graphene-MoS_2/gold, etc. hybrid film-like architectures (Fig. 7). It is proven that the intensity of SPP on gold film surface can be effectively strengthened by the seamlessly integrated graphene layer. When graphene and gold are in contact, the work function difference between the two materials (4.5 eV for graphene and 5.54 eV for gold) causes electrons flow from graphene to gold to equilibrate the Fermi levels (Giovannetti et al. 2008; Khomyakov et al. 2009). As a result, the electron density at gold film surface increases as the graphene becomes p-type doped. Therefore, a stronger SPP so as a higher sensitivity can be achieved.

Even though the 2D material/metal film hybrid structure had been widely proposed on the prism-based SPR configuration, systematical analysis and experimental

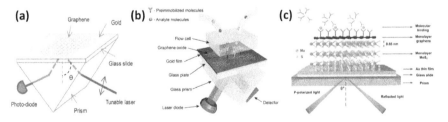

Fig. 7 The prism-based SPR configurations with hybrid plasmonic structures of **a** single-layer graphene/gold (Adapted with permission from Salihoglu et al. 2012); **b** multilayer graphene/Py/gold (Adapted with permission from Stebunov et al. 2015); **c** graphene-MoS2/gold (Adapted with permission from Zeng et al. 2015)

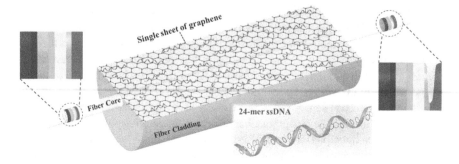

Fig. 8 The schematic illustration of proposed hybrid graphene-on-gold SPR sensor. The nucleobases of target ssDNA molecules can form stable π–stacking interaction with the honeycomb arrange carbon atoms of graphene (Adapted with permission from Zhang et al. 2017)

demonstration of integrating such hybrid plasmonic structure with flexible waveguides such as optical fibers were rare. As a proof of concept, a graphene-on-gold hybrid structure is proposed to be seamlessly integrated with a side-polished optical fiber, purposing to demonstrate that the 2D material/metal hybrid structures could achieve enhanced plasmonic biosensing performance on flexible waveguide platforms. As illustrated in Fig. 8, the exposed evanescent field of guided core mode interacts with the graphene-on-gold structure deposited at the surface of polished facet of optical fiber, leading to strong SPP-biomolecules interaction. Meanwhile, the single graphene layer functions as excellent surface functionalization of the thin gold film. Since the SPP at gold film surface exponentially decays with the penetration depth, the thickness of surface functionalization is a crucial factor that affects sensitivity. The graphene layer as thin as 0.34 nm could hardly compromise the SPR sensitivity (Kim et al. 2013). Moreover, the carbon atoms of graphene arranged in honeycomb format can easily form π–stacking interaction with the aromatic rings commonly existed in biomolecules (Song et al. 2010). Hence, it facilitates effective adsorption of target biomolecules such as ssDNA, providing high sensitivity and low LOD.

Simulation can verify the SPP enhancement capability of the additional graphene sheet on conventional gold film coated side-polished fiber. The inset of Fig. 9a plots the whole electrical field distribution of guided core mode in the fiber as well as the SPP on the side-polished facet. The magnified field distribution of the SPP at the gold/graphene surface is shown in Fig. 9a. As expected, introducing single or multiple graphene layers can effectively enhance the SPP intensity on the thin gold film surface, which is benefited from the electrons transfer as explained above. Another interesting finding in the simulation is that bilayer or multilayer graphene slightly compromise the SPP intensity compared with the single-layer graphene. This is due to the electrons' energy loss induced by the increase of graphene layers (Kim et al. 2013). Therefore, with the SPP intensity boosted by ~30.2%, single graphene layer most enhances the plasmonic sensing behavior. The experimental results further verify the graphene-on-gold hybrid structure can effectively improve the plasmonic

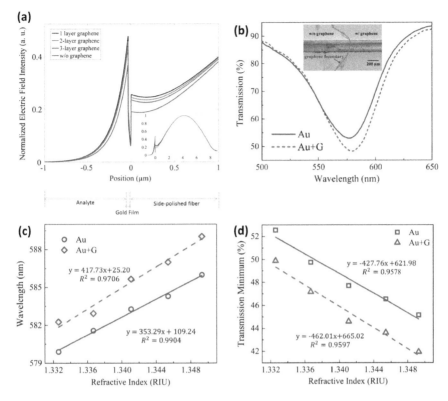

Fig. 9 **a** The comparison of electric field intensities when the SPP is excited by bare thin gold film, single-layer graphene/gold, 2-layer graphene/gold and 3-layer graphene/gold. (Inset) The electric field distribution over the entire fiber-optic graphene/gold hybrid structure. **b** The comparison of fiber transmission spectra with and without single graphene layer. (Inset) The microscopic view of the single graphene layer transferred on the side-polished fiber. **c** The comparison of sensitivities with and without graphene transfer for wavelength interrogation. **d** The comparison of sensitivities with and without graphene transfer for intensity interrogation (Adapted with permission from Zhang et al. 2017)

sensing behavior. Figure 9b compares the resonant peaks of the conventional thin gold film coated side-polished fiber and the graphene-on-gold hybrid structure integrated side-polished fiber when both sensing configurations are immersed in DI water. The inset of Fig. 9b shows the microscopic view of the boundary of transferred graphene on the thin gold film coated side-polished fiber facet. The graphene-on-gold hybrid structure corresponds to a deeper resonant peak, indicating a stronger SPP intensity, which matches well with the simulation. As a result, the hybrid graphene-on-gold structure provides enhanced plasmonic sensing performance in both wavelength and intensity interrogations (Fig. 9c, d).

The biosensing capability of the proposed plasmonic hybrid SPR configuration can be validated by detecting ssDNA concentration. ssDNA quantization provides biomedical significance in gene expression, DNA sequencing and polymerase chain

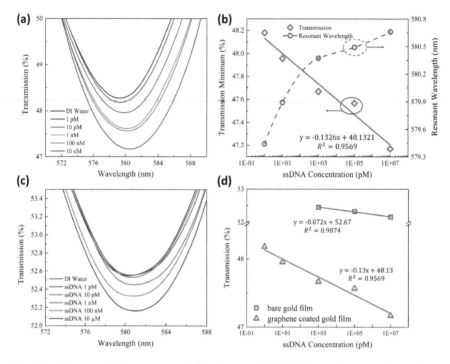

Fig. 10 a The transmission spectra variation of proposed hybrid plasmonic sensor along with the increase of ssDNA concentration. **b** The transmission minimum and resonant wavelength variations along with log-scale ssDNA concentration increase. **c** The transmission spectra variation along with the increase of ssDNA concentration if there is no graphene transferred on gold film. **d** The comparison of sensitivities to ssDNA concentration of fiber-optic plasmonic sensors with and without graphene layer (Adapted with permission from Zhang et al. (2017)

reaction (PCR) (Bhat et al. 2010). Figure 10a shows the magnified SPR peaks of the biosensing platform with the incrementing ssDNA concentration. This can be explained by Eq. (1) that the surrounding refractive index of the plasmonic architecture is increased due to the efficient adsorption of ssDNA molecules on graphene surface via π–stacking interaction. Also, the SPP evanescent field is scattered by the bonding of ssDNA molecules, which further induces transmission loss thereby a deeper SPR peak. The LOD of the biosensor to ssDNA molecules is as small as 1 pM based on the distinguishable enhancement of SPR peak (the red curve of Fig. 10a). The sensitivity of biosensor can be determined by the linear response of transmission minimum against the log-scale pM ssDNA concentration from 1 pM to 10 μM as plotted in Fig. 10b. To experimentally verify that the biosensing performance is improved by the additional graphene layer, a conventional thin gold film-based side-polished fiber-optic SPR sensor is prepared and applied to measure the same ssDNA solutions. As shown in Fig. 10c, the LOD increases to 1 nM which is less sensitive than the hybrid plasmonic structure. The comparison of sensitivities corresponding

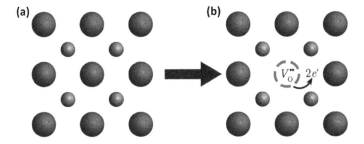

Fig. 11 The doping of free electrons by inducing oxygen vacancies in TMO lattice. **a** The pristine TMO lattice. **b** Electrons doped TMO lattice. Red spheres: oxygen anions. Yellow spheres: metal cations (Adapted with permission from Lounis et al. 2014)

to the two structures in Fig. 10d can obviously indicate that the graphene-on-gold hybrid structure can effectively improves the sensitivity almost twice higher.

Although the 2D material/metal hybrid structures facilitate remarkable light–matter interaction in plasmonic sensing, the intrinsic SPP of most common 2D materials (e.g., graphene and MX_2) located at MIR range is hardly accessible for practical applications. Therefore, an alternative class of 2D plasmonic material, heavily doped ultrathin TMOs, have captured research attention in recent years aiming for manipulating the intrinsic plasmonics of 2D materials with exceptional field confinement and in situ plasmonic tunability in the frequently used visible and NIR optical window (Alsaif et al. 2014; Cheng et al. 2014, 2016; Liu et al. 2017a). To realize SPP in visible or NIR frequencies, sufficient free carrier concentration must be achieved in 2D materials. The unique character of outer–d valence electrons enables TMOs to achieve sufficient free carriers doping via ionic intercalation. Taking the most representative TMOs molybdenum trioxide (MoO3) and tungsten oxide (WO3) as examples, free electrons can be abundantly doped by introducing oxygen vacancies in the TMO lattice. As shown in Fig. 11, an oxygen atom acts as a dianion when it is removed from lattice so that two free electrons are contributed to the conduction band (Manthiram and Alivisatos 2012; Lounis et al. 2014). Therefore, the plasmonic behavior of 2D TMOs can be easily tuned by manipulating the oxygen vacancies. So far, the tuneable plasmonics of heavily doped MoO_3 nanoflakes in visible or NIR region has been most widely studied, yet the exploration on integrating such emerging 2D materials with optical devices especially the highly integrated waveguide-based sensing devices is very limited.

Driven by the purpose of investigating the potential of 2D TMOs on highly integrated plasmonic devices, a biosensor based on a microfiber functionalized with α-MoO3 nanoflakes is developed and validated by BSA molecules detection. As shown in Fig. 12, few-layer α-MoO3 nanoflakes are synthesized by the liquid phase exfoliation method (Balendhran et al. 2013) and then heavily doped with free electrons via an H^+ intercalation process. After doping, sub-stoichiometric α-MoO$_{3-x}$ nanoflakes solution with strong SPP at NIR region is formed. Immobilized on the surface of the 2-μm-diameter tapered portion of microfiber, the MoO$_{3-x}$ nanoflakes

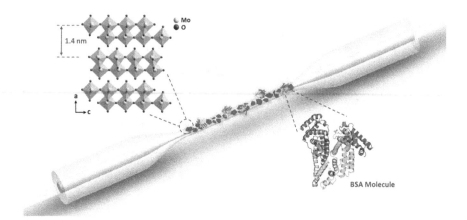

Fig. 12 The schematic illustration of heavily doped MoO_{3-x} nanoflakes-based hybrid fiber-optic plasmonic biosensor. (Inset 1) The crystal structure of α-MoO_3 lattice. (Inset 2) Molecular structure of BSA (Adapted with permission from Zhang et al. 2018)

not only induce sensitive resonance dip on the microfiber transmission spectrum but also provide excellent affinity to negatively charged biomolecules such as BSA, contributing to promising LOD and sensitivity.

The 2D morphology of the synthesized α-MoO_3 nanoflakes can be verified by TEM and AFM. The TEM shown in Fig. 13a clearly shows the flake-like shape of as-prepared α-MoO_3 samples. As the thickness of double-layer α-MoO_3 is 1.4 nm (the inset of Fig. 13), the synthesized nanoflakes consist of two such double-layer planner units based on the AFM assessment as shown in Fig. 13b. The H^+ inter-calation (i.e., the doping of free electrons) can be realized by adding $NaBH_4$ into the MoO_3 nanoflakes solution. During the intercalation process, the solution color gradually evolves from colorless to Prussian blue indicating the pristine MoO_3 is gradually reduced to sub-stoichiometric MoO_{3-x} (Fig. 13c). As a result, the absorption spectrum of MoO_3 nanoflakes solution is also tailored during the electrons doping. Pristine MoO_3 only introduces absorption at UV wavelengths, which is due to the large bandgap of 3.2 eV (the black curve in Fig. 13d) (Cheng et al. 2015). After electrons are increasingly doped, a distinct absorption peak appears and enhances at 700–800 nm range, in the meantime, undergoes a blueshift. This phenomenon can be explained by Drude model that the plasma frequency is inversely correlated to electron density (Liu et al. 2017c). A more intuitive way to prove the formation of sub-stoichiometric MoO_{3-x} is to characterize via X-ray photoelectron spectroscopy (XPS). In XPS measurements, pristine MoO_3 only possesses two binding energy peaks at 233.1 eV and 236.2 eV, corresponding to $Mo^{6+}3d_{5/2}$ and $Mo^{6+}3d_{5/2}$, respectively (Fig. 13e). After doping, two additional peaks corresponding to Mo^{5+} at 231.9 eV and 235.1 eV appear and coexist with the original two Mo^{6+} peaks (Fig. 13f) (Prabhakaran et al. 2016), which originate from the leftover electrons due to the removal of oxygen atoms from MoO_3 lattice.

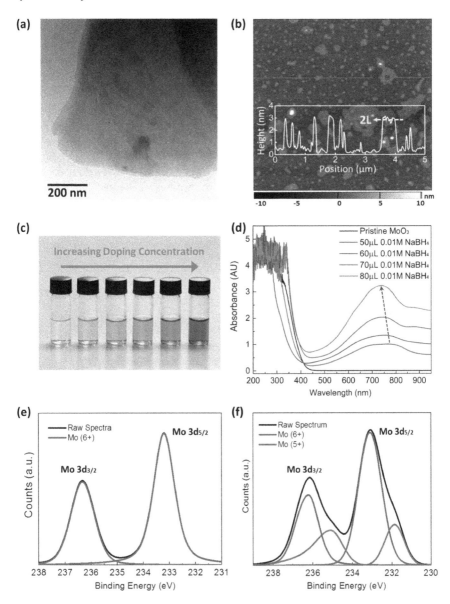

Fig. 13 **a** The TEM image of flake-like α-MoO$_3$ sample. **b** The characterization of the thickness of α-MoO$_3$ sample using AFM. **c** The color evolvement of MoO$_3$ nanoflakes solution as the doping concentration increases. **d** The evolvement of absorption spectrum of MoO$_{3-x}$ nanoflakes solution as the doping concentration increases. **e** The XPS spectrum of pristine MoO$_3$. **f** The XPS spectrum of highly doped MoO$_{3-x}$ (Adapted with permission from Zhang et al. 2018)

The MoO_{3-x} nanoflakes can be stably immobilized on the microfiber surface via electrostatic interaction. Since MoO_{3-x} is positively charged (Balendhran et al. 2013), the microfiber surface functionalized with evenly distributed negative charges (e.g., self-assembled poly(allylamine) (PAA)/poly(styrene sulfonate) (PSS) bilayer) applies strong attraction to the nanoflakes. Similarly, the immobilized positively charged MoO_{3-x} nanoflakes on microfiber surface can effectively attract negatively charged target molecules, such as BSA (Balendhran et al. 2013). Dye labeled BSA molecules are adopted to verify the effectiveness of electrostatic interaction-based target molecule adsorption as well as fiber surface functionalization. Figure 14a shows the fluorescent microscope views of four MoO_{3-x} nanoflakes deposited microfibers after immersing in different concentrations of dye labeled BSA solutions. It is obvious that fiber brightens as the BSA concentration increases. Also, the even brightness on fiber surface implies the uniformity of adsorbed BSA molecules so as the MoO_{3-x} nanoflakes.

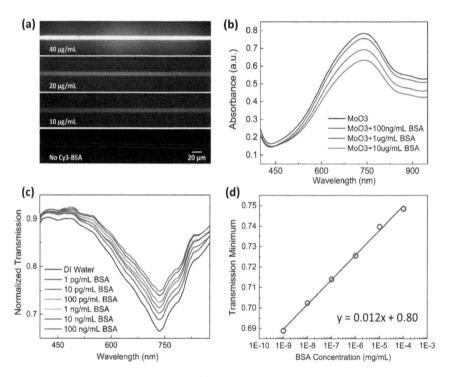

Fig. 14 **a** The fluorescent microscopic view of MoO_{3-x} nanoflakes functionalized microfibers coated with different concentrations of dye labeled BSA molecules. **b** The absorption spectra of MoO_{3-x} nanoflakes solutions mixed with different BSA concentrations. **c** Transmission spectrum variation of proposed hybrid plasmonic biosensor along with increasing BSA concentration. **d** The linear increase of transmission minimum against log-scale BSA concentration (Adapted with permission from Zhang et al. 2018)

The binding of negatively charged BSA molecules on MoO_{3-x} nanoflakes surface impacts the plasmonic behavior. When MoO_{3-x} nanoflakes suspensions mix with different concentrations of BSA solution, the absorption peak of MoO_{3-x} weakens as the BSA concentration increases (Fig. 14b). This is due to the free electrons at MoO_{3-x} surface are repelled by the negatively charged BSA molecules, resulting in the reduced free electron density involved in the plasmonic resonance (Alsaif et al. 2014, 2016; Liu et al. 2017b). Therefore, fiber-optic sensor based on MoO_{3-x} nanoflakes shows a unique characteristic that the resonance peak on fiber transmission spectrum gradually shallows along with the increasing concentration of target BSA (Fig. 14c), while other plasmonic sensors are on the contrary. Profited from the full utilization of high aspect ratio of 2D MoO_{3-x}, a LOD of BSA as low as 1 pg/mL is achieved. Moreover, the transmission minimum of plasmonic resonance peak provides linear response to the log-scale BSA concentration (Fig. 14d).

- Fiber-Optic Localized Surface Plasmon Resonance Sensors

Different from the propagating SPP at thin metal film surface, the resonant electron oscillation induced by light interacting with a metallic nanoparticle is nonpropagating due to the particle size restriction. Therefore, it is called localized SPR (LSPR). LSPR can be excited when the oscillation frequency of nanoparticle electron cloud matches with the frequency of incident light (Fig. 15) (Mayer and Hafner 2011; Peiris et al. 2016). A proper model to understand how incident light is scattered and absorbed by a nanoparticle with diameter much smaller than the wavelength is the Mie theory. The Mie theory constructs a model to deduce the extinction cross-section of nanoparticle based on the assumption that the nanoparticle is a homogeneous conducting sphere:

$$\sigma_{ext} = 9\left(\frac{\omega}{c}\right)(\varepsilon_{dielectric})^{\frac{3}{2}} V \frac{\varepsilon''_{metal}}{\left(\varepsilon'_{metal} + 2\varepsilon_{dielectric}\right)^2 + \left(\varepsilon''_{metal}\right)^2} \qquad (2)$$

where V is the volume of nanoparticle, and ε'_{metal} and ε''_{metal} are the real and the imaginary parts of metal dielectric function respectively in the Drude model (Parkins

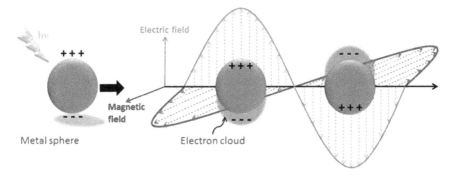

Fig. 15 The schematic diagram of localized surface plasmon resonance (Adapted with permission from Peiris et al. 2016)

et al. 1981):

$$\varepsilon'_{metal} = 1 - \frac{\omega_p^2}{(\omega^2 + \gamma^2)} \tag{3}$$

$$\varepsilon''_{metal} = \frac{\omega_p^2 \gamma}{(\omega^2 + \gamma^2)\omega} \tag{4}$$

where γ is the damping of electron oscillation and ω_p is the bulk plasma frequency. More detailed definitions of γ and ω_p can be found in (Luther et al. 2011). Since LSPR operation frequencies are generally within the visible and NIR optical windows where $\gamma \ll \omega_p$, Eq. (3) can be simplified as:

$$\varepsilon'_{metal} = 1 - \frac{\omega_p^2}{\omega^2} \tag{5}$$

Based on Eq. (2), the resonance is satisfied (i.e., the extinction cross-section is maximum) when $\varepsilon'_{metal} = -2\varepsilon_{dielectric}$. The LSPR resonant frequency is thereby expressed as:

$$\omega_{LSPR} = \frac{\omega_p}{\sqrt{2\varepsilon_{dielectric} + 1}} \tag{6}$$

Furthermore, for dielectric medium, $\varepsilon_{dielectric} = n^2_{dielectric}$. Therefore, the refractive index of ambient dielectric medium of nanoparticle impacts the LSPR resonant wavelength:

$$\lambda_{LSPR} = \lambda_p \sqrt{2n^2_{dielectric} + 1} \tag{7}$$

Similar with SPR-based fiber-optic sensing platforms, LSPR-based fiber-optic devices have also captured intensive research attention. Various fiber structures such as microfiber, cascaded unclad fiber, fiber endface have been integrated with silver or gold nanoparticles and shown promising plasmonic sensing performance (Fig. 16). To achieve efficient selectivity to analyte molecules, it is necessary to apply surface functionalization on metallic nanoparticles. Taking the most widely employed gold nanoparticle (AuNP) as an example, many effective functionalization strategies have been proven such as biomolecule coating, ligand substitution, polymer deposition (Bolduc and Masson 2011; Mieszawska et al. 2013; Tadepalli et al. 2015; Oliverio et al. 2017). However, when sensitivity is the crucial factor of plasmonic sensors, it is critical to keep the surface functionalization as thin as possible. Due to the evanescent nature of surface plasmon, thick surface functionalization considerably compromises the plasmon–matter interaction. For instance, a study has compared the LODs and sensitivities of two functionalization strategies with thicknesses of 4.24 nm and 0.96 nm, respectively, and shown that surface functionalization thinner

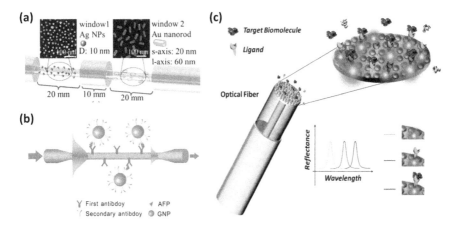

Fig. 16 The LSPR devices based on various optical fiber structures such as **a** cascaded unclad fiber (Adapted with permission from Lin et al. 2012); **b** microfiber (Adapted with permission from Li et al. 2014); **c** optical fiber endface (Adapted with permission from Ricciardi et al. 2015)

than 1 nm significantly improves the sensing performance (Tadepalli et al. 2015). In such cases, macrocyclic supramolecules has shown potential to meet the challenges of achieving both sub-nanometer functionalization thickness and target molecules recognition.

Benefited from their macrocyclic cavities, macrocyclic supramolecules like cyclodextrins (CDs), cucurbiturils, pillararenes, calixarenes show excellent molecular recognition capability by the host–guest interaction. The host–guest interaction is noncovalent interaction like van der Vaals attraction, electrostatic force, hydrophobic association. between macrocyclic supramolecules and certain guest molecules to form inclusion complexations (Qu et al. 2015). Encouragingly, it is proven that host–guest interaction is a more effective target molecule recognition and adsorption mechanism compared with the conventional biomolecule-ligand binding (Yang et al. 2014). Another advantage of macrocyclic supramolecules being surface functionalization of metallic nanoparticles is the heights of their macrocyclic cavities are normally less than 1 nm (Lagona et al. 2005; Yu et al. 2015; Xie et al. 2016), which facilitates sensitive molecular detection as discussed above. In addition, the macrocyclic supramolecules also eliminate the cytotoxicity of nanoparticles that is often favored in biomedical sensing applications. In recent years, studies have been carried out to achieve functionalizing metallic nanoparticles with macrocyclic supramolecules during the synthesis of nanoparticles instead of through post-processing surface modification. Hence, the nanoparticle formation and the surface functionalization could be realized in a more efficient one-step process. In most of these attempts (Fig. 17), however, harsh reducing reagents such as thiols, NaBH$_4$, NaOH have to be introduced which violates the purpose of achieving biocompatibility in many LSPR biosensing applications (Li et al. 2013; Sánchez et al. 2013; Bhoi et al. 2016). Therefore, inspired by the method proposed by Zhao et al. (2016), where CDs act as both reducing and capping agent for AuNP synthesis, a microfiber-based

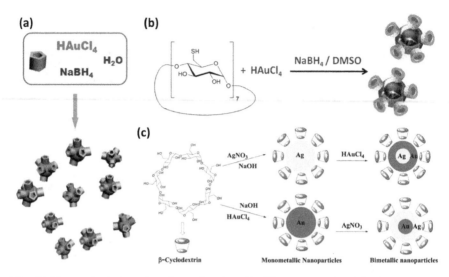

Fig. 17 The synthesis formulas of **a** carboxylatopillar[5]arene capped AnNPs (Adapted with permission from Li et al. (2013). Further permissions related to the material excerpted should be directed to the ACS); **b** CD capped AuNPs (Adapted with permission from Sánchez et al. 2013); **c** CD capped silver nanoparticles, AuNPs and Ag$_{core}$-Au$_{shell}$/Au$_{core}$-Ag$_{shell}$ bimetallic nanoparticles (Adapted with permission from Bhoi et al. 2016)

LSPR biosensor is developed to comprehensively investigate the plasmonic sensing potential of one-step synthesized macrocyclic supramolecules decorated metallic nanoparticles.

Figure 18 illustrates the configuration of LSPR biosensor based on a microfiber integrated with one-step synthesized β-CD-capped AuNPs. Functioning as both the reducing and capping agent, β-CD facilitates the AuNP formation and forms biocompatible functionalization layer via the conjunction of carboxyl groups and gold surface. More detailed synthesis procedure of β-CD-capped AuNPs can be found in (Zhang et al. 2019). The evanescent field of guide mode in microfiber leaks out at the tapered portion and interacts with immobilized AuNPs at fiber surface, leading to strong LSPR peak on the transmission spectrum. Cholesterol, the guest molecule of β-CD, is employed to validate the biosensing performance of proposed LSPR device. The sterol groups of cholesterol molecules can tightly fit into the β-CD macrocyclic cavity, meanwhile forming stable host–guest interaction through the hydrophobic associations (Yu et al. 2006; Christoforides et al. 2018).

The as-synthesized β-CD-capped AuNPs show good uniformity in particle size. As seen from the TEM image in Fig. 19a, the AuNPs possess spherical structures with diameters range from ~18 to ~21 nm. The dynamic light scattering (DLS) characterization of the AuNPs further verifies the observation (Zhang et al. 2019). The absorption peak of β-CD-capped AuNP solution is shown in Fig. 19b. The resonance band with linewidth as narrow as 47 nm also indicates the monodispersity of the particles, which is comparable with that of conventionally synthesized AuNPs.

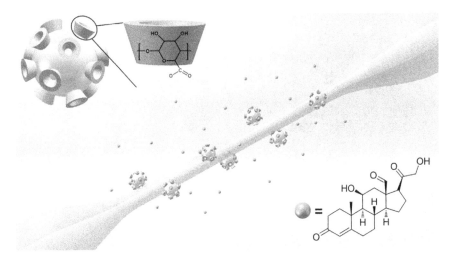

Fig. 18 The schematic illustration of the β-CD-capped AuNP-based fiber-optic biosensor. (Inset1): The conjunction between β-CD molecule and AuNP surface. (Inset2): The molecular structure of cholesterol (Adapted with permission from Zhang et al. 2019)

Based on the proton nuclear magnetic resonance (^{1}H NMR) spectrum (Fig. 19c), it can be concluded that the macrocyclic structures of β-CDs are preserved on the AuNPs (Schneider et al. 1998). Figure 19d shows the FTIR spectra comparison of β-CDs before and after the formation of AuNPs. The FTIR profiles are similar except that the hydroxyl group becomes much narrower, indicating the hydroxyl groups in β-CDs mainly contribute to reducing Au^{3+} ions to Au^0 atoms.

The as-synthesized AuNPs are negatively charged; hence, it can be stably immobilized on positively charged microfiber surface (e.g., functionalize the fiber surface with homogeneous PAA layer) via electrostatic attraction. As shown in Fig. 20a, b, the prepared microfiber is 4 μm in diameter and decorated with evenly distributed AuNPs. The attached AuNPs induce a deep resonance band centered at 530.7 nm on the microfiber transmission spectrum. When the fiber-optic sensing device is sequentially immersed in cholesterol solutions with concentrations range from 5 aM to 0.5 μM, the LSPR resonance band gradually deepens along with the increasing cholesterol concentration meanwhile the resonant wavelength shifts from 530.7 to 531.4 nm (Fig. 20c). Such ultralow LOD of 5 aM is profited from the highly efficient host–guest interaction between β-CD and cholesterol. The transmission minimum of the resonance band can be taken as the sensing parameter and provides linear response to the log-scale cholesterol concentration (Fig. 20d).

The selectivity of the proposed biosensor to cholesterol is validated by an interference study, where common interfering substances in human serum such as glutamic acid, cysteine, ascorbic acid, dopamine and human serum albumin (HSA) are introduced. Figure 20e shows the real-time average transmission intensity within 530–535 nm of microfiber when the interfering substances are introduced during the detection of cholesterol. It is clear that the β-CD-capped AuNP-based fiber-optic

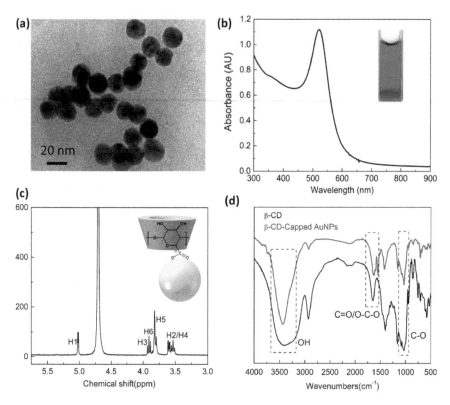

Fig. 19 **a** The TEM images of as-synthesized β-CD-capped AuNPs. **b** The absorption peak of β-CD-capped AuNP solution. **c** ^{1}H NMR spectrum of synthesized β-CD-capped AuNPs. **d** The FTIR spectra comparison between pristine β-CD and β-CD-capped AuNPs (Adapted with permission from Zhang et al. 2019)

sensor only responses to cholesterol molecules but not interfered by other substances. To further validate the cholesterol recognition capability of the proposed sensor, recovery experiments are also carried out to evaluate the accuracy of detecting real human serum samples diluted by a factor of 10^{14} and spiked with different cholesterol concentrations. As summarized in Table 1, the measurement of cholesterol concentration in unspiked human serum sample is 4.23 mM. The measurement of the same sample using commercial blood cholesterol monitor is 4.35 mM, which indicates the proposed fiber-optic biosensor is reliable. In addition, the recoveries of the spiked samples are 105.2–112.2%, which is also within a satisfactory range, further verifies the accuracy of proposed sensor. Therefore, it indicates the great plasmonic sensing potential of highly integrated fiber-optic sensors based on the macrocyclic supramolecules modified metallic nanoparticles.

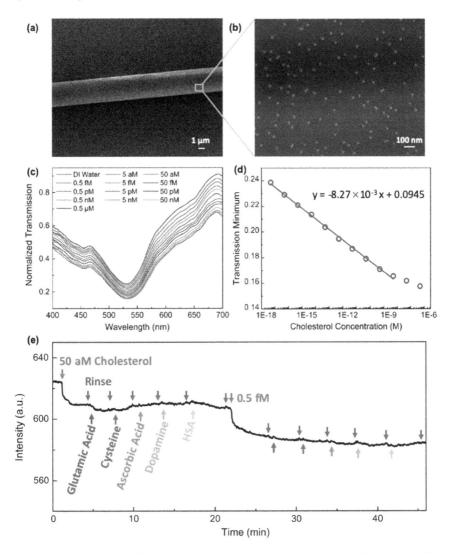

Fig. 20 **a** The SEM image of 4-μm-diameter microfiber. **b** The distribution of β-CD-capped AuNPs on the microfiber surface. **c** Transmission spectrum variation of microfiber-based hybrid plasmonic biosensor along with increasing cholesterol concentration. **d** The linear decrease of transmission minimum against log-scale cholesterol concentration. **e** The real-time average transmission intensity within 530–535 nm of microfiber when the interfering substances are introduced during cholesterol detection (Adapted with permission from Zhang et al. 2019)

Table 1 Recovery results of detecting cholesterol in human serum samples

Sample	Added (aM)	Found[a] (aM)	Recovery (%)
Human serum (male)	0.0	42.3 ± 2.8	–
	50.0	94.9 ± 6.7	105.2
	100.0	154.5 ± 16.5	112.2

[a]The values are mean of 4 independent experiments ± standard deviation

Reprinted with permission from Zhang et al. (2019)

3 Conclusion

Coupled with the advances of specialty fibers and functional nanomaterials, fiber-optic hybrid plasmonic sensors can deliver more promising sensing capability. As covered in this chapter, proper design of MOFs with high birefringence provides wide possibilities in highly integrated microfluidic sensing devices with improved measurement accuracy and stability. Profited from 2D material-based hybrid plasmonic structures, fiber-optic plasmonic sensors can deliver more promising sensing capability. The exceptional surface-to-volume ratio, near-field confinement and in situ plasmonic properties tunability of 2D materials facilitate further enhancement of plasmon–matter interaction so as the sensitivity and LOD. In addition, the development of supramolecular chemistry brings new solutions in LSPR nanoparticles surface functionalization, leading to excellent target molecule selectivity via host–guest interaction. Given the numerous possibilities in optical design and hybrid plasmonic architectures construction, fiber-optic hybrid plasmonic sensors possess vast potential in various sensing scenarios with distinct advantages of high sensitivity, flexibility, miniaturization and high degree of integration.

References

M.M.Y.A. Alsaif, K. Latham, M.R. Field, D.D. Yao, N.V. Medhekar, G.A. Beane, R.B. Kaner, S.P. Russo, J.Z. Ou, K. Kalantar-Zadeh, Tunable plasmon resonances in two-dimensional molybdenum oxide nanoflakes. Adv. Mater. **26**(23), 3931–3937 (2014). https://doi.org/10.1002/adma.201306097

M.M.Y.A. Alsaif, M.R. Field, T. Daeneke, A.F. Chrimes, W. Zhang, B.J. Carey, K.J. Berean, S. Walia, J. Van Embden, B. Zhang et al., Exfoliation solvent dependent plasmon resonances in two-dimensional sub-stoichiometric molybdenum oxide nanoflakes. ACS Appl. Mater. Interfaces **8**(5), 3482–3493 (2016). https://doi.org/10.1021/acsami.5b12076

S. Balendhran, S. Walia, M. Alsaif, E.P. Nguyen, J.Z. Ou, S. Zhuiykov, S. Sriram, M. Bhaskaran, K. Kalantar-Zadeh, Field effect biosensing platform based on 2D α-MoO3. ACS Nano **7**(11), 9753–9760 (2013). https://doi.org/10.1021/nn403241f

S. Bhat, N. Curach, T. Mostyn, G.S. Bains, K.R. Griffiths, K.R. Emslie, Comparison of methods for accurate quantification of DNA mass concentration with traceability to the international system of units. Anal. Chem. **82**(17), 7185–7192 (2010). https://doi.org/10.1021/ac100845m

V.I. Bhoi, S. Kumar, C.N. Murthy, Cyclodextrin encapsulated monometallic and inverted core–shell bimetallic nanoparticles as efficient free radical scavengers. New J. Chem. **40**(2), 1396–1402 (2016). https://doi.org/10.1039/c5nj02511g. http://xlink.rsc.org/?DOI=C5NJ02511G

O.R. Bolduc, J.F. Masson, Advances in surface plasmon resonance sensing with nanoparticles and thin films: nanomaterials, surface chemistry, and hybrid plasmonic techniques. Anal. Chem. **83**(21), 8057–8062 (2011). https://doi.org/10.1021/ac2012976

H. Cheng, T. Kamegawa, K. Mori, H. Yamashita, Surfactant-free nonaqueous synthesis of plasmonic molybdenum oxide nanosheets with enhanced catalytic activity for hydrogen generation from ammonia borane under visible light. Angew. Chem. Int. Ed. **53**(11), 2910–2914 (2014). https://doi.org/10.1002/anie.201309759

H. Cheng, X. Qian, Y. Kuwahara, K. Mori, H. Yamashita, A plasmonic molybdenum oxide hybrid with reversible tunability for visible-light-enhanced catalytic reactions. Adv. Mater. **27**(31), 4616–4621 (2015). https://doi.org/10.1002/adma.201501172

H. Cheng, M. Wen, X. Ma, Y. Kuwahara, K. Mori, Y. Dai, B. Huang, H. Yamashita, Hydrogen doped metal oxide semiconductors with exceptional and tunable localized surface plasmon resonances. J. Am. Chem. Soc. **138**(29), 9316–9324 (2016). https://doi.org/10.1021/jacs.6b05396

E. Christoforides, A. Papaioannou, K. Bethanis, Crystal structure of the inclusion complex of cholesterol in β-cyclodextrin and molecular dynamics studies. Beilstein J. Org. Chem. **14**, 838–848 (2018). https://doi.org/10.3762/bjoc.14.69

G. Giovannetti, P.A. Khomyakov, G. Brocks, V.M. Karpan, J. Van Den Brink, P.J. Kelly, Doping graphene with metal contacts. Phys. Rev. Lett. **101**(2) (2008). https://doi.org/10.1103/physrevlett.101.026803

A. Hassani, M. Skorobogatiy, Design criteria for microstructured-optical-fiber-based surface-plasmon-resonance sensors. J. Opt. Soc. Am. B **24**(6), 1423 (2007). https://doi.org/10.1364/josab.24.001423. https://www.osapublishing.org/abstract.cfm?URI=josab-24-6-1423

J. Jana, M. Ganguly, T. Pal, Enlightening surface plasmon resonance effect of metal nanoparticles for practical spectroscopic application. RSC Adv. **6**(89), 86174–86211 (2016). https://doi.org/10.1039/c6ra14173k

L. Ju, B. Geng, J. Horng, C. Girit, M. Martin, Z. Hao, H.A. Bechtel, X. Liang, A. Zettl, Y.R. Shen et al., Graphene plasmonics for tunable terahertz metamaterials. Nat. Nanotechnol. **6**(10), 630–634 (2011). https://doi.org/10.1038/nnano.2011.146

P.A. Khomyakov, G. Giovannetti, P.C. Rusu, G. Brocks, J. Van Den Brink, P.J. Kelly, First-principles study of the interaction and charge transfer between graphene and metals. Phys. Rev. B Condens. Matter Mater. Phys. **79**(19), 1–12 (2009). https://doi.org/10.1103/PhysRevB.79.195425

J.A. Kim, T. Hwang, S.R. Dugasani, R. Amin, A. Kulkarni, S.H. Park, T. Kim, Graphene based fiber optic surface plasmon resonance for bio-chemical sensor applications. Sens. Actuators B Chem. **187**, 426–433 (2013). https://doi.org/10.1016/j.snb.2013.01.040

F.H.L. Koppens, D.E. Chang, F.J. García De Abajo, Graphene plasmonics: a platform for strong light-matter interactions. Nano Lett. **11**(8), 3370–3377 (2011). https://doi.org/10.1021/nl201771h

J. Lagona, P. Mukhopadhyay, S. Chakrabarti, L. Isaacs, The cucurbit[n]uril family. Angew. Chem. Int. Ed. Engl. **44**(31), 4844–4870 (2005). https://doi.org/10.1002/anie.200460675. http://www.ncbi.nlm.nih.gov/pubmed/16052668

H. Li, D.X. Chen, Y.L. Sun, Y.B. Zheng, L.L. Tan, P.S. Weiss, Y.W. Yang, Viologen-mediated assembly of and sensing with carboxylatopillar[5]arene-modified gold nanoparticles. J. Am. Chem. Soc. **135**(4), 1570–1576 (2013). https://doi.org/10.1021/ja3115168. Direct link: https://pubs.acs.org/doi/abs/10.1021/ja3115168

K. Li, G. Liu, Y. Wu, P. Hao, W. Zhou, Z. Zhang, Gold nanoparticle amplified optical microfiber evanescent wave absorption biosensor for cancer biomarker detection in serum. Talanta **120**, 419–424 (2014). https://doi.org/10.1016/j.talanta.2013.11.085

H.-Y. Lin, C.-H. Huang, C.-C. Huang, Y.-C. Liu, L.-K. Chau, Multiple resonance fiber-optic sensor with time division multiplexing for multianalyte detection. Opt. Lett. **37**(19), 3969 (2012). https://doi.org/10.1364/OL.37.003969

W. Liu, Q. Xu, W. Cui, C. Zhu, Y. Qi, Surface plasmon resonance very important paper CO_2-assisted fabrication of two-dimensional amorphous molybdenum oxide nanosheets for enhanced plasmon resonances. Angew. Int. Ed. Chem. **450052**(6), 1600–1604 (2017a). https://doi.org/10.1002/anie. 201610708

W. Liu, Q. Xu, W. Cui, C. Zhu, Y. Qi, CO_2-assisted fabrication of two-dimensional amorphous molybdenum oxide nanosheets for enhanced plasmon resonances. Angew. Chem. Int. Ed. **56**(6), 1600–1604 (2017b). https://doi.org/10.1002/anie.201610708

X. Liu, J.-H. Kang, H. Yuan, J. Park, S.J. Kim, Y. Cui, H.Y. Hwang, M.L. Brongersma, Electrical tuning of a quantum plasmonic resonance. Nat. Nanotechnol. (2017c). https://doi.org/10.1038/ nnano.2017.103. http://www.nature.com/doifinder/10.1038/nnano.2017.103

S.D. Lounis, E.L. Runnerstrom, A. Llordés, D.J. Milliron, Defect chemistry and plasmon physics of colloidal metal oxide nanocrystals. J. Phys. Chem. Lett. **5**(9), 1564–1574 (2014). https://doi. org/10.1021/jz500440e

X. Luo, T. Qiu, W. Lu, Z. Ni, Plasmons in graphene: recent progress and applications. Mater. Sci. Eng. R Rep. **74**(11), 351–376 (2013). https://doi.org/10.1016/j.mser.2013.09.001

J.M. Luther, P.K. Jain, T. Ewers, A.P. Alivisatos, Localized surface plasmon resonances arising from free carriers in doped quantum dots. Nat. Mater. **10**(5), 361–366 (2011). https://doi.org/10. 1038/nmat3004. http://www.nature.com/doifinder/10.1038/nmat3004

K. Manthiram, A.P. Alivisatos, Tunable localized surface plasmon resonances in tungsten oxide nanocrystals. J. Am. Chem. Soc. **134**(9), 3995–3998 (2012). https://doi.org/10.1021/ja211363w

K.M. Mayer, J.H. Hafner, Localized surface plasmon resonance sensors. Chem. Rev. **111**(6), 3828–3857 (2011). https://doi.org/10.1021/cr100313v

A.J. Mieszawska, W.J.M. Mulder, Z.A. Fayad, D.P. Cormode, Multifunctional gold nanoparticles for diagnosis and therapy of disease. Mol. Pharm. **10**(3), 831–847 (2013). https://doi.org/10.1021/ mp3005885

M. Oliverio, S. Perotto, G.C. Messina, L. Lovato, F. De Angelis, Chemical functionalization of plasmonic surface biosensors: a tutorial review on issues, strategies, and costs. ACS Appl. Mater. Interfaces **9**(35), 29394–29411 (2017). https://doi.org/10.1021/acsami.7b01583

G.R. Parkins, W.E. Lawrence, R.W. Christy, Intraband optical conductivity (T) of Cu, Ag, and Au: contribution from electron-electron scattering. Phys. Rev. B **23**(12), 6408–6416 (1981). https:// doi.org/10.1103/PhysRevB.23.6408

S. Peiris, J. McMurtrie, H.-Y. Zhu, Metal nanoparticle photocatalysts: emerging processes for green organic synthesis. Catal. Sci. Technol. **6**(2), 320–338 (2016). https://doi.org/10.1039/c5c y02048d. http://xlink.rsc.org/?DOI=C5CY02048D

Y. Peng, J. Hou, Z. Huang, Q. Lu, Temperature sensor based on surface plasmon resonance within selectively coated photonic crystal fiber. Appl. Opt. **51**(26), 6361–6367 (2012). https://doi.org/ 10.1364/ao.51.006361. http://www.ncbi.nlm.nih.gov/pubmed/22968275

V. Prabhakaran, B.L. Mehdi, J.J. Ditto, M.H. Engelhard, B. Wang, K.D.D. Gunaratne, D.C. Johnson, N.D. Browning, G.E. Johnson, J. Laskin, Rational design of efficient electrode–electrolyte interfaces for solid-state energy storage using ion soft landing. Nat. Commun. **7**, 11399 (2016). https:// doi.org/10.1038/ncomms11399. http://www.nature.com/doifinder/10.1038/ncomms11399

D.H. Qu, Q.C. Wang, Q.W. Zhang, X. Ma, H. Tian, Photoresponsive host-guest functional systems. Chem. Rev. **115**(15), 7543–7588 (2015). https://doi.org/10.1021/cr5006342

A. Ricciardi, A. Crescitelli, P. Vaiano, G. Quero, M. Consales, M. Pisco, E. Esposito, A. Cusano, Lab-on-fiber technology: a new vision for chemical and biological sensing. Analyst **140**(24), 8068–8079 (2015). https://doi.org/10.1039/c5an01241d. http://xlink.rsc.org/?DOI=C5AN01241D

D. Rodrigo, O. Limaj, D. Janner, D. Etezadi, F.J. García De Abajo, V. Pruneri, H. Altug, Mid-infrared plasmonic biosensing with graphene. Science **349**(6244), 165–168 (2015). https://doi. org/10.1126/science.aab2051

S. Roh, T. Chung, B. Lee, Overview of the characteristics of micro- and nano-structured surface plasmon resonance sensors. Sensors **11**(2), 1565–1588 (2011). https://doi.org/10.3390/s11020 1565

O. Salihoglu, S. Balci, C. Kocabas, Plasmon-polaritons on graphene-metal surface and their use in biosensors. Appl. Phys. Lett. **100**(21) (2012). https://doi.org/10.1063/1.4721453

A. Sánchez, P. Díez, R. Villalonga, P. Martínez-Ruiz, M. Eguílaz, I. Fernández, J.M. Pingarrón, Seed-mediated growth of jack-shaped gold nanoparticles from cyclodextrin-coated gold nanospheres. Dalton Trans. **42**(39), 14309–14314 (2013). https://doi.org/10.1039/c3dt51368h

H.-J. Schneider, F. Hacket, V. Rüdiger, H. Ikeda, NMR studies of cyclodextrins and cyclodextrin complexes. Chem. Rev. **98**(5), 1755–1786 (1998). https://doi.org/10.1021/cr970019t. http://pubs.acs.org/doi/abs/10.1021/cr970019t

T. Schuster, N. Neumann, C. Schäffer, Miniaturized fiber-optic surface-plasmon-resonance sensor, p. 77530W (2011). https://doi.org/10.1117/12.888684. http://proceedings.spiedigitallibrary.org/proceeding.aspx?doi=10.1117/12.888684

Y. Shevchenko, T.J. Francis, D.A.D. Blair, R. Walsh, M.C. Derosa, J. Albert, In situ biosensing with a surface plasmon resonance fiber grating aptasensor. Anal. Chem. **83**(18), 7027–7034 (2011). https://doi.org/10.1021/ac201641n

B. Song, D. Li, W. Qi, M. Elstner, C. Fan, H. Fang, Graphene on Au(111): a highly conductive material with excellent adsorption properties for high-resolution bio/nanodetection and identification. ChemPhysChem **11**(3), 585–589 (2010). https://doi.org/10.1002/cphc.200900743

Y.V. Stebunov, O.A. Aftenieva, A.V. Arsenin, V.S. Volkov, Highly sensitive and selective sensor chips with graphene-oxide linking layer. ACS Appl. Mater. Interfaces **7**(39), 21727–21734 (2015). https://doi.org/10.1021/acsami.5b04427

S. Tadepalli, Z. Kuang, Q. Jiang, K.-K. Liu, M.A. Fisher, J.J. Morrissey, E.D. Kharasch, J.M. Slocik, R.R. Naik, S. Singamaneni, Peptide functionalized gold nanorods for the sensitive detection of a cardiac biomarker using plasmonic paper devices. Sci. Rep. **5**, 16206 (2015). https://doi.org/10.1038/srep1620610.1038/srep16206. http://www.nature.com/srep/2015/151110/srep16206/full/srep16206.html

J. Xie, T. Zuo, Z. Huang, L. Huan, Q. Gu, C. Gao, J. Shao, Theoretical study of a novel imino bridged pillar[5]arene derivative. Chem. Phys. Lett. **662**, 25–30 (2016). https://doi.org/10.1016/j.cplett.2016.09.010

Y.W. Yang, Y.L. Sun, N. Song, Switchable host-guest systems on surfaces. Acc. Chem. Res. **47**(7), 1950–1960 (2014). https://doi.org/10.1021/ar500022f

X. Yang, Y. Lu, M. Wang, J. Yao, An exposed-core grapefruit fibers based surface plasmon resonance sensor. Sensors (Switzerland) (2015). https://doi.org/10.3390/s150717106

Y. Yu, C. Chipot, W. Cai, X. Shao, Molecular dynamics study of the inclusion of cholesterol into cyclodextrins. J. Phys. Chem. B **110**(12), 6372–6378 (2006). https://doi.org/10.1021/jp056751a

X. Yu, Y. Zhang, S. Pan, P. Shum, M. Yan, Y. Leviatan, C. Li, A selectively coated photonic crystal fiber based surface plasmon resonance sensor. J. Opt. **12**, 015005 (2009). https://doi.org/10.1088/2040-8978/12/1/015005

G. Yu, K. Jie, F. Huang, Supramolecular amphiphiles based on host−guest molecular recognition motifs. Chem. Rev. **115**(15), 7240–7303 (2015)

S. Zeng, S. Hu, J. Xia, T. Anderson, X.-Q. Dinh, X.-M. Meng, P. Coquet, K.-T. Yong, Graphene–MoS2 hybrid nanostructures enhanced surface plasmon resonance biosensors. Sens. Actuators B Chem. **207**, 801–810 (2015). https://doi.org/10.1016/j.snb.2014.10.124. http://linkinghub.elsevier.com/retrieve/pii/S0925400514013367

N.M.Y. Zhang, D.J.J. Hu, P.P. Shum, Z. Wu, K. Li, T. Huang, L. Wei, Design and analysis of surface plasmon resonance sensor based on high-birefringent microstructured optical fiber. J. Opt. (United Kingdom) (2016). https://doi.org/10.1088/2040-8978/18/6/065005

N.M.Y. Zhang, K. Li, P.P. Shum, X. Yu, S. Zeng, Z. Wu, Q.J. Wang, K.T. Yong, L. Wei, Hybrid graphene/gold plasmonic fiber-optic biosensor. Advanced Materials Technologies (2017). https://doi.org/10.1002/admt.201600185

N.M.Y. Zhang, K. Li, T. Zhang, P. Shum, W. Zhe, W. Zhixun, N. Zhang, J. Zhang, T. Wu, L. Wei, Electron-rich two-dimensional molybdenum trioxides for highly integrated plasmonic biosensing. ACS Photonics (2018). https://doi.org/10.1021/acsphotonics.7b01207

N.M.Y. Zhang, M. Qi, W. Zhixun, W. Zhe, M. Chen, K. Li, P. Shum, L. Wei, One-step synthesis of cyclodextrin-capped gold nanoparticles for ultra-sensitive and highly-integrated plasmonic biosensors. Sens. Actuators B Chem. (2019). https://doi.org/10.1016/j.snb.2019.01.166

Y. Zhao, Z. Deng, J. Li, Photonic crystal fiber based surface plasmon resonance chemical sensors. Sens. Actuators B Chem. **202**, 557–567 (2014). https://doi.org/10.1016/j.snb.2014.05.127

Y. Zhao, Y. Huang, H. Zhu, Q. Zhu, Y. Xia, Three-in-one: sensing, self-assembly, and cascade catalysis of cyclodextrin modified gold nanoparticles. J. Am. Chem. Soc. **138**(51), 16645–16654 (2016). https://doi.org/10.1021/jacs.6b07590

Microstructured Fibers for Sensing

Nan Zhang, Georges Humbert, and Zhifang Wu

Abstract Microstructured optical fibers (MOFs), which have a holey structure in the cladding/core region, exhibit enhanced sensing sensitivity and performance for liquid/gas samples. In MOFs, the presence of sensing samples in the holey cladding/core region increases mode-field overlap and effective interaction length between the samples and the optical signals, resulting in a deep modulation on the optical signal. Moreover, in places of a bulky chamber for hosting liquid/gas samples in conventional fiber-based sensing configurations, the tiny voids in MOFs save the volume of sensing samples and avoid contaminations, making the sensing scheme more compact for in-line sensing applications. In this chapter, we first introduce the structures of MOFs and lightwave guiding mechanisms in MOFs, including index-guiding mechanism, photonic bandgap guiding mechanism, and antiresonance guiding mechanism. Then, we present MOF fabrication methods for different fiber structures and materials. Last but not least, several kinds of MOF-incorporated sensing configurations, including fiber gratings, Fabry–Pérot interferometers, Mach–Zehnder interferometers, and Sagnac interferometers, and surface-enhanced Raman scatterings, are discussed with theoretical analysis and cutting-edge achievements in a few application scenarios.

N. Zhang (✉)
Shenzhen JPT Opto-Electronics Co., Ltd, Tellhow Industrial Park,
Hi-tech Industrial Area, Guanlan Town, Shenzhen 518110, China
e-mail: zhangnan@jptoe.com

G. Humbert
XLIM Research Institute - UMR 7252 CNRS, University of Limoges,
123 Avenue Albert Thomas, Limoges Cedex, France
e-mail: georges.humbert@xlim.fr

Z. Wu
Fujian Key Laboratory of Light Propagation and Transformation,
College of Information Science and Engineering, Huaqiao University,
Xiamen 361021, China
e-mail: zfwu@hqu.edu.cn

© Springer Nature Singapore Pte Ltd. 2020
L. Wei (ed.), *Advanced Fiber Sensing Technologies*,
Progress in Optical Science and Photonics 9,
https://doi.org/10.1007/978-981-15-5507-7_3

Keywords Photonic crystal fibers · PCF fabrication methods · Long period fiber grating · Fiber interferometer · Fiber surface-enhanced Raman scattering sensing

1 Introduction

Optical fiber sensing networks have distinct features of high robustness, immunity to electromagnetic interference, small size, good repeatability, and enabling remote sensing configurations (Consales et al. 2012). In particular, optical fibers are made of silica. It is a dielectric material and has high melt temperature (~1700 °C), making optical fibers immune to the ambient electromagnetic signals and harsh environment. By possessing a small cross section of hundred micrometers and a low transmission loss, optical fibers show great flexibility and benefits in constructing networks for remote sensing applications. Moreover, the advanced fiber fabrication technology leads to good consistency in the fiber geometry over long length and helps to save cost for mass production. Owing to these distinguished features, fiber-based sensing platforms have attracted intensive research and exploration, and rapidly being applied widely from environment monitoring to biological detections.

Conventional single-mode fibers (SMFs) and multimode fibers (MMFs) have a germanium-doped silica fiber core surrounded by a pure-silica cladding. The refractive index (RI) of fiber core is larger than that of fiber cladding, and light guidance in the fiber core relies on total internal reflection (TIR). Though SMFs and MMFs have been successfully applied to temperature (Peng et al. 2016), strain (Zhu et al. 2012), curvature (Mao et al. 2014), and RI (Rong et al. 2012) analysis, the extremely weak light–matter interaction property hampers their spectral response's sensitivity to ambient variations, hence, making the sensing scheme exhibits either low sensing sensitivity or redundant in fiber length. A few techniques have been applied to conventional fibers to enhance light–matter interaction, such as fiber gratings, microbubbles, microfibers, and side-polished fibers. These microstructures on fibers have low fabrication reproducibility and are fragile, causing difficulties in integration and instability of sensing networks. Besides, all-solid fiber geometry requires external chambers for liquid flow, which compromises compactness of the systems.

To address these challenges and enhance light–matter interaction, MOFs are proposed and have been investigated for sensing applications in the past two decades. MOFs, mainly refer to photonic crystal fibers (PCFs), enable light trapping in the solid/hollow fiber core through modified TIR, photonic bandgap (PBG) or antiresonance etc. by the periodic microstructures in the fiber cladding region. The scanning electron microscope (SEM) images of a typical pure-silica solid-core PCF and a PBG fiber are shown in Fig. 1a, b, respectively. Because the main motivation when the idea of PCFs was first proposed was to fabricate a hollow-core fiber that confines light through PBG effect enabled by a cross-sectionally periodic photonic crystal structure, the term PCF was introduced and remains in use for the whole class of microstructured fibers, either with or without any kind of "crystal" structure.

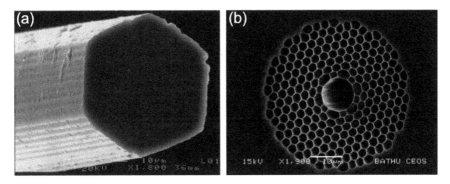

Fig. 1 SEM photograph of **a** the first solid-core index-guiding PCF (Adapted with permission from Knight et al. 1996); and **b** a typical hollow photonic bandgap PCF (Adapted with permission from Lynch-Klarup et al. 2013)

The holey structure of PCFs provides inherent advantages in improving the compactness and performance of fiber-based sensors. In particular, firstly, the air channels can host the liquid/gas media. And, as the air channels generally have a diameter of a few micrometers or tens of micrometers, relatively small sample volume is needed for completing the sensing/detection. Secondly, the large hollow core or the air channels tightly surrounding the solid fiber core enable a large overlap between the sample in the channels and the electrical field of the light beam, thus to dramatically enhance the modulation capability to the light signal, resulting in a better sensing resolution. Thirdly, the unprecedented long effective interaction length enabled by the air channels running along the entire fiber length can contribute to improve the limit of detection in absorption-based sensing configurations. Besides, PCF-based sensors are especially suitable for in-line real-time sensing applications in which the liquid/gas samples continuously flow through the sensing region via inlet and outlet, and the spectrum is recorded and analyzed in real time.

Over the past few decades, the main research interests of PCFs-based sensors has been focused on increasing sensing sensitivity, constructing in-line sensing configurations, and expanding the application scenarios by exploring new PCF structures and integrating other sensing techniques like fiber gratings and interferometers. In terms of exploring new fiber structure, great efforts have been put on enlarging evanescent field in the air channels and easy accessibility of liquid/gas samples to the internal air channels. For example, suspended-core fiber, with 3 larger air channels surrounding a small solid-core region (as shown in Fig. 3a), has been intensively explored for high-sensitivity molecular detection because of its extremely large evanescent power ratio in the air channels. The simple wheel-like structure makes the lateral access of liquid easy as well. The integration with other techniques enables higher sensitivity sensing, and wider application scenarios like RI sensing, or extracting structural fingerprinting information of biological molecules.

In this chapter, we first give a brief introduction to the fundamentals of PCFs, including lightwave guiding mechanisms, the key geometric parameters that affects its optical properties, and advanced fiber fabrication methods. Then, we will summarize a few most compact and highly sensitive liquid/gas sensing configurations that adopt PCFs, including fiber gratings, fiber interferometers, and surface-enhanced Raman scattering, and their cutting-edge achievements in versatile sensing scenarios.

2 Guiding Mechanisms of PCFs

Lightwave guidance mechanisms in the core region of PCFs are mainly classified as index guiding, PBG guiding, and antiresonance guiding.

- Index guiding

In index-guiding PCFs, light is confined in the core area through modified TIR. TIR is a phenomenon that the total amount of the incident light energy is reflected back at the boundary between two different media. It only occurs when the lightwave travels in a high RI medium approaches the other medium with a lower RI at an angle of incidence larger than the critical angle. Normally, index-guiding PCFs are made of pure silica. The RI difference between the core region and the cladding region is obtained by creating periodic holey structures in the cladding region to reduce the averaged RI of the cladding region. A typical structure of an index-guiding PCF is illustrated in Fig. 2. The presence of air holes decreases the overall RI value of the cladding region, because air has a low RI value (1.0) than that of silica (e.g., 1.444 at the wavelength of 1550 nm). The pitch of the air channels (distance between two adjacent air channels) is represented by symbol Λ, and the diameter of air hole

Fig. 2 A schematic illustration of an index-guiding PCFs made of pure silica. The green color represents the silica material, and the white regions are air

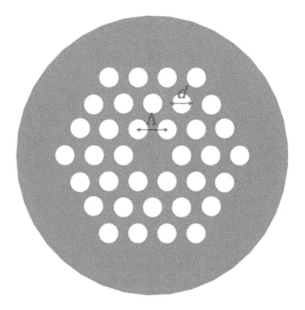

is indicated by d. Simulation results suggest that a smaller pitch and a larger air hole diameter can help to increase the evanescent power fraction in the air channels located in the innermost ring. And, increasing the number of rings helps to reduce the fiber transmission loss.

The mode number analysis of index-guiding PCFs can refer to that of step-index fibers. The number of transverse modes (M) that are supported in the step-index conventional fiber core can be theoretically predicted by the following equations:

$$V = \tfrac{2\pi}{\lambda} a \sqrt{n_{\text{core}}^2 - n_{\text{cladding}}^2} \tag{1}$$

$$M = \tfrac{V^2}{4} \tag{2}$$

where a is the radius of the fiber core, λ is the wavelength of the light wave, and n_{core} and n_{cladding} refer to the RIs of the fiber core and cladding regions, respectively. V number is the normalized frequency parameter. When $V < 2.405$, step-index fibers only support single-mode propagation, which is the fundamental mode. For a typical PCF structure which has periodic air holes in the cladding, the RI of the cladding can be estimated through an effective index model (Knight et al. 1998). In the effective index model, the RI of the cladding is defined according to the propagation constant of the fundamental space-filling mode of the infinite cladding, and d/Λ play an important role in determining the number of modes. An effective V value (V_{eff}) for a PCF can be defined as (Birks et al. 1997; Vengsarkar et al. 1996):

$$V_{\text{eff}} = \tfrac{2\pi}{\lambda} a \sqrt{(n_{\text{core}}^2 - n_{\text{a}}^2)F} \tag{3}$$

where, n_{a} is the RI of air ($n_{\text{a}} = 1$), and $F = d/\Lambda$ is the air filling ratio.

- PBG guiding

There is another kind of PCFs in which the fiber core exhibits a smaller RI than that of the cladding region. Obviously, TIR does not work in this situation. In this kind of PCF, light of a certain band of wavelength is confined in the low index core by the PBG effect of the photonic crystal structure in the cladding region. Therefore, it is categorized as PBG fibers. Usually, PBG effect only guarantees relatively narrow transmission bandwidth. The first hollow-core photonic bandgap structure fiber was fabricated in 1998. However, the guided mode in this fiber was observed to be evanescent in the central air channel. To support stable light propagation in the central air core, numerical calculations reveal that a large air filling ratio and a small pitch in the surrounding photonic crystal cladding structure are necessary (Cregan 1999). Later, in 1999, R. F. Cregan in the University of Bath successfully fabricated the first photonic bandgap guiding fiber that supported single-mode propagation in the air core (Cregan 1999). In this fiber, the large hollow core was obtained by removing 7 capillaries in the center, and the air filling ratio and pitch in the cladding are ~39% and 4.9 μm, respectively. However, the researchers only receive 35% of an input laser energy after propagating through a 4 cm long fiber sample due to low input coupling efficiency and the scattering loss resulted from surface roughness of the inner ring of cladding. In 2005, an ultimate low-loss photonic bandgap fiber (1.2 dB/km) was obtained by reducing surface roughness (Roberts et al. 2005).

In addition, Bragg fibers and all-solid photonic bandgap fibers are also proposed and investigated for exploring new optical properties. Bragg fibers are usually made of polymer and soft glasses instead of silica. Light is confined in the large hollow core by the surrounding multilayer reflector consist of alternating high index (soft glasses) thin films and low index films (polymers). A Bragg fiber that is used for CO_2 laser beam delivery (Temelkuran et al. 2002) is shown in Fig. 3b.

All-solid photonic bandgap fiber structure is realized by replacing the air holes of index-guiding PCFs with high index material (larger than core), as shown in Fig. 3c. Compared with holey structured PBG fiber, all-solid PBG fibers are easier to fabricate, and it shows better coupling performance especially when splicing with standard SMFs or MMFs.

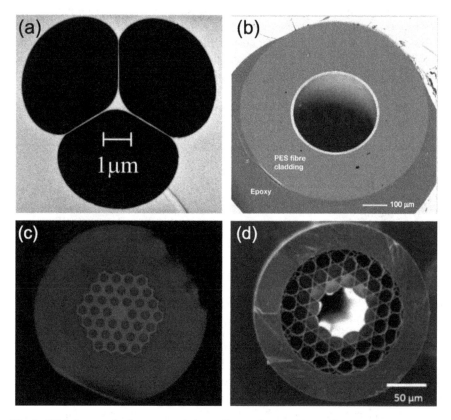

Fig. 3 SEM photograph of **a** a suspended core fiber (Adapted with permission from Monro et al. 2010); **b** a Bragg fiber (Adapted with permission from Temelkuran et al. 2002); **c** all-solid PCF (Adapted with permission from Wang et al. 2015); and **d** Kagome-like hollow-core PCF (Adapted with permission from Subramanian et al. 2017)

- Antiresonance

The first hollow-core PCF capable of transmitting light over a broad range of optical frequencies was demonstrated in 2002 (Benabid et al. 2002). The cladding structure of this fiber consists of thin (nanometer scale) silica webs arranged in a Kagome lattice instead of conventional triangular hollow-core PCF structures reported previously. But numerical simulation reveals that the cladding structure does not support any PBGs. Further research progress on broadband-guiding hollow-core PCF indicates that: 1. The geometry of the cladding is not the crucial factor in exhibiting broadband guidance in the Kagome-like hollow-core PCFs; 2. in contrast to hollow-core PBG fibers, the transmission loss of this kind of fiber is independent of the number of cladding layers; and 3. the geometry of the first silica layer surrounding the hollow core is essential for determining the performance of the broadband-guiding hollow-core PCFs (Markos et al. 2017). There are a few other types of hollow-core PCF structures that support broadband spectrum guidance, such as negative-curvature hollow-core PCFs or single-ring hollow-core PCFs. According to current research, broadband-guiding hollow-core PCFs does not show a unique cladding design and neither has an accepted guidance mechanism. It may be composed of a Kagome-like (Fig. 3d), a honey-comb lattice or a reduced or even single layer of annular tubes (Pryamikov et al. 2011). The guidance mechanisms in broadband-guiding hollow-core PCFs are also still a subject of study or debate, commonly perceived as inhibited-coupling or antiresonance guiding.

3   PCF Fabrication Technology

The fabrication of PCFs is proved far from easy. It took a long journey to succeed until the year of 1996 when Prof. J. C. Knight and his colleagues reported that they successfully fabricated the first silica-air PCF in the University of Southampton by means of capillaries stacking (Knight et al. 1996). This PCF has a solid core surrounded by periodic air arrays in the cladding region, and it confines light through TIR (index guiding) the same as the conventional SMFs and MMFs. Soon after, with the successful experience of fabricating index-guiding PCFs, the first PBG PCF was achieved in 1998.

The fabrication process of a fiber normally involves two basic steps: preform preparation and drawing process with a fiber drawing tower. The preform has the same structure and RI profile as the final fiber sample, but larger in size. The diameter of the cross section of a fiber preform is around a few centimeters.

Depending on different fiber structures and materials, a few techniques have been practiced to prepare the large fiber preform, such as stacking (Russell 2003), extrusion (Kumar et al. 2002), and rolling (Temelkuran et al. 2002).

- Stack-and-draw method

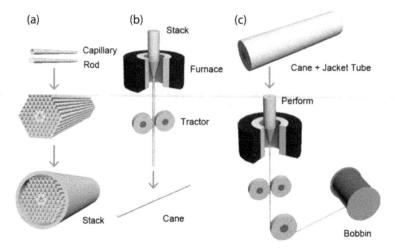

Fig. 4 A brief schematic diagram of stack-and-draw method for the fabrication of PCFs. **a** Assembling silica capillaries and rods into macroscopic preform stack. **b** The large preform stack is drawn into canes with a diameter of a few millimeters. **c** The final thermal draw step to achieve PCFs with a diameter of a few hundred micrometers (Adapted with permission from Ma et al. 2017)

In terms of fabricating silica PCFs preform, the most commonly used method is capillary/rod stacking. Capillaries are small silica tubes that have the same air filling ratio with that of the PCF cladding region. They are generally 1 m long with an outer diameter of a few millimeters and obtained from drawing large silica tube/rods. For preparing a preform with stacking method, first, capillaries and rods are stacked following the designed photonic crystal pattern. Then, the stack is inserted into a larger silica tube to provide an outer protecting silica layer. Next, the preform will be mounted in the preform holder on the top of a fiber drawing tower and drawn to be thinner cans/fibers when fed into a high-temperature furnace (operating at above 1700 °C). Figure 4 shows the procedure of a fiber drawing process.

Because the thermal expansion coefficients of silica and air are quite different and the draw-down ratio from the preform to fiber is large, to prevent the breaking of preform or damage of the fiber structure, the complete drawing process normally involves two steps:

1. Feed the preform into the high-temperature furnace and achieve a smaller-sized cane which has an outer diameter of a few millimeters.
2. Feed the cane into the furnace again and draw to obtain regular fibers following adding an extra silica jacket to achieve thicker outer cladding region.

To obtain a perfect fiber structure, the drawing process is very challenging. The feeding speed (speed of the preform/cane, V_{seed}) and drawing speed (speed of the fiber, V_{fiber}) of the preform together determine the draw-down ratio of the drawing process by the following relationship:

$$V_{\text{seed}} R_{\text{seed}}^2 = V_{\text{fiber}} R_{\text{fiber}}^2 \qquad (4)$$

where R_{seed} and R_{fiber} are the radii of the seed preform/cane and the produced cane/fiber, respectively. Moreover, to perfectly maintain the fiber structure, the furnace temperature, the pressure in the air channels, and vacuum environment in some circumstances need to be precisely managed. In particular, for holey structured PCFs, when the temperature goes too high, the air in the channels expand rapidly and may destroy the thin silica webs in the microstructure. Inversely, the fiber is easy to break in a low temperature applied. The temperature applied to the preform/cane is crucial. It relies on both of the temperature of furnace and the feeding speed of the preform as fast feeding speed reduce heat accumulation in the fiber. The pressure applied in the air channel helps to maintain the fiber structure, and the vacuum applied to the preform tube in some circumstances is intended to suppress the interstitial holes in the joints of silica web.

Therefore, to obtain a fine fiber structure, those parameters including furnace temperature, the pressure applied, feeding speed of preform, and drawing speed need to be precisely adjusted. And this requires a few rounds of trials and improvements. Besides, the large surface area of capillaries and rods in the preform stack could introduce contamination easily. But this method of making preform allows large flexibility in fiber structure design.

- Extrusion

Extrusion method can be applied to realize silica-air preforms that are not readily achievable by stacking capillaries and rods. In the preform fabrication process, a molten glass is penetrated through by a die having the desired patterns of holes. Therefore, it is especially suitable for fabricating fiber preform made of low melting temperature materials, for example, chalcogenide glass, polymers, or those materials which are not readily available in tube form. Then, the bulky glass preform is drawn using a thermal-drawing tower which is the same to the one shown in Fig. 4b. This kind of fabrication method allows the fabrications of any regular or irregular structured fibers. The first non-silica glass PCF using extrusion method is made of glass SF57 (Kiang et al. 2002). It exhibits a "wheel"-like structure with 3 large air channels surrounding a small solid core, and observed to be single mode over a broad wavelength range.

Extrusion provides a controlled and reproducible method for fabricating holey preform with complex structure and good surface quality. By broadening the applicable range of materials and structure designs, extruded holey fibers offer a wider range of optical properties compared with the stacking method. Moreover, as fewer interfaces are involved, extrusion shows the potential to minimize the ultimate transmission loss by avoiding contamination.

- Rolling method

Rolling is perfectly suitable for the fabrication of fiber with periodic annular layers structure, like Bragg fibers or omnidirectional fibers. In this method, multilayers of polymer-based mirrors are rolled up to present a preform with a hollow core and alternate high refractive index and low index in the cladding region. The index variation in the radial direction provides the light guidance capability in the hollow fore region through photonic bandgap mechanism.

This method is widely used in fabricating preform that combines polymer and glass materials (Temelkuran et al. 2002). And the wavelength scalability (the control over transmission band through the fiber's geometrical parameters) is available by changing the thickness of the thin layers. A. M. Stolyarov et al. demonstrated an omni-directional PBG fiber that confines the visible wavelength for trace vapor sensing (Stolyarov et al. 2012). This fiber was thermally drawn from a preform comprised of 15 bilayers of chalcogenide glass (As_2S_3) and polyetherimide (PEI) and a cladding made of PEI.

4   Processing for in-Line Access

The in-line accessibility to the PCF channels is important for prompting the use of PCF for sensing applications. Because it speeds up getting sensing results, ensures the reusability of fiber samples, and also eliminates the measurement error introduce when realign a new fiber sample. A considerable part of PCF-based in-line sensing configuration is constructed in this form that: The light beam is coupled into the fiber core through precise position alignment (free space coupling), while both of the PCF fiber ends are housed by microfluidic reservoirs, as illustrated in Fig. 5a. The liquid flow through the PCFs is forced to continue by the pressure difference between the two reservoirs. In this way, simultaneous liquid flow in all the air channels of PCF and light coupling to the fiber core are realized. This configuration can also be applied to selective liquid filling into the air channels in PCFs, with a pretreatment to block the rest of the air channels. However, the precise free space light coupling requires high mechanical robustness and might be subject to the pressure perturbation in the microfluidic reservoir. There is another attempt to insert and splice a 20-μm-width C-shaped fiber between the PCF and SMFs to create liquid inlet and outlet, as shown in Fig. 5b. The splicing connection is much more robust, but this method requires precise manipulation on very tiny C-shaped fiber, which is difficult. Besides, laser micromachining or coupled with hydrogen fluoride (HF) etching can drill a hole from the side of the PCF to enable the liquid flow into the air channels. This method makes it possible to choose any drilling positions, but it may destroy the fiber structure, introduce contamination, and most importantly only the outer rings of air channels are accessible.

In addition to the post-processing methods, researchers also have proposed novel fiber structures to enable convenient liquid access to the air channels in PCFs. In 2011, H. W. Lee et al. in Prof. P. Russell's group proposed a step-index fiber with a parallel hollow micro-channel (Lee et al. 2011) to provide convenient liquid insertion and avoid complex selective channel introduction process. Also, a side-channel PCF (SC-PCF) is proposed to provide easy access to the lateral large side channel and

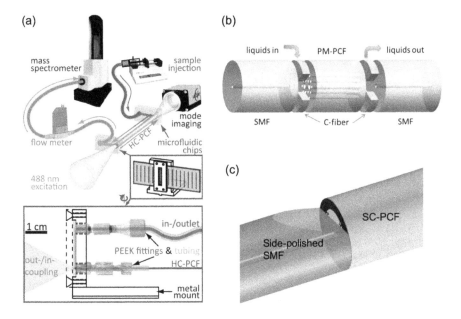

Fig. 5 A schematic diagram showing the methods to enable simultaneous light coupling and in-line liquid access to PCFs: **a** free space coupling (Adapted with permission from Unterkofler et al. 2012); **b** inserting a C-shaped fiber between SMF and PCF (Adapted with permission from Wu et al. 2013); and **c** splicing the SC-PCF to side-polished SMF (Adapted with permission from Zhang et al. 2016a, b)

large evanescent power in the side channel for improving sensing performance. The structure of SC-PCF is elaborately designed to have a larger channel in one side of the PCF, so that the simultaneous light coupling and liquid access can be enabled through splicing it with a side-polished SMF (as shown in Fig. 5c). The side-polished SMF is manufactured by polishing a glass cuboid with a bare SMF buried in its groove with wax, and its produced geometric parameters are well designed to expose the side channel to the external environment only.

5 Sensing Applications

PCF-based fiber sensing configurations have been applied to physical sensing, chemical sensing, and biomedical sensing, owing to its diversity in optical waveguide properties. Among these, aqueous sensing is particularly attractive as the voids in PCF are natural liquid carrier. The analysis of aqueous/gaseous samples can be done through tracing a few parameters: absorption loss of light beam it introduced, RI, photoluminescence phenomenon, Raman spectrum, etc. Measuring the RI of aqueous sample is a very important method to analyze the presence of analytes in the aqueous solutions and its concentrations. It is a label-free detection method and is easy to implement and commonly realized through monitoring the resonant wavelength shift of a long period grating (LPG) or interferometry since the RI of liquid in the sensing region has an impact on the effective refractive index (ERI) of the transverse modes and the distribution of the modal energy. With the increasing demands of specific detection of chemical and biological molecules, some researchers have explored specific detection on DNA or RNAs by adding preparations on fiber/core surface in RI sensors. Another technique, surface enhance Raman scattering (SERS), also has attracted large interests, for its capability of revealing fingerprinting information of the molecule with high sensitivity by directly analyzing the Raman spectrum collected from the fiber ends.

In this section, we will focus on LPG fiber sensors and interferometric fiber sensors, as well as SERS fiber sensors which particularly interest the current research and also make up a large proportion of PCF based sensors.

• Long period gratings fiber sensors

LPG is a structure of periodic RI variation along the fiber axial direction. The structure has a period in the sub-millimeter scale and is able to realize mode energy coupling between two co-propagating transverse modes at a certain resonant wavelength (λ_{res}) when satisfy the phase matching condition:

$$n_{eff}^1 - n_{eff}^n = \frac{\lambda_{res}}{\delta} \qquad (5)$$

where, n_{eff}^1 and n_{eff}^n are the ERIs of the coupling transverse modes. δ is the pitch of the grating structure. Therefore, if a gating pitch has been chosen to realize the energy coupling from fundamental mode to higher-order core mode or cladding modes, there will be a series of low power bands in the transmission spectrum as high-order modes experience larger transmission loss.

According to Eq. (5), we can clearly see that with a certain pitch, the resonant dips are dependent on the difference between the ERIs of the two coupling modes which can be affected by the RI of liquid filling in the channels of PCFs. In this way, the variation in the RI of the liquid sample can be monitored through tracing the shift of LPG dip wavelength.

Fiber LPGs can be fabricated through several methods, for example, point-by-point laser writing method (Wang et al. 2006), mechanical pressure (Hu et al. 2012a, b), and electric arc technique (Humbert et al. 2003). Among of these, CO_2 laser point-by-point imprinting method is the most commonly used one. The CO_2 laser irradiation imposes thermal effect on fiber surface and can affect the RI of the PCF via three mechanisms:

1. Release stress introduced during the fiber fabrication process
2. Cause slight deformation on the fiber surface
3. Introduce stretch force when heating up the silica fiber.

With the combination of a computer-controlled precise moving stage, CO_2 laser point-by-point imprinting method enables great flexibility in fabricating LPGs with different pitch, grating numbers, and writing position to control over the modulation depth and resonant wavelength of the LPGs. The grating obtained with laser inscription is permanent, so it can be established for long time and stable sensing applications.

PCF LPGs sensing configurations provide reliable RI measurements with high sensitivity. Attempts have been made to inscribe LPG on various kinds of PCFs. It is reported that fiber gratings on the large-mode-area PCF are able to deliver a high RI sensitivity of 1500 nm/RIU near a RI value close to water (1.33), which is nearly 2 order of magnitude higher than that can be obtained with SMF LPG (Rindorf and Bang 2008). In this work, the authors fabricate an LPG on a large-mode-area PCF which has a solid core and an air channel lattice cladding with the pitch and the ratio of air hole diameter to pitch being 7.12 μm and 0.478, respectively. The resonance occurs at the wavelength of 1050 nm with a grating period of 580 μm and period number of 60. The RI sensitivity is characterized by tracing the blue shift amount of the resonant wavelength when tuning the RI of the methanol solution infiltrated in the channels through temperature.

In-line access of liquid into the air channels of PCF LPGs enables direct measurements of RI sensitivity through dynamic liquid infiltration. An LPG inscribe on a side-channel PCF (SC-PCF) designed to enable direct and highly sensitive RI measurement (Zhang et al. 2016a, b). The SC-PCF enables a large side channel adjacent to the solid core for fast liquid infiltration of the test sample and large evanescent power. The LPG on SC-PCF has a grating period of 153 μm and a period number of 50, resulting in the modal energy coupling between LP_{01} and LP_{11} mode. The resonance dip in the transmission spectrum is centered at 1520 nm with a full bandwidth at half maximum (FWHM) of ~21 nm and an extinction ratio of ~7 dB. The RI sensitivity is characterized to be 1145 nm/RIU over the RI range of 1.3330–1.3961 through the infiltration of RI solutions (sodium chloride solutions with different concentrations) into the large side channel adjacent to the solid core. This continuous-flow configuration makes this sensing platform very promising for real-time monitoring.

In addition to non-specific RI sensing, PCF-based LPG sensors are also reported for specific biological sensing application. In 2006, L. Rindorf et al. demonstrated the use of a PCF LPG to detect the thickness of poly-L-lysine and double-stranded DNA immobilized on the inner wall of the air channels through electrostatic (Rindorf et al. 2006). Because the test molecules are deposited on the side of the holes where they interact with intense evanescent wave, the sensor is regarded more robust to environmental temperature variation and the influence of the liquid in the air holes. In 2011, Z. He et al. reported an endless single-mode PCF LPG sensor that realized specific goat anti-mouse IgG (H + L) antibodies. It, for the first time, realized a PCF LPG that carry out multilayer biomolecular binding events in the air channel of PCF cladding (He et al. 2011) and each step of binding event on the wall of channels cam be monitored through the shift of LPG resonant wavelength. Though the sensing/detection of biomolecules with fiber LPGs is essentially attributed to the modulation of sample RI to the resonant wavelength, immobilization of multilayer biomolecular binding method eliminates the perturbation from other molecules.

PCF LPG sensors are compact in configuration and offer relatively high sensitivity. Taking advantages of the liquid voids in PCF, it is very promising in serving as a platform for real-time and online measurements, as well as for biological and chemical sensing which may require multilayer treatment on the silica surface for specific identification.

- Interferometer sensors

In optical fibers, interference between two propagation modes can be realized in a single fiber or in two separate fibers as long as the two modes travel over different optical paths. Assume the two interference beams are split from a single beam, which means that there is no phase difference at the very beginning. After passing through different optical paths, the recombined output intensity I can be predicted by the following equation:

$$I = I_1 + I_2 + 2\sqrt{I_1 I_2} \cos(\Delta \psi) \tag{6}$$

$$\Delta \psi = \frac{2\pi}{\lambda}(n_1 l_1 - n_2 l_2) \tag{7}$$

where I_1 and I_2 are the intensities of the two interference beams at the position of recombination, respectively. n_1, l_1 and n_2, l_2 are ERIs of the two modes and the fiber lengths they pass through. $\Delta \psi$ represents the optical phase difference (OPD) between the two beams, and λ is the wavelength. From the equation, we can identify that the output intensity shows periodic resonant dips/peaks in the transmission spectrum. Because OPD is relevant to ERIs of the propagation modes, the resonance dips will shift in response to the RI variation in the sensing region.

PCF-based interferometric RI sensing configurations includes Fabry–Pérot inter-ferometers (FPI), Mach–Zehnder interferometers (MZI), and Sagnac interferom-eter (SI). In particular, a concave-core PCF constructed FPI sensor is reported to exhibit an RI sensitivity of 1635.62 nm/RIU (Tian et al. 2016). The FP sensor is built by splicing a segment of concave-core PCF with a SMF, and the shallow concave-core of the fiber form the FP cavity directly. J. J. Hu et al. demonstrate a highly sensitive PCF MZI by splicing both ends of a PCF with SMFs using fusion arc splicing technique. The air channels of the PCF are completely collapsed in both splicing region, serving as beam splitter and beam combiner to excite mode coupling to/from high-order cladding modes (Hu et al. 2012a, b). Since the evanescent field of the high-order cladding mode extends to the external medium, the RI of the external medium affects the ERI of the cladding mode, whereas not that of the core modes. The RI variation will induce OPD change and lead to variation in the transmission spectrum. The RI sensitivity was demonstrated to be 320 nm/RIU over the biosensing RI range (1.33–1.34) by immersing the PCF in different RI solutions. In 2013, T. Han et al. reported a selective-filling PCF-based Sagnac interferometer (Han et al. 2013). The two channels in the innermost ring of the fiber cladding are selectively filled with liquids that has a higher RI index than silica to guide light by a combination of index-guiding and bandgap guiding. It is characterized to enable an RI sensitivity of 63,882 nm/RIU by heating the fiber with a temperature chamber. C. Wu reports an in-line optofluidic polarization maintaining PCF-based SI sensor. It allows simultaneous light injection into the fiber core and liquids flow in and out by inserting C-shaped fibers between PCF and SMFs (Wu et al. 2014). The schematic diagram is shown in Fig. 5b. By changing the RI solutions (sodium chloride solutions) that flows through the fiber from the opening, its RI sensitivity is measured to be 8699 nm/RIU.

In 2018, N. Zhang et al. presented a side-channel PCF SI for ultra-sensitive chemical and biological analysis (Zhang et al. 2018). The geometry of the side-channel PCF is shown in Fig. 6a. To increase the fraction of evanescent power in the large side channel and maintain a relatively small insertion loss when spliced with SMFs, the pitch and the ratio of air hole diameter to the pitch are fabricated to be 6 μm and 0.6, respectively. To enable lateral liquid access to the large side channel for the convenience of liquid flow in and out, the side-channel PCF is spliced with a side-polished SMF to create an opening. The sensing scheme is constructed in this way to support specific chemical and biological analysis that require multilayer treatments on the silica surface. Figure 6c shows the resonance in the transmission spectra when the side channel is empty and filled with water. RI sensitivity is measured to be 2849 nm/RIU. Specific sensing performance with side-channel PCF Sagnac interferometer is characterized using human cardiac troponin T (cTnT) protein detection.

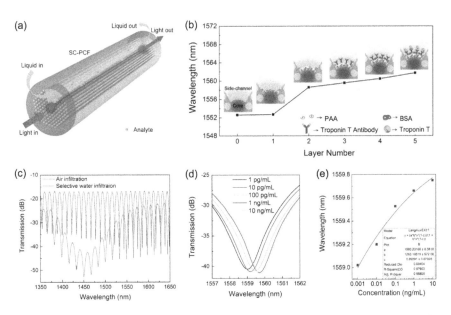

Fig. 6 **a** Schematic diagram of the sensing configuration; **b** resonance dips recorded along with the multilayer deposition; **c** transmission spectrum of the side-channel PCF interferometer; **d** resonant wavelength shift corresponding to different antibody concentration, and **e** their correlation relationship (Zhang et al. 2018). Reproduced by permission of The Royal Society of Chemistry

Specific cTnT detection needs a series immobilization of OH^- group, a monolayer of poly(allylamine), active cTnT antibody, and cTnT antibody in sequence. Before each immobilization step, distilled water or phosphate-buffered saline (PBS) is used to rinse the residuals from the previous step. And resonance wavelength variations are recorded in real time after each binding step of the layers, as recorded in Fig. 6b. Eventually, cTnT solutions are infiltrated in an ascending order of concentration from 1 pg/mL, and the shift of resonance dips is plotted in Fig. 6d, e, revealing a limit of detection of 1 ng/mL.

Overall, PCF-based interferometers exhibit characteristics of easy fabrication, simple structures, high integration level, and high sensitivity. Similar to PCF-based LPG sensors, they deliver a highly efficient solution to chemical and biological analysis for environmental monitoring or biological/medical diagnosis.

- Surface-enhanced Raman scattering sensing

Raman spectrum reveals the information of a molecule in vibrational mode level which relies on the molecule's constituent atoms, chemical bind strengths, and their structural arrangements. As each kind of molecules has its unique structures, the Raman spectrum is specific to one kind of molecules. Surface-enhanced Raman scattering (SERS) is a technique that amplifies Raman signal of the molecules thus to dramatically improve the limit of detection of test molecules. This amplification is realized by an intense electromagnetic field on a roughed noble metal surface that stems from the excitation of certain special coupling of surface plasmon. The intensity of SERS signal ($P^{SERS}(v_S)$) obeys the following equation:

$$P^{SERS}(v_S) = N' \cdot \sigma_{inc}^{R} \cdot |A(v_L)|^2 \cdot |A(v_S)|^2 \cdot I(v_L) \tag{8}$$

where $A(v_L)$ and $A(v_S)$ are the enhancement factors of excitation laser electromagnetic field and scattered Raman electromagnetic field, respectively. σ_{inc}^{R} is the effective Raman cross section. N' is the total number of the molecule that contribute to the Raman signal. $I(v_L)$ is the intensity of excitation laser. SERS is a label-free method that can identify the test molecules with its fingerprinting information; thus, it has attracted large interest in sensing applications with both index-guiding PCFs and bandgap guiding PCFs.

The performance of PCF-based SERS sensors is largely affected by the fiber structure and morphology of the noble metals. In particular, fiber structure affects mode-field overlap in the air channels and minimum confinement loss of the guided modes, and the morphology of the noble metal determines the enhancement factor of both excitation laser electromagnetic field and the scattered Raman electromagnetic field. Compared with solid-core index-guiding PCFs, hollow-core PCFs can obtain Raman signals with low silica background, because the laser beam directly interacts with analyte in the hollow core where both excitation laser and Raman signal are guided. But the narrow transmission window coming along with the bandgap confinement hinders their use in broad Raman scattering spectra.

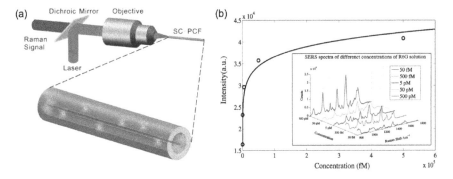

Fig. 7 **a** Schematic diagram of backscattered SERS sensing scheme. **b** The Raman signal intensity corresponding to the R6G concentration (Zhang et al. 2016a, b)

K. Khaing Oo et al. compared a few PCF structures for SERS applications, including nonlinearity PCFs, steering-wheel PCFs, and suspended-core PCF. The experimental results suggest that the PCF microstructure has a significant effect on the evanescent wave-based sensing and detection (Khaing Oo et al. 2010). And a best limit of detection of 10^{-10} M Rhodamine 6G is achieved with the suspended-core fiber which exhibits the largest fraction of evanescent power in the sensing region. In the experiments, the gold nanoparticles are immobilized on the silica surface of the cladding air channels to provide electromagnetic field amplification. In 2016, the geometry of a side-channel PCF was well defined for SERS sensing applications. The fiber geometry enables large evanescent power in the air channels adjacent to the solid fiber core and a low transmission loss at the SERS excitation wavelength of 632.8 nm with a pitch of 3 μm and an air holes diameter of 2.8 μm in the cladding region. For the SERS measurements, the schematic diagram is shown in Fig. 7a. The excitation laser source at 632.8 nm is coupled into the fiber core from one end of the fiber, and the backscattered enhanced Raman signal is collected and analyzed from the same fiber end. Both the performance under full-filling and selective-filling of the air channels conditions are characterized with a mixture solution of gold nanoparticles and Rhodamine 6G molecule infiltrated. The full-filing configuration reveals a better limit of detection of 50 fM Rhodamine 6G solution. The intensity of the SERS signal under full-filling configuration is plotted in Fig. 7b. In addition, the accumulative phenomenon of Raman signal along the SERS-active fiber length is also investigated, suggesting that properly increasing fiber length is an effective way to further enhance the SERS intensity and thus further reduce the limit of detection.

Compared with conventional fiber-based and substrate-based counterparts, of which the performance is limited by small SERS-active region and weak excitation electromagnetic field in the sensing region, PCF-based SERS sensing platform shows great potentials in constructing high-performance SERS sensing system. By synergistically enabling flexible design of optical waveguide properties and liquid transmission cell in a single fiber, PCFs can contribute to reaching ultra-low limit of detection and more opportunities for diverse and high-performance sensing applications, such as single molecular detection on cells and clinical diagnosis in liquid samples like urine and blood.

6 Conclusion

In this chapter, we focus on microstructured fibers for sensing applications. By manipulating the materials and fiber structures, light propagation in the PCF fiber core can be realized through index-guiding mechanism, PBG guiding mechanism, and antiresonance guiding mechanism. Coupled with advanced fiber fabrication technology, it allows great flexibility in the design of optical waveguide properties for versatile sensing applications. Hollow-core photonic bandgap fibers offer extremely large mode-field-overlap, but the light transmission loss induced by scattering by liquid is relatively high and the transmission window is relatively narrow. Though, in solid-core index-guiding fibers, the proportion of the evanescent field that extends to the sensing channels is relatively small, it exhibits stable light transmission window and low transmission loss which benefit a stable, robust platform for wide range applications. LPGs, interferometers, and SERS sensing platforms that take advantages of PCF properties show great potentials in in-line and real-time liquid RI sensing, specific chemical and biological molecule identification.

References

F. Benabid, J.C. Knight, G. Antonopoulos, P.S.J. Russell, Stimulated Raman scattering in hydrogen-filled hollow-core photonic crystal fiber. Science **298**(5592), 399–402 (2002). https://doi.org/10.1126/science.1076408

T.A. Birks, J.C. Knight, P.S. Russell, Endlessly single-mode photonic crystal fiber. Opt. Lett. **22**(13), 961–963 (1997). http://doi.org/10.1364/OL.22.000961

M. Consales, M. Pisco, A. Cusano, Lab-on-fiber technology: a new avenue for optical nanosensors. Photonic Sens. **2**(4), 289–314 (2012). https://doi.org/10.1007/s13320-012-0095-y

R.F. Cregan, Single-mode photonic band gap guidance of light in air. Science **285**(5433), 1537–1539 (1999). https://doi.org/10.1126/science.285.5433.1537

T. Han, Y. Liu, Z. Wang, J. Guo, Z. Wu, S. Wang et al., Unique characteristics of a selective-filling photonic crystal fiber Sagnac interferometer and its application as high sensitivity sensor. Opt. Express **21**(1), 122–128 (2013). https://doi.org/10.1364/OE.21.000122

Z. He, F. Tian, Y. Zhu, N. Lavlinskaia, H. Du, Long-period gratings in photonic crystal fiber as an optofluidic label-free biosensor. Biosens. Bioelectron. **26**(12), 4774–4778 (2011). https://doi.org/10.1016/j.bios.2011.05.048

D.J.J. Hu, J.L. Lim, M. Jiang, Y. Wang, F. Luan, P.P. Shum et al., Long period grating cascaded to photonic crystal fiber modal interferometer for simultaneous measurement of temperature and refractive index. Opt. Lett. **37**(12), 2283–2285 (2012a). https://doi.org/10.1364/OL.37.002283

D.J.J. Hu, J.L. Lim, M.K. Park, L.T.H. Kao, Y. Wang, H. Wei, W. Tong, Photonic crystal fiber-based interferometric biosensor for streptavidin and biotin detection. IEEE J. Sel. Top. Quantum Electron. **18**(4), 1293–1297 (2012b). https://doi.org/10.1109/JSTQE.2011.2169492

G. Humbert, A. Malki, S. Février, P. Roy, J.L. Auguste, J.M. Blondy, Long period grating filters fabricated with electric arc in dual concentric core fibers. Opt. Commun. **225**, 47–53 (2003). https://doi.org/10.1016/j.optcom.2003.07.007

M.K. Khaing Oo, Y. Han, J. Kanka, S. Sukhishvili, H. Du, Structure fits the purpose: photonic crystal fibers for evanescent-field surface-enhanced Raman spectroscopy. Opt. Lett. **35**(4), 466–468 (2010). https://doi.org/10.1364/OL.35.000466

K.M. Kiang, K. Frampton, T.M. Monro, R. Moore, J. Tucknott, D.W. Hewak et al., Extruded singlemode non-silica glass holey optical fibres. Electron. Lett. **38**(12), 546 (2002). https://doi.org/10.1049/el:20020421

J.C. Knight, T.A. Birks, P.S.J. Russell, D.M. Atkin, All-silica signal-mode optical fiber with photonic crystal cladding. Opt. Lett. **21**(19), 1547–1549 (1996)

J.C. Knight, T.A. Birks, P.S.J. Russell, J.P. de Sandro, Properties of photonic crystal fiber and the effective index model. **15**(3), 748–752 (1998). https://doi.org/10.1364/JOSAA.15.000748

V.V.R. Kumar, A. George, W. Reeves, J. Knight, P. Russell, F. Omenetto, A. Taylor, Extruded soft glass photonic crystal fiber for ultrabroad supercontinuum generation. Opt. Express **10**(25), 1520–1525 (2002). https://doi.org/10.1364/OE.10.001520

H.W. Lee, M.A. Schmidt, P. Uebel, H. Tyagi, N.Y. Joly, M. Scharrer, P.S.J. Russell, Optofluidic refractive-index sensor in step-index fiber with parallel hollow micro-channel. Opt. Express **19**(9), 8200–8207 (2011). https://doi.org/10.1364/OE.19.008200

K.E. Lynch-Klarup, E.D. Mondloch, M.G. Raymer, D. Arrestier, F. Gerome, F. Benabid, Supercritical xenon-filled hollow-core photonic bandgap fiber. Opt. Express **21**(11), 13726 (2013). https://doi.org/10.1364/OE.21.013726

J. Ma, H.H. Yu, X. Jiang, D.S. Jiang, High-performance temperature sensing using a selectively filled solid-core photonic crystal fiber with a central air-bore. Opt. Express **25**(8), 9406 (2017). https://doi.org/10.1364/oe.25.009406

L. Mao, P. Lu, Z. Lao, D. Liu, J. Zhang, Highly sensitive curvature sensor based on single-mode fiber using core-offset splicing. Opt. Laser Technol. **57**, 39–43 (2014). https://doi.org/10.1016/j.optlastec.2013.09.036

C. Markos, J.C. Travers, A. Abdolvand, B.J. Eggleton, O. Bang, Hybrid photonic-crystal fiber. Rev. Mod. Phys. **89**(4), 1–55 (2017). https://doi.org/10.1103/RevModPhys.89.045003

T.M. Monro, S. Warren-Smith, E.P. Schartner, A. Franois, S. Heng, H. Ebendorff-Heidepriem, S. Afshar, Sensing with suspended-core optical fibers. Opt. Fiber Technol. **16**(6), 343–356 (2010). https://doi.org/10.1016/j.yofte.2010.09.010

Z. Peng, L. Wang, H. Yan, Research on high-temperature sensing characteristics based on modular interference of single-mode multimode single-mode fiber, in *Proceedings of SPIE*, vol. 10025, ed. by T. Liu, S. Jiang, R. Landgraf (2016), p. 1002519. https://doi.org/10.1117/12.2245525

A.D. Pryamikov, A.S. Biriukov, A.F. Kosolapov, V.G. Plotnichenko, S.L. Semjonov, E.M. Dianov, Demonstration of a waveguide regime for a silica hollow-core microstructured optical fiber with a negative curvature of the core boundary in the spectral region >3.5 μm. Opt. Express **19**(2), 1441 (2011). https://doi.org/10.1364/OE.19.001441

L. Rindorf, O. Bang, Highly sensitive refractometer with a photonic-crystal-fiber long-period grating. Opt. Lett. **33**(6), 563–565 (2008). https://doi.org/10.1364/OL.33.000563

L. Rindorf, J.B. Jensen, M. Dufva, L.H. Pedersen, P.E. Høiby, O. Bang, Photonic crystal fiber long-period gratings for biochemical sensing. Opt. Express **14**(18), 8224–8231 (2006). https://doi.org/10.1364/OE.14.008224

P.J. Roberts, F. Couny, H. Sabert, B.J. Mangan, D.P. Williams, L. Farr et al., Ultimate low loss of hollow-core photonic crystal fibres. Opt. Express **13**(1), 236–244 (2005). https://doi.org/10.1364/OPEX.13.000236

Q. Rong, X. Qiao, R. Wang, H. Sun, M. Hu, Z. Feng, High-sensitive fiber-optic refractometer based on a core-diameter-mismatch mach-zehnder interferometer. IEEE Sens. J. **12**(7), 2501–2505 (2012). https://doi.org/10.1109/JSEN.2012.2194700

P. Russell, Photonic crystal fibers. Science **299**(5605), 358–362 (2003). https://doi.org/10.1126/science.1079280

A.M. Stolyarov, A. Gumennik, W. McDaniel, O. Shapira, B. Schell, F. Sorin et al., Enhanced chemiluminescent detection scheme for trace vapor sensing in pneumatically-tuned hollow core photonic bandgap fibers. Opt. Express **20**(11), 12407 (2012). https://doi.org/10.1364/OE.20. 012407

K. Subramanian, I. Gabay, A. Shadfan, M. Pawlowski, Y. Wang, T. Tkaczyk, A. Ben-Yakar, A Kagome fiber-based, high energy delivery laser scalpel system for ultrafast laser microsurgery, in *2016 Conference on Lasers and Electro-Optics, CLEO 2016*, vol. 10066, ed. by T.P. Ryan (2017), p. 100660U. https://doi.org/10.1117/12.2253446

B. Temelkuran, S.D. Hart, G. Benoit, J.D. Joannopoulos, Y. Fink, Wavelength-scalable hollow optical fibres with large photonic bandgaps for CO_2 laser transmission. Nature **420**, 650–653 (2002). https://doi.org/10.1038/nature01275

J. Tian, Z. Lu, M. Quan, Y. Jiao, Y. Yao, Fast response Fabry-Perot interferometer microfluidic refractive index fiber sensor based on concave-core photonic crystal fiber. Opt. Express **24**(18), 20132–20142 (2016). https://doi.org/10.1364/OE.24.020132

S. Unterkofler, R.J. McQuitty, T.G. Euser, N.J. Farrer, P.J. Sadler, P.S.J. Russell, Microfluidic integration of photonic crystal fibers for online photochemical reaction analysis. Opt. Lett. **37**(11), 1952–1954 (2012). https://doi.org/10.1364/OL.37.001952

A.M. Vengsarkar, P.J. Lemaire, J.B. Judkins, V. Bhatia, T. Erdogan, J.E. Sipe, Long-period fiber gratings as band-rejection filters. J. Lightwave Technol. **14**(1), 58–64 (1996). https://doi.org/10. 1109/50.476137

Y.P. Wang, D.N. Wang, W. Jin, Y.J. Rao, G.D. Peng, Asymmetric long period fiber gratings fabricated by use of CO_2 laser to carve periodic grooves on the optical fiber. Appl. Phys. Lett. **89**(15), 19–21 (2006). https://doi.org/10.1063/1.2360253

L. Wang, D. He, S. Feng, C. Yu, L. Hu, J. Qiu, D. Chen, Phosphate ytterbium-doped single-mode all-solid photonic crystal fiber with output power of 13.8 W. Sci. Rep. **5**, 10–13 (2015). https:// doi.org/10.1038/srep08490

C. Wu, M.-L.V. Tse, Z. Liu, B.-O. Guan, C. Lu, H.-Y. Tam, In-line microfluidic refractometer based on C-shaped fiber assisted photonic crystal fiber Sagnac interferometer. Opt. Lett. **38**(17), 3283–3286 (2013). https://doi.org/10.1364/OL.38.003283

C. Wu, M.-L.V. Tse, Z. Liu, B.-O. Guan, A.P. Zhang, C. Lu, H.-Y. Tam, In-line microfluidic integration of photonic crystal fibres as a highly sensitive refractometer. The Analyst **139**(21), 5422–5429 (2014). https://doi.org/10.1039/C4AN01361A

N. Zhang, G. Humbert, Z. Wu, K. Li, P.P. Shum, N.M.Y. Zhang et al., In-line optofluidic refractive index sensing in a side-channel photonic crystal fiber. Opt. Express **24**(24), 419–424 (2016a). https://doi.org/10.1364/OE.24.027674

N. Zhang, G. Humbert, T. Gong, P.P. Shum, K. Li, J.-L. Auguste et al., Side-channel photonic crystal fiber for surface enhanced Raman scattering sensing. Sens. Actuators B Chem. **223**, 195–201 (2016b). https://doi.org/10.1016/j.snb.2015.09.087

N. Zhang, K. Li, Y. Cui, Z. Wu, P.P. Shum, J.-L. Auguste et al., Ultra-sensitive chemical and biological analysis via specialty fibers with built-in microstructured optofluidic channels. Lab Chip **18**(4), 655–661 (2018). https://doi.org/10.1039/C7LC01247K

T. Zhu, D. Wu, M. Liu, D.W. Duan, In-line fiber optic interferometric sensors in single-mode fibers. Sensors (Switzerland) **12**(8), 10430–10449 (2012). https://doi.org/10.3390/s120810430

Optical Microfiber Sensors

Kaiwei Li, Jiajia Wang, and Tuan Guo

Abstract Optical microfiber is a class of specialty fibers, which is featured with wavelength scale diameters. With such small dimensions, the optical microfiber offers large fractions of evanescent fields and high surface field intensities, making it highly sensitive to disturbances in the surrounding medium. Thus, the optical microfiber is an ideal building block for high-performance photonics sensing devices. In this chapter, recent progress in optical microfiber-based sensors is reviewed. It starts with a brief introduction of the fundamental optical properties of optical microfibers and the well-developed fabrication techniques. Then, a brief summarization of the well-established microfiber-based refractive index sensing schemes is given, including working principles and sensing performances. Following this section, the latest progress on new effects and strategies for sensing enhancement is reviewed. In the last, the conclusions and an outlook are presented.

Keywords Optical microfiber · Optical fiber sensor · Modal interferometer · Biochemical sensor

1 Introduction

Optical microfibers have been gaining tremendous attention in a wide range of research fields ever since its first discovery (Tong et al. 2003). With diameters close to the wavelength of light, these tiny optical fibers cannot confine the guided light tightly inside the fiber, like their telecommunication counterparts. Instead, the optical

K. Li (✉) · T. Guo
Institute of Photonics Technology, Jinan University, Guangzhou, China
e-mail: kaiweili@jnu.edu.cn

T. Guo
e-mail: tuanguo@jnu.edu.cn

J. Wang
College of Agricultural Equipment Engineering, Henan University of Science and Technology, Luoyang, China
e-mail: johnnyjiajia@163.com

© Springer Nature Singapore Pte Ltd. 2020
L. Wei (ed.), *Advanced Fiber Sensing Technologies*,
Progress in Optical Science and Photonics 9,
https://doi.org/10.1007/978-981-15-5507-7_4

microfibers show a strong evanescent field, which means that a substantial amount of the guiding light can enter and interact with the surrounding medium. Besides, the optical microfibers also offers several unique properties, including small footprint, low optical loss, high nonlinearity, and high mechanical flexibility. All these superior properties make the optical microfiber a promising platform for high efficient light-matter interactions. Applications in fields ranging from optical sensing (Chen et al. 2019) to ultrafast laser (Luo et al. 2015b), nonlinear optics (Vienne et al. 2008), and optomechanics (Zhang et al. 2020) have been explored.

In particular, the small geometry size and the strong evanescent filed make the optical microfibers an ideal platform to perform sensing on the nanoscale (Tong et al. 2004). Any environmental parameter variations, external forces, object movements, as well as the chemical reaction that can alter the optical guiding property of the microfiber, can be easily detected. As such a perspective filed, the optical microfiber sensors have witnessed prosperity during the last decades, along with the improvement of the fabrication techniques. Multifarious novel microfiber sensors with unique advantages of ultra-small footprint, high sensitivity, and fast response have been explored. These sensing devices include microfiber modal interferometers (Yadav et al. 2014), Mach–Zehnder interferometers (MZI) (Hu et al. 2012), microfiber couplers (Tazawa et al. 2007), microfiber resonators (Sumetsky et al. 2006), microfiber gratings (Ran et al. 2011; Wang et al. 2017), and nanomaterial decorated microfibers (Yu et al. 2017; Zhou et al. 2019). These versatile sensors have found applications in temperature sensing, pressure sensing, strain sensing, flow rate sensing, refractive index (RI) sensing, as well as biochemical sensing. In this chapter, the recent progress in microfiber sensors is reviewed, with an emphasis on the basic sensing principles of the well-developed microfiber-based sensing devices and the newly developed strategies for sensing enhancement.

2   Basics of Optical Microfibers

The typical construction of an optical microfiber is depicted in Fig. 1, where the uniform microfiber lies in the middle region connects with two standard optical fibers through two conical tapers. When the standard optical fibers are tapered into microwires with diameters around the operation wavelength, the original core reduces

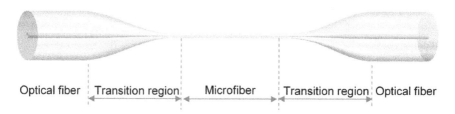

Optical fiber ┊ Transition region┊   Microfiber   ┊ Transition region┊ Optical fiber

Fig. 1 Schematic diagram of the optical microfiber

to below 100 nm and becomes negligible. Thus, the whole microfiber works as the core, and the environmental medium (air/liquid) whose RI usually is lower than the fiber works as the new cladding.

2.1 Optical Properties

2.1.1 Few-Mode/Single-Mode Operation

As the diameter of the microfiber is on the scale of the wavelength, only a limited number of guided optical modes can be supported. The number of the guided modes are determined by the V number, which is defined as (Tong et al. 2003):

$$V = \frac{\pi D}{\lambda} \sqrt{n_{co}^2 - n_{cl}^2} \tag{1}$$

where D, λ, n_{co}, and n_{cl} are the fiber diameter, the wavelength in vacuum, core RI, and cladding RI, respectively. The critical condition for single-mode operation is $V <$ 2.405 when all the high-order modes are cut off (Tong et al. 2003). The propagation constant, effective RI (ERI), and the spatial field distribution of the modes can be obtained by solving the Helmholtz equations (Tong et al. 2003). The modal dispersion curve showing the ERI of the supported modes in optical microfibers in water is depicted in Fig. 2. It is evident that when the diameter of the fiber is smaller than 2 μm, the single-mode operation can be achieved.

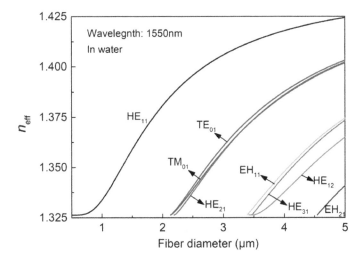

Fig. 2 Modal dispersion of an optical microfiber with a water cladding

2.1.2   Strong Evanescent Filed

The unique and attractive optical property of the optical microfiber is the strong evanescent field. It means that a substantial portion of the guided optical power can enter the surrounding medium and propagate along the fiber. The modal field distributions of the HE_{11} mode for optical microfibers with a diameter of 2 μm and 0.8 μm are shown in Fig. 3. It is evident that the 2-μm-thick microfiber only possesses a small portion of the evanescent field, whereas for the 0.8-μm-thick nanofiber, more than half of the optical power is guided outside the fiber. The evanescent power ratio of the HE_{11} mode for microfibers with different diameters is displayed in Fig. 4a. With the enhanced evanescent field, the penetration depth also increases significantly. For nanofibers thinner than 0.5 μm, the penetration depth can reach several microns, which is far more significant than the small value of half-wavelength, for the conventional evanescent field in a prism configuration. The strong evanescent

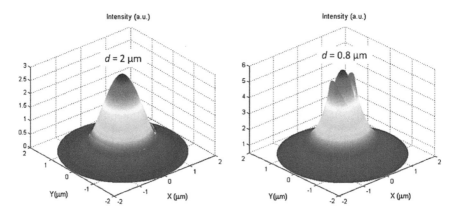

Fig. 3   Modal field distribution of the fundamental HE_{11} mode for optical microfibers (in water)

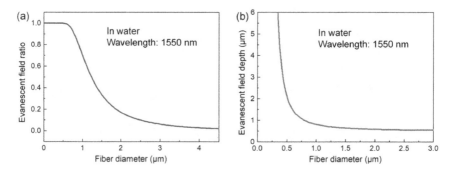

Fig. 4   a Evanescent power ratio of the fundamental HE_{11} mode. **b** The penetration depth of the evanescent field of the optical microfiber

field provides a perfect platform for light-matter interactions and further serves as essential building blocks for super sensitive sensing devices.

2.1.3   Low Optical Loss

The insertion loss is also an essential parameter for the microfiber. A low optical loss facilitates the application of the optical microfiber in the field of lasers, nonlinear optics, and sensing. The optical loss mainly comes from the surface roughness-induced scattering. Utilizing the heating and pulling method, microfibers with surface roughness as low as sub-nanometer can be achieved, and the insertion loss can be as low as 0.06 dB (Hoffman et al. 2014), which can be almost ignored.

2.2   Fabrication Techniques

In order to manufacture optical microfibers with well-defined geometries and optical characteristics in a scalable manner, several methods have been developed. The most common fabrication techniques are the so-called heating and pulling method (Felipe et al. 2012) and the chemical etching method (Li et al. 2018d). As the dimension of these tiny microfibers is generally below 5 μm, less than 4% of the original diameter of the 125-μm-thick standard telecommunication optical fibers, the precise control of the diameter can be challenging. Therefore, in-line monitoring methods have been developed to monitor the diameter or the optical spectra of the fabricated microfibers during the fabrication process. With such thin diameters, the microfibers usually have low mechanical strength and are quite fragile. Thus, proper packaging is essential to keep the microfiber safe and stable.

2.2.1   Heating and Pulling Method

The typical fabrication process of the heating and pulling technique is displayed in Fig. 5a. Firstly, a section of standard optical fiber with the polymer coating removed in the central part is mounted on two translation stages and fixed using fiber holders. Secondly, a heating source is applied to the silica fiber to keep it above the glass transition temperature. Then, the fiber is pulled by the two translation stages, and the dimension of the melting central region reduces gradually as the two translation stages move apart. Figure 5b shows a micrograph of a typical optical microfiber fabricated by this method. The fabrication process usually takes about several minutes. The diameter of the microfiber can be controlled by adjusting the pulling velocity and the pulling duration. The width of the heating source determines the length of the uniform waist region. Various heating elements have been adopted, including oxyhydrogen flame (Felipe et al. 2012), ceramic heater (Ma et al. 2016), and the CO_2 laser beam (Xuan et al. 2009). Typically, the oxyhydrogen flame and the ceramic heater that

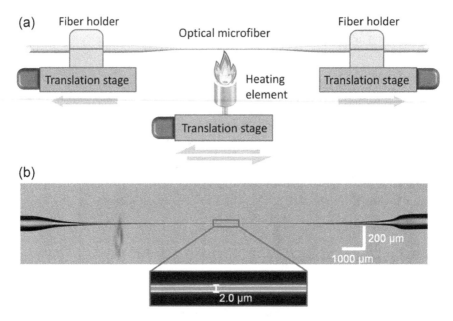

Fig. 5 **a** Schematic diagram of the heating and pulling fabrication method. **b** A micrograph of an optical microfiber fabricated by the heating and pulling method. Adapted with permission Zhang et al. (2018a)

provide millimeter-long heating regions are favorable for fabricating adiabatic optical microfibers. The CO_2 laser beam provides a heating width of as small as several hundreds of microns and is suitable for producing short and abrupt tapers. So as to obtain optical microfibers with the desired length, a scanning heating source is usually used, which can extend the effective width of the heating source. The most significant merit of the heating and pulling method is the sub-nanometer-scale surface roughness, which renders the microfiber with ultra-low optical loss.

2.2.2 Chemical Etching Method

The chemical etching technique is illustrated in Fig. 6a. Typically, a section of standard single-mode optical fiber with the desired length is stripped off the protective polymer coating and cleaned with acetone. Then, the bare silica fiber is placed inside the fluidic channel of a specially designed silicon chip and fixed by glue. Afterward, the hydrofluoric acid etching solution was added into the fluid cell, and the etching process begins. The diameter of the fabricated microfiber can be tuned by adjusting the etching duration. In order to achieve a proper diameter, a two-step etching strategy is adopted. In the first step, the course etching using high-concentration hydrofluoric acid is applied to the fiber. It takes about 50 min to decrease the 125 μm diameter to about 5 μm. In the second step, low-concentration hydrofluoric acid is added into

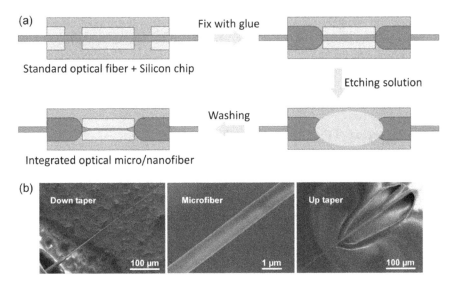

Fig. 6 **a** Schematic diagram of the chemical etching method. **b** An SEM image of an optical microfiber fabricated using the chemical etching method. Adapted with permission Li et al. (2018d)

the fluid cell to fine-etch the 5-μm-thick microfiber to the target dimension (Li et al. 2014). A typical image of a resultant sub-wavelength optical microfiber is depicted in Fig. 6b. The whole manufacture procedure takes about several hours, which is much longer than the heating and pulling method. However, the productivity can be easily improved by a parallel fabrication strategy (Li et al. 2018d). The transmission loss of the fabricated fibers is usually more significant than those fabricated by the heating and pulling method. However, one distinct advantage of the etching method is that the microfluidic channel for the fabrication can readily serve as channels for the infiltration of bio-samples. Thus, a further packaging procedure is eliminated.

2.2.3 In-Line Monitoring Technologies

The in-line monitoring systems can facilitate the fabrication process to enhance the capability in controlling the dimension and optical properties of the fabricated microfiber in a real-time manner. The widely adopted approach is connecting the two ends of the optical fiber to a light source and a detector, respectively. Thus, the optical loss and the output spectra can be measured in real-time, and the fabrication process can be terminated once the desired parameter is reached. This in-line monitoring system has been applied to both the heating and pulling method and the chemical etching method (Li et al. 2014). In order to control the diameter and the optical properties more precisely, advanced in-line measuring techniques have been developed. The cutoff of a high-order mode occurs at a determined diameter. Thus, by monitoring the cutoff effect of high-order modes and the time interval between two drops, the diameter of the nanofiber can be precisely determined in real-time during the pulling process (deviation <5 nm) (Xu et al. 2017a). An in-line monitoring technique utilizing the harmonic generation was also developed to measure the diameter of the nanofiber during the tapering process. The manufacturing deviation can be lower than 2% (Wiedemann et al. 2010).

2.2.4 Packaging of the Optical Microfibers

The convenient and safe handling of the fabricated microfibers is an essential prerequisite to the practical applications. In order to protect the fiber from mechanical disturbs and airborne dust, different approaches have been developed by groups around the world that are specialized in optical microfibers. For example, the fabricated microfiber can be embedded inside low-refractive-index substrates such as PDMS (Polynkin et al. 2005), Teflon (Xu et al. 2007), and silica aerogels (Xiao et al. 2011) to maintain the low optical loss and enhance the portability and long-term stability. For biosensing and gas sensing applications, the microfiber can be fixed inside a microfluidic channel (Li et al. 2014) or sealed inside silica tubes (Mao et al. 2018) that provide input and output ports to make the sensor accessible to the samples under test.

3   Optical Microfiber-Based Sensing Devices

Owing to the unique advantages of strong evanescent field and mechanically flexibility that are not available in standard telecommunication optical fibers, the optical microfiber sensors have witnessed a flourishing development ever since its invention. Various microfiber-based sensing devices employing diverse mechanisms have been explored, which show considerable promise in fields ranging from fundamental physical parameter measurement to practical biochemical analysis. The number of published research works in this field keeps increasing substantially, and a number of reviews are available. In this section, a brief summarization of the well-established sensing devices with a focus on their schemes and sensing mechanisms is provided to give the readers a general concept of how these sensors work. Then, the latest progress in new effects and strategies for sensing enhancement is reviewed.

3.1   Basic Sensing Devices

The optical interferometric sensors probably account for the largest group of the microfiber-based RI sensors. For a specific guided mode in the microfiber with a determined optical path, a small variation in the RI of the surrounding medium can induce a phase change. Thus, by introducing multi-modes into a single microfiber device or incorporating an additional reference arm, interferometers can be constructed. Based on the structural characteristics, these interferometers can be classified into the following categories: the modal interferometers (Zhang et al. 2018a) and the Sagnac interferometers (Sun et al. 2014). Apart from the interferometers, optical microfiber based resonators and gratings also have been widely explored.

3.1.1 Modal Interferometers

The simplest form of the modal interferometer is the biconical tapered microfiber (Fig. 7a). It consists of a uniform waist region that can support multimode propagation and two abrupt tapered regions, which can couple a portion of the fundamental core mode into high-order modes in the microfiber region. The fundamental HE_{11} mode and the high-order mode possess different propagation constants and ERIs so that a phase difference can be formed between the two modes. The high-order mode can be HE_{12} mode or other high-order modes. The transmission of a microfiber interferometer can be calculated through

$$I = I_1 + I_2 + 2\sqrt{I_1 I_2} \cos\left(\frac{2\pi L \Delta n_{\text{eff}}}{\lambda}\right) \tag{2}$$

where I_1 and I_2 represent the optical power of two interference modes. L, λ, and Δn_{eff} denote the lengths of the optical microfiber, the operating wavelength, and the difference between the refractive index of the two modes, respectively. With this

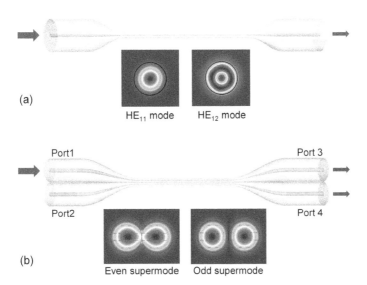

Fig. 7 a Schematic diagram of a biconical tapered microfiber-based modal interferometer. **b** Schematic diagram of an optical microfiber coupler

compact sensing device, physical or chemical changes that can alter the length or the mode property of the microfiber can be easily detected.

The optical microfiber coupler (OMC) is a traditional fiber-optic device, which has been extensively employed in fiber-optic networks, fiber lasers, and fiber-optic imaging. The tapered fiber coupler also serves as a promising candidate for sensing. The OMC is consists of two input ports, two output ports, two transition regions, and a uniform waist region where two microfibers are placed in parallel, as shown in Fig. 7b. The underlying operating mechanism is the interference between the even supermode and the odd supermode. When light is injected from port 1, the output power at port 3 and 4 can be obtained as (Yang et al. 1998).

$$P_3 = P_1 \cos^2 \left(\frac{2\pi L \left(n_{\text{eff}}^{\text{even}} - n_{\text{eff}}^{\text{odd}} \right)}{\lambda} \right) \tag{3}$$

$$P_4 = P_1 \sin^2 \left(\frac{2\pi L \left(n_{\text{eff}}^{\text{even}} - n_{\text{eff}}^{\text{odd}} \right)}{\lambda} \right) \tag{4}$$

where $n_{\text{eff}}^{\text{even}}$ and $n_{\text{eff}}^{\text{odd}}$ are the ERIs of the even supermode and odd supermode, respectively. L and λ denote the coupling length and the working wavelength.

Due to the periodical coupling between the even and odd supermodes, sine wave-like transmission spectra can be obtained from the two output ports. When it is used as a sensor, the variation of the environmental parameter can affect the coupling and eventually can be analyzed through the transmission spectra. The microfiber coupler sensor regularly exhibits equivalent sensing performances to the microfiber interferometric sensors (Tazawa et al. 2007).

3.1.2 Sagnac Interferometers

The microfiber-based Sagnac interferometer usually employs the highly birefringent optical microfibers with none circular cross-sections (Fig. 8a). In this sense, the two orthogonally polarization states of the fundamental mode experience different propagation constants and effective RIs. The birefringent optical microfibers are conventionally incorporated into a Sagnac loop to form the Sagnac interference between the two orthogonally polarized optical beams, which are also named the fast beam and the slow beam (Fig. 8b). To date, microfiber Sagnac interference with ellipse cross-sections (Sun et al. 2014), rectangular cross-sections (Li et al. 2011), H-shaped cross-sections (Xuan et al. 2010), and D-shaped cross-sections have been demonstrated (Fig. 8c) (Luo et al. 2016). The transmission of the Sagnac interferometer is obtained as (Li et al. 2011)

$$T = P_0 \sin^2 \left(\frac{\pi B L}{\lambda} \right) \tag{5}$$

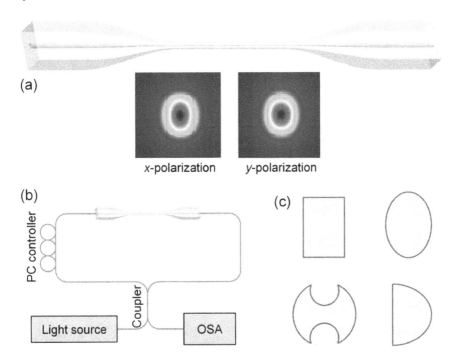

Fig. 8 **a** Schematic diagram of a highly birefringent optical microfiber. **b** Schematic diagram of the Sagnac interferometer. **c** Cross-sections of typical highly birefringent optical microfibers (rectangular shape, oval shape, H shape, and D shape)

where $B = n_{\text{eff}}^x - n_{\text{eff}}^y$ denotes the birefringence of a guided mode, with n_{eff}^x and n_{eff}^y represent the ERIs of the x-polarized mode and y-polarized mode, respectively.

3.1.3 Optical Microfiber Resonators

The optical microfiber resonators can be formed by microfiber loops (Fig. 9a) (Guo and Tong 2008) or microfiber knot resonators (Xiao and Birks 2011), and microfiber coil resonator (Fig. 9b) (Xu et al. 2007) that are wrapped around a rod. Theses resonators can confine the light in a marginal volume and significantly enhance the evanescent field, which can significantly promote the sensing capability. The quality factor of microfiber resonators can reach as high as 10^6 (Xiao and Birks 2011). Although the sensitivity of these sensors is relatively low compared to microfiber interferometers and microfiber couplers, the detection limit can be several orders of magnitude lower, considering the high-quality factor. Numerous progress has been made both in exploring new resonator schemes and improving the quality factor. Despite the RI sensing performance, these sensors are seldomly employed for biochemical sensing. The relatively low mechanical robustness may limit this. Advanced packaging techniques are expected to promote practice applications.

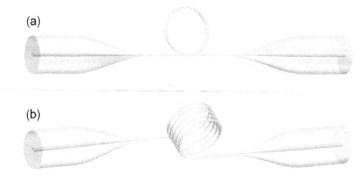

Fig. 9 Schematic diagram of **a** optical microfiber ring resonator. **b** optical microfiber coil resonator

Fig. 10 Schematic diagram of the optical microfiber grating

3.1.4 Optical Microfiber Gratings

Fiber-optic gratings are the most classic fiber sensors for strain and temperature measurements, which have found wide applications in the fields of railway monitoring, structure health monitoring, and oil pipeline inspections. Thanks to the pronounced evanescent field, microfiber gratings (Fig. 10) can extend their application to RI sensing and biochemical detection. These gratings are routinely inscribed in the microfiber using the ArF excimer laser (Ran et al. 2011) or the femtosecond laser (Fang et al. 2010). Microstructure-patterned microfiber gratings fabricated by the focused ion beam (FIB) milling (Ding et al. 2011) and the Rayleigh-Plateau instability (Li et al. 2017) have also been reported. Based on the period of the grating, microfiber grating can be divided into two classes: the microfiber Bragg grating (Kou et al. 2012) and the long-period microfiber grating (Li et al. 2017). The merits of the microfiber gratings are the possibility of automatic production with high uniformity. However, the relatively low sensitivity remains a critical problem which needs to be overcome.

3.2 New Effects and Strategies for Sensing Enhancement

3.2.1 Dispersion Turning Point

Searching for new strategies to enhance the performance of existing sensors is a permanent theme. For microfiber RI sensors, the highest sensitivity is always achieved when the RI of the surrounding medium approaches that of the fiber itself. This can be explained by the fact that the evanescent field becomes pronounced when the external RI comes close to that of the fiber, which can enhance light-matter interaction. However, in practical applications, most of the samples are with low RI. For example, in biosensing, biomolecules are always given in the form of water solutions whose RI is quite close to that of water (around 1.33–1.36). Also, for gas sensing and gas-phase biomarker detections, the working medium usually shows an extremely low RI of ~1. Therefore, it is of great practical importance to develop RI sensors that can achieve high sensitivity in the low RI region.

By carefully designing the parameters of an OMC, it can work around the modal dispersion turning point, which can significantly improve the sensitivity in a given RI range (Li et al. 2016, 2018b, c). As shown in Fig. 11a, the OMC mainly relies on the interference of the even supermode and the odd supermode. The RI sensitivity can be calculated through (Li et al. 2018c):

$$S_{\text{RI}} = \frac{\partial \lambda_N}{\partial n} = \frac{\lambda_N}{n_g^{\text{even}} - n_g^{\text{odd}}} \frac{\partial \left(n_{\text{eff}}^{\text{even}} - n_{\text{eff}}^{\text{odd}}\right)}{\partial n} \tag{6}$$

where λ_N represents the wavelength of Nth dip on the interference spectrum. n_g^{even} and n_g^{odd} denotes the ERI of the even and odd supermodes, and can be obtained by $n_g = n_{\text{eff}} - \lambda_N \partial (n_{\text{eff}}) / \partial \lambda_N$. $n_{\text{eff}}^{\text{even}}$ and $n_{\text{eff}}^{\text{odd}}$ are the ERIs of the two modes, respectively. It is evident from Eq. (6) that when the group index difference $g = n_g^{\text{even}} - n_g^{\text{odd}}$

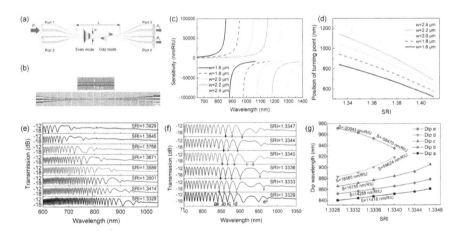

Fig. 11 **a** Schematic diagram of an OMC with the interference between the even mode and odd mode. **b** The microscopic view of an OMC with a waist width of 1.8 μm. **c** The calculated sensitivity of near the dispersion turning point for OMCs with different waist width. **d** Positions of the dispersion turning point for OMCs ($w = 1.6$–2.4 μm) as SRI increases from 1.3329 to 1,4100. **e** Variation of the transmission spectrum of the OMC along with increasing SRI from 1.3329 to 1.3929. **f** spectral responses of the OMC in the RI range around 1.3329. **g** The sensitivities of five dips closest to the dispersion turning point. Adapted with permission Li et al. (2018c)

approaches zero, the sensitivity can be significantly enhanced. Numerical simulation results (Fig. 11c) indicate that OMCs with a width ranging from 1.6 to 2.4 μm exhibit dispersion turning points in the wavelength range of 600–1400 nm, along with extremely high sensitivities of above 10,000 nm/RIU. When the waist width of the microfiber coupler is appropriately selected, the dispersion turning points can be readily obtained in a vast range of SRIs, ranging from as low as 1.0 for the gas medium (Li et al. 2018b) to liquid samples as high as 1.44 (Li et al. 2018c). Dual-peaks/dips interference characteristic near the dispersion turning point, which is different from conventional modal interferometers, was discovered and investigated experimentally (Fig. 11f). An exceptional sensitivity as high as 59,624 nm/RIU was achieved (Fig. 11g). In particular, OMCs with sub-micron waist width show clear dispersion turning points in the gaseous environment, and ultra-sensitive gas RI detection has been demonstrated (Li et al. 2018b).

Analogous dispersion turning point and ultra-high sensitivities have also been demonstrated in optical microfiber modal interferometers, which mainly rely on the interference of the HE_{11} and the HE_{12} modes (Luo et al. 2015a; Zhang et al. 2018a). By optimizing the waist diameter, dispersion turning points can be achieved in a vast RI range from 1.0 to 1.44. However, one problem with the tapered microfiber modal interferometer is that high-quality interference fringes are challenging to achieve experimentally. This is because the high-order HE_{12} mode is excited by the LP_{01} core mode of the standard optical fiber at the down taper. The excitation ratio usually is relatively low, and other high-order modes are also excited simultaneously. This newly discovered sensing effect is not limited to microfiber couplers and biconical tapered optical microfibers. It may be further extended to other microfiber-based sensing devices.

Leveraging on this well demonstrated OMC sensor, ultra-sensitive detection of the cardiac troponin I (cTnI) biomarker has realized (Zhou et al. 2018). The measuring setup is shown in Fig. 12a. The packaging of the OMC is shown in Fig. 12b. The OMC biosensor is fabricated through the heating and pulling method, followed by surface modification with site-specific cTnI antibodies via covalent binding. The results in Fig. 12c, d demonstrate the remarkable performance of this sensor for cTnI biomarker detection. A limit of detection (LOD) as low as 2 fg/mL was achieved, which is much low than previously reported fiber-optic biosensors.

3.2.2 Vernier Effect

The Vernier effect is adapted from the Vernier caliper, which is used for length measurement with enhanced resolution. This effect can be applied to fiber-optic sensors with enhanced performance. To realize the Vernier effect, one should obtain and superpose two paths of resonances or interferences with identical but unequal free spectral range (FSR). Such structures could be easily constructed using the optical microfibers. To date, the Vernier effect has been demonstrated within cascaded microfiber ring resonators and the OMCs.

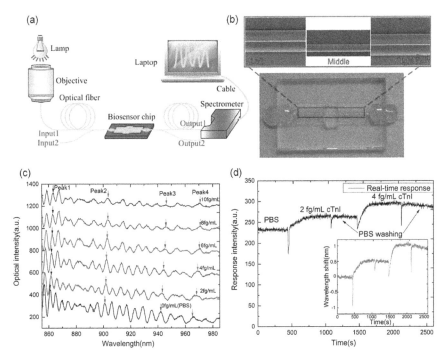

Fig. 12 **a** Schematics of the experimental setup for the OMC biosensor. **b** Scanning Electron Microscopy images of the OMC and a photograph of the sensor chip. **c** Spectral responses of the OMC (waist width: ~1.0 μm) to cardiac troponin I biomarkers in PBS buffers with concentrations of 2 fg/mL, 4 fg/mL, 6 fg/mL, 8 fg/mL, 10 fg/mL, respectively. **d** Real-time response of the sensor to PBS, 2 fg/mL, and 4 fg/mL. Adapted with permission Zhou et al. (2018)

By cascading a θ-shaped microfiber resonator (Fig. 13a) with an additional fiber Fabry–Perot interferometer, the Vernier effect can be generated when the FSR of the resonator and the interference are identical. The experimental measuring setup is shown in Fig. 13b. Benefiting from the Vernier effect, the RI sensitivity of the combined structure is m times higher than the sensitivity of the singular θ-shaped microfiber resonator. Moreover, this sensor is highly tunable, as the researchers have demonstrated in Fig. 13c, d: when the cavity length of the θ-shaped microfiber resonator is adjusted from 9.4 to 8.7 mm, the RI sensitivity can be widely tuned from 311.77 to 2460.07 nm/RIU (Xu et al. 2017b).

The OMC is a naturally birefringent waveguide, which can be utilized to compose the Vernier effect without incorporating an additional structure (Li et al. 2018a). The basic principle is shown in Fig. 14a. Due to the modal birefringence, two paths of orthogonally polarized interferences that vary slightly can be achieved using a single OMC. Figure 14b shows the microscopic images of an OMC with a waist width of 3.2 μm, together with the transmission spectra in both air and water. The pronounced envelopes in the spectra result from the Vernier effect. Benefitting from the Vernier effect, this sensor shows an improvement of one order of magnitude in

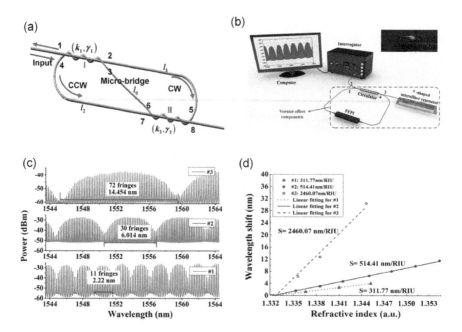

Fig. 13 **a** Schematic diagram of the θ-shaped ring resonator. **b** The experimental setup of the θ-shaped microfiber resonator cascaded with the Fabry–Perot interferometer. **c** Vernier spectra with different periods. **d** The wavelength shifts of three θ-shaped microfiber resonators to SRI. Adapted with permission Xu et al. (2017b)

RI sensing and the highest sensitivity of 35,823.3 nm/RIU (Fig. 14c, d). Biosensing application was also performed for cardiac troponin T (cTnT) biomarker detection, and a detection limit of 1 ng/mL was achieved (Fig. 14e). Compared with other Vernier effect enhanced fiber-optic sensors that utilize cascaded schemes, this sensor is simple and compact in structure and shows high stability. Although the Vernier effect has only been demonstrated in microfiber ring resonators and OMCs, it is applicable to other microfiber sensors to achieve improved performance.

3.2.3 Integration with Functional Nanomaterials

The integration of functional nanomaterials with optical microfibers can bring about novel optical properties and provide a versatile platform for various sensing applications. Multifarious nanomaterials with diverse optical, optoelectronic, and photothermal properties can be readily incorporated on the surface of optical microfibers via electrostatic interactions or covalent binding to compose novel sensors (Fig. 15). By decorating silica nanoparticles on the surface of a tapered multimode optical fiber, a Mie scattering enhanced RI sensor can be constructed. The measured sensitivity in terms of transmitted optical intensity raises two orders

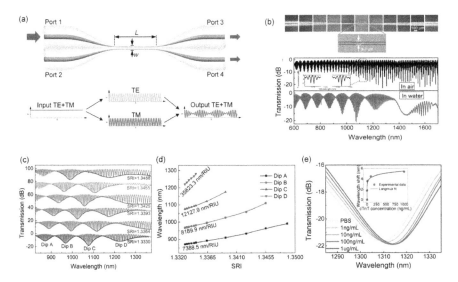

Fig. 14 Vernier effect in OCMs for enhanced sensing. **a** Basic sensing scheme and mechanism. **b** An OMC with a waist width of 3.2 μm, and the transmission spectra in both air and water, with the Vernier effect. **c** Spectral response of the sensor to SRI. **d** The dependence of wavelength shift on SRI. **e** Spectral response of the sensor for cardiac troponin T biomarker detection. Adapted with permission Li et al. (2018a)

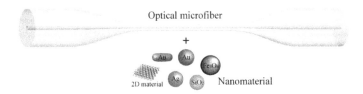

Fig. 15 The basic concept of integration of functional nanomaterials with the optical microfibers

of magnitude for a 2.8-μm-thick microfiber after modified by silica nanospheres (Liu et al. 2012). Decorating the microfiber surface with plasmonic nanoparticles can make a compact sensing probe with enhanced light-matter interactions (Zhou et al. 2019). The strong evanescent field can exit the localized surface plasmon resonance of the plasmonic nanoparticles and confine the light in narrow spaces of about tens of nanometers over the surface of the nanoparticles. Thus, ultra-high sensitivities can be achieved.

To this end, understanding the optical interaction between the guided light with the nanomaterials is the key to unlocking the potential. Researches have built a model to analyses the strength of the interaction on the single-nanoparticle level (Fig. 16a) by taking the plasmonic gold nanosphere as an example (Li et al. 2018d). The proportional optical power encountered by a single nanoparticle is a valid indicator of the strength of light-nanoparticle interaction. The numerical result in Fig. 16b shows that,

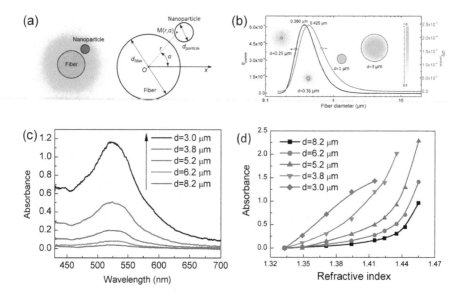

Fig. 16 **a** Schematic diagram of the model. **b** Calculated $\eta_{particle}$ and $d\eta_{particle}$ of a 20 nm-sized nanoparticle located on the surface of an optical microfiber as a function of fiber diameter (wavelength: 520 nm). **c** Absorption spectra of optical microfibers with different diameters as gold nanoparticle binding reaches saturation. **d** RI sensing performances of optical microfiber-based LSPR sensors with different fiber diameters. Adapted with permission Li et al. (2018d)

when the diameter of the microfiber decreases from 10 μm to sub-micron, the proportional optical power $\eta_{particle}$ encountered by a 20-nm gold nanosphere dramatically improves and reaches a peak value at $d = 0.36$ μm. However, when it falls below 0.36 μm, the proportional optical power near fiber surfaces is relatively reduced. Thus, generally, thinner optical microfibers are preferable to achieve better sensing performance as long as $d > 0.36$ μm. This tendency was proved experimentally using gold nanosphere decorated optical microfibers with different diameters, as shown in Fig. 16c, d.

Based on the preliminary analysis, a highly integrated microfiber LSPR sensor decorated with high quality and monodisperse β-cyclodextrin-capped gold nanoparticles (β-CD-capped AuNPs) has been constructed (Fig. 17a, b). The bi-functional gold nanoparticles, which were synthesized in a one-step facile and eco-friendly process, can effectively deliver promising localized surface plasmon resonance properties. Moreover, the β-CD molecule can also work as the receptor for cholesterol, which is of great diagnostic significance for various disorders, including coronary heart disease, atherosclerosis, hypertension, and nephrosis. The proposed biosensor reaches an ultra-low cholesterol LOD of 5 aM (shown in Fig. 17c), which is the most sensitive among the state-of-the-art cholesterol detections (Zhang et al. 2019).

Two-dimensional (2D) plasmonic nanomaterials facilitate exceptional light-matter interaction and enable in situ plasmon resonance tunability. By introducing heavily doped MoO3-x nanoflakes that provide strong plasmon resonance located

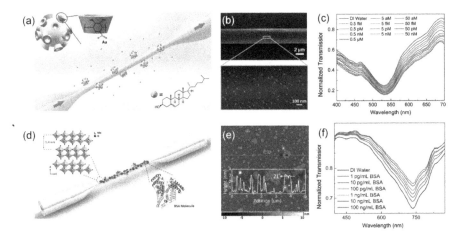

Fig. 17 **a** Schematic diagram of the integrated microfiber LSPR biosensor capped with β-CD-capped AuNPs. **b** SEM images showing the profile of the optical microfiber and the nanoparticles on the fiber surface. **c** Spectral responses of the sensor to cholesterol samples. Adapted with permission Zhang et al. (2019). **d** Schematic diagram of the microfiber optic biosensor integrated with two-dimensional molybdenum trioxides. **e** Atomic force microscope (AFM) image of the two-dimensional molybdenum trioxides. **f** Spectral responses of the sensor to BSA samples. Adapted with permission Zhang et al. (2018b)

at ~735 nm onto the surface of an optical microfiber, a new sensing platform was constructed (Fig. 17d). The MoO3-x nanoflakes (Fig. 17e) show an excellent affinity to negatively charged biomolecules. A LOD of 1 pg/mL was experimentally demonstrated for the detection of bovine serum albumin (Fig. 17f). It proved the feasibility and prospects of employing 2D plasmonic materials in highly integrated devices compliant with frequently used and a cost-effective optical system (Zhang et al. 2018b).

Aside from working as the sensing element, the nanomaterials can also function as the signal indicator and amplifier in optical microfiber biosensors (Li et al. 2014). The primary sensing scheme of a microfiber biosensor is illustrated in Fig. 18a, in which a sandwiched immunoassay method was adopted. First, a single layer of AFP specific first antibodies was coated on the surface of the optical microfiber. Then, the samples were injected, and the first antibodies captured the biomarkers. Finally, the secondary antibody-coated gold nanoparticles are injected and bind to the surface of the fiber through the antibody-antigen conjugations. The captured gold nanoparticles that laid inside the evanescent field of the microfiber can induce significant loss. The loss is proportional to the quantity of the nanoparticles, and hence the number of the captured biomarkers. This sensing strategy was experimentally verified using a 1-μm-thick optical microfiber and 40-nm gold nanoparticles as the label. The concentration of AFP biomarker was successfully detected with a LOD of 2 ng/mL in bovine serum (Fig. 18b–e). The advantages of this biosensor are simple detection scheme, fast response time, the immunity to other irrelevant proteins in complex

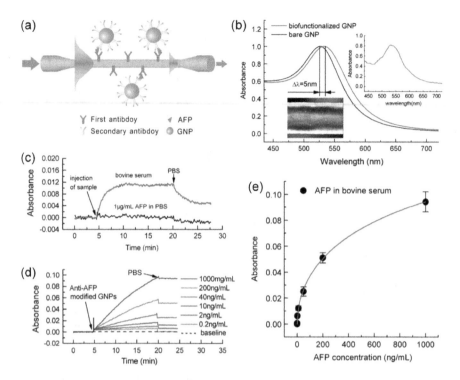

Fig. 18 a Schematic diagram of the microfiber biosensor that utilizes the gold nanoparticles as the signal indicator and amplifier. **b** Absorption spectra of gold nanoparticle solution and biofunctionalized gold nanoparticle solution. The insets show the absorption spectra of the secondary antibody functionalized gold nanoparticle binding to a biomarker coated microfiber and an SEM image of the fiber decorated with GNPs. **c, d** Real-time responses of the sensor to AFP biomarkers that are spiked in bovine serum. **e** Calibration curve of the biosensor as a function of analyte target concentration in bovine serum. Adapted with permission Li et al. (2014)

samples, which might make this biosensor a promising platform for clinical cancer diagnosis and prognosis.

4 Conclusion

This chapter reviewed the fundamental optical properties, fabrications techniques, the well-developed sensing schemes as well as the latest progress on new effects and strategies for enhanced sensing. The optical microfiber, with the unique property of strong evanescent field and compact micron-scale size, can serve as building blocks to construct versatile sensing schemes with unprecedented sensitivities. On the one hand, new sensing schemes with better performances will be continuously developed; on the other hand, the integration of functional nanomaterials with the microfiber has opened new opportunities and will bring prosperity to this field. The optical microfiber sensors are also facing some challenges, which have blocked the path to practical applications. One challenge is the batch production of optical microfibers with high uniformities. The other is effective packaging, which can protect the microfiber-based devices from external continents and provide long-term stability. It is believed that remarkable progress will be made in the near future to address the challenges mentioned above.

References

J.H. Chen, D.R. Li, F. Xu, J. Light. Technol. **37**, 2577 (2019)

M. Ding, M.N. Zervas, G. Brambilla, Opt. Express **19**, 15621 (2011)

X. Fang, C.R. Liao, D.N. Wang, Opt. Lett. **35**, 1007 (2010)

A. Felipe, G. Espíndola, H.J. Kalinowski, J.A.S. Lima, A.S. Paterno, Opt. Express **20**, 19893 (2012)

X. Guo, L. Tong, Opt. Express **16**, 14429 (2008)

J.E. Hoffman, S. Ravets, J.A. Grover, P. Solano, P.R. Kordell, J.D. Wong-Campos, L.A. Orozco, S.L. Rolston, AIP Adv. **4**, 067124 (2014)

Z. Hu, W. Li, Y. Ma, L. Tong, Opt. Lett. **37**, 4383 (2012)

J.L. Kou, M. Ding, J. Feng, Y.Q. Lu, F. Xu, G. Brambilla, Sensors (Switzerland) **12**, 8861 (2012)

J. Li, L.-P. Sun, S. Gao, Z. Quan, Y.-L. Chang, Y. Ran, L. Jin, B.-O. Guan, Opt. Lett. **36**, 3593 (2011)

K. Li, G. Liu, Y. Wu, P. Hao, W. Zhou, Z. Zhang, Talanta **120**, 419 (2014)

K. Li, T. Zhang, G. Liu, N. Zhang, M. Zhang, L. Wei, Appl. Phys. Lett. **109**, 101101 (2016)

B. Li, J. Chen, F. Xu, Y. Lu, Opt. Express **25**, 4326 (2017)

K. Li, N. Zhang, N.M.Y. Zhang, W. Zhou, T. Zhang, M. Chen, L. Wei, Sensors Actuators, B Chem. **275**, 16 (2018a)

K. Li, N. Zhang, N.M.Y. Zhang, G. Liu, T. Zhang, L. Wei, Opt. Lett. **43**, 679 (2018b)

K. Li, N.M.Y. Zhang, N. Zheng, T. Zhang, G. Liu, L. Wei, J. Light. Technol. **36**, 2409 (2018c)

K. Li, W. Zhou, S. Zeng, Sensors (Switzerland) **18**, 3295 (2018d)

G. Liu, Y. Wu, K. Li, P. Hao, P. Zhang, M. Xuan, IEEE Photonics Technol. Lett. **24**, 658 (2012)

H. Luo, Q. Sun, X. Li, Z. Yan, Y. Li, D. Liu, L. Zhang, Opt. Lett. **40**, 5042 (2015a)

Z.-C. Luo, M. Liu, Z.-N. Guo, X.-F. Jiang, A.-P. Luo, C.-J. Zhao, X.-F. Yu, W.-C. Xu, H. Zhang, Opt. Express **23**, 20030 (2015b)

H. Luo, Q. Sun, Y. Li, D. Liu, L. Zhang, IEEE Sens. J. **16**, 4793 (2016)

Y. Ma, G. Farrell, Y. Semenova, B. Li, J. Yuan, X. Sang, B. Yan, C. Yu, T. Guo, Q. Wu, Opt. Laser Technol. **78**, 101 (2016)

Y. Mao, Y. Zhang, R. Xue, Y. Liu, K. Cao, S. Qu, Appl. Opt. **57**, 1061 (2018)

P. Polynkin, A. Polynkin, N. Peyghambarian, M. Mansuripur, Opt. Lett. **30**, 1273 (2005)

Y. Ran, Y.-N. Tan, L.-P. Sun, S. Gao, J. Li, L. Jin, B.-O. Guan, Opt. Express **19**, 18577 (2011)

M. Sumetsky, Y. Dulashko, J.M. Fini, A. Hale, D.J. DiGiovanni, J. Light. Technol. **24**, 242 (2006)

L.-P. Sun, J. Li, S. Gao, L. Jin, Y. Ran, B.-O. Guan, Opt. Lett. **39**, 3531 (2014)

H. Tazawa, T. Kanie, M. Katayama, Appl. Phys. Lett. **91**, 113901 (2007)

L. Tong, R.R. Gattass, J.B. Ashcom, S. He, J. Lou, M. Shen, I. Maxwell, E. Mazur, Nature **426**, 816 (2003)

L. Tong, J. Lou, E. Mazur, Opt. Express **12**, 1025 (2004)

G. Vienne, Y. Li, L. Tong, P. Grelu, Opt. Lett. **33**, 1500 (2008)

C. Wang, L. Zeng, Z. Li, D. Li, J. Raman Spectrosc. **48**, 1040 (2017)

U. Wiedemann, K. Karapetyan, C. Dan, D. Pritzkau, W. Alt, S. Irsen, D. Meschede, Opt. Express **18**, 7693 (2010)

L. Xiao, T.A. Birks, Opt. Lett. **36**, 1098 (2011)

L. Xiao, M.D.W. Grogan, W.J. Wadsworth, R. England, T.A. Birks, Opt. Express **19**, 764 (2011)

F. Xu, P. Horak, G. Brambilla, Opt. Express **15**, 9385 (2007)

Y. Xu, W. Fang, L. Tong, Opt. Express **25**, 10434 (2017a)

Z. Xu, Y. Luo, D. Liu, P.P. Shum, Q. Sun, Sci. Rep. **7**, 1 (2017b)

H. Xuan, W. Jin, M. Zhang, Opt. Express **17**, 21882 (2009)

H. Xuan, J. Ju, W. Jin, Opt. Express **18**, 3828 (2010)

T.K. Yadav, R. Narayanaswamy, M.H. Abu Bakar, Y.M. Kamil, M.A. Mahdi, Opt. Express **22**, 22802 (2014)

S.W. Yang, T.L. Wu, C.W. Wu, H.C. Chang, J. Light. Technol. **16**, 691 (1998)

C. Yu, Y. Wu, X. Liu, F. Fu, Y. Gong, Y.J. Rao, Y. Chen, Sensors Actuators. B Chem. **244**, 107 (2017)

N.M.Y. Zhang, K. Li, N. Zhang, Y. Zheng, T. Zhang, M. Qi, P. Shum, L. Wei, Opt. Express **26**, 29148 (2018a)

N.M.Y. Zhang, K. Li, T. Zhang, P. Shum, Z. Wang, Z. Wang, N. Zhang, J. Zhang, T. Wu, L. Wei, ACS Photonics **5**, 347 (2018b)

N.M.Y. Zhang, M. Qi, Z. Wang, Z. Wang, M. Chen, K. Li, P. Shum, L. Wei, Sensors Actuators, B Chem. **286**, 429 (2019)

Q. Zhang, R. Zhai, S. Yang, S. Yang, Y. Li, ACS Photonics (2020). https://doi.org/10.1021/acsphotonics.9b01560

W. Zhou, K. Li, Y. Wei, P. Hao, M. Chi, Y. Liu, Y. Wu, Biosens. Bioelectron. **106**, 99 (2018)

N. Zhou, P. Wang, Z.X. Shi, Y.X. Gao, Y.X. Yang, Y.P. Wang, Y. Xie, D.W. Cai, X. Guo, L. Zhang, J.R. Qiu, L.M. Tong, Opt. Express **27**, 8180 (2019)

Fiber-Based Infrasound Sensing

Shun Wang, Wenjun Ni, Liang Zhang, Ping Lu, Yaowen Yang, and Lei Wei

Abstract With the maturity of optical fiber sensing technology, its related applications have penetrated into various fields. Fiber-based acoustic sensors (FAS) are one of the most important research fields in fiber optic sensors. So far, it has been widely applied in natural disaster warning, medical diagnosis, geological exploration, and even battlefields. In recent years, natural disasters such as earthquakes have occurred frequently, and modern naval warfare technology has developed rapidly. Due to the special demand on acoustic detection in these fields, FAS have developed in the direction of high precision, low frequency, low cost, and miniaturization. Aiming at the low frequency or infrasound detection requirements for special applications and taking the advantages of fiber interferometric sensing structure, this chapter proposes a composite diaphragm-type fiber (external Fabry–Perot interferometer) EFPI infrasound sensor. The circular composite film is formed by combining a polymer film and an aluminum foil; then, an EFPI interferometer structure is formed between the

S. Wang (✉)
Wuhan Institute of Technology, Wuhan, China
e-mail: 18164113661@163.com

W. Ni · L. Wei
School of Electrical and Electronic Engineering, Nanyang Technological
University, Singapore, Singapore
e-mail: wenjun.ni@ntu.edu.sg

L. Wei
e-mail: wei.lei@ntu.edu.sg

L. Zhang
Shenzhen Institutes of Advanced Technology, Chinese Academy of Sciences, Shenzhen, China
e-mail: liang.zhang@siat.ac.cn

P. Lu
Huazhong University of Science and Technology, Wuhan, China
e-mail: pluriver@mail.hust.edu.cn

Y. Yang
School of Civil and Environmental Engineering, Nanyang Technological
University, Singapore, Singapore
e-mail: cywyang@ntu.edu.sg

© Springer Nature Singapore Pte Ltd. 2020
L. Wei (ed.), *Advanced Fiber Sensing Technologies*,
Progress in Optical Science and Photonics 9,
https://doi.org/10.1007/978-981-15-5507-7_5

composite film and the end face of the fiber FC joint. By theoretically simulating and optimizing the materials, sizes, and structures of the transducer, the relevant parameters are selected according to the detecting requirements. Then, the fiber EFPI infrasound sensor is well fabricated and packaged. In the simulated infrasound field environment, a comparison test was performed with a standard B&K infrasound sensor. The experimental results show an acoustic sensitivity of up to -138.3 dB re 1 V/μPa ($\sim$121 mV/Pa) in the infrasound frequency range of 1–20 Hz, which is higher than the commercial acoustic sensor used for comparison.

Keywords Optical fiber sensing · Fiber-based infrasound sensor · External Fabry-Perot interferometer

1 Introduction

Fiber-based acoustic sensors (FAS), also known as fiber optic microphones, are an important branch of fiber optic sensors (Teixeira et al. 2014; Wild et al. 2008). FAS technology combines the advantages of acoustic wave detection and analysis technology with optical fiber sensing. In recent years, it has been widely used in many fields such as seismic wave detection, fiber optic hydrophone, non-destructive testing, industrial process control, and structural health monitoring (SHM) (Le Pichon et al. 2006; Barruol et al. 2006; Cusano et al. 2006; Huke et al. 2013; Tsuda et al. 2010). The detection and monitoring of sound waves in low frequency and even infrasound bands is of great significance in some special fields. For example, volcanoes, avalanches, lightning, and nuclear explosions are accompanied by the generation of low-frequency sound waves. This low-frequency sound wave has strong penetrating power, small attenuation, and long propagation distance. If such low frequency or infrasound can be detected, early warning of such incidents may be achieved, thereby greatly reducing the economic and property losses caused by such incidents (Matoza et al. 2011; Green et al. 2011; Sindelarova et al. 2009; Morrissey et al. 2008). Therefore, in recent years, fiber-based infrasound sensing technology has gradually become a research hot spot for researchers and engineers.

According to the detection principle, fiber-based infrasound sensing technology reported in recent years can be divided into three types: capacitive, piezoelectric, and optical. The optical fiber-based infrasound sensor (FIS) can be divided into the following types according to the sensing principle: wavelength type, intensity type, laser type, phase type, polarization type, etc. FIS has many advantages over the first two electrical types: high sensitivity, high signal-to-noise ratio (SNR), large dynamic range, anti-electromagnetic interference, and capable to work in complex environments such as high temperature and high pressure. FIS will be an inevitable trend in the future due to its excellent characteristics. Besides, the wavelength-type FIS is suitable for detecting static sound pressure signals, but not suitable for detecting dynamic signals, and the size is often large since the wavelength-type device (such

as FBG) is large itself; The intensity-type FIS has a simple structure, easy to manufacture, convenient in demodulation, but it is susceptible to power source jitter and
environmental interference. The fiber laser-type FIS can achieve a high SNR and good
stability, but its structure is complex and the cost is relatively high. The sensitivity
of the phase-type FIS can achieve a very high level with and compact structure size,
but phase noise may be introduced which does harm to sound detection. Overall,
the corresponding type of FIS can be applied according to the actual application
requirements.

Low-frequency infrasound measurement is an important mean to warn of natural
disasters, military applications, etc. (D'spain et al. 1991; Wooler and Crickmore
2007). A series of FIS methods have been proposed in the prior art: wavelength type,
intensity type, and laser type, but it is difficult to break their lowest limit of their
frequency detection 20 Hz. The reason can be mainly attributed to two aspects: First,
the sensitivity that can be achieved by the above non-interference methods is limited,
while the disturbance caused by the infrasound signal is usually very weak, so it is
difficult to meet the requirement. On the other hand, in the free field for infrasound
measurement, extremely large transducer size is required but that is clearly not readily
achievable in the laboratory. Therefore, when the low-frequency infrasound coupling
cavity sound source (pressure field) is used to simulate the infrasound environment,
the sensing head needs to be placed inside the coupling cavity, so that a smaller
sensing device is required to meet the requirements. That is to say, from the actual
situation, it is necessary to design an interference-type sensor with high sensitivity
and small size to realize the infrasound sensing measurement.

Therefore, combined with the actual needs, using experimental conditions such
as laboratory infrasound source, processing unit, and detection environment, we
design and fabricate a fiber optic external Fabry–Perot interferometer (EFPI)-type
FIS based on the composite film. And the infrasound test was performed with good
experimental results. This chapter firstly introduces the basic theory of optical fiber
EFPI infrasound sensor and determines the design standards based on the principle
of infrasound detection and experimental conditions. On this basis, the parameters
in terms of structure, materials, and size of the sensor are optimized and processed.
Infrasound measurement experiments were performed in combination with experimental instruments and test environments to achieve high-sensitivity (~121 mV/Pa)
infrasound sensing application in the infrasound frequency range (1–20 Hz).

2   Main Body Text

2.1   Basic Theory of Acoustics

Acoustics is a branch of physics, an ancient and developing discipline. Along with
the continuous development of science and technology, acoustic research has gradually penetrated into other various natural science fields. Besides, it is promoting

the germination and growth of many marginal disciplines, and also organically combining with other disciplines for common development, for example, surface acoustic wave devices, medical acoustics, molecular acoustics, cryogenic acoustics, photo-acoustics and sonoluminescence, and thermo-acoustics. Modern acoustics will become more and more important in various industrial and agricultural and service industries. Research on acoustics is committed to improving people's working environment and productivity.

Sound is a mechanical fluctuation that propagates in gaseous, liquid, and solid matter. The study on sound is to study about its production, transmission, reception, action, and reproduction processing. In the spread process of the sound, it also causes property change in terms of the optical, electromagnetic, mechanical, chemical, and human psychological and physiological properties, which in turn affect the spread itself. Therefore, the scope of acoustic research is extremely wide and there are many branches.

It is well known that the vibration of an object often leads to the generation of sound. For example, when the guitar string vibrates, it will produce a pleasant music sound. When the speaker's acoustic paper vibrates, it can make a sound, and hitting the glass will make a "beep" sound. Then, how does the vibration of the object spread into the human ear and we feel the sound? This process actually contains two aspects: First, how the vibration of the object spreads to the human ear and causes the tympanic membrane inside the human ear to vibrate; second, how the vibration of the tympanic membrane makes us feel the sound. The latter question belongs to the category of physiology and it will not be discussed in this chapter. Here, we only focus on the first aspect of how the vibration of an object propagates through the medium.

When an object vibrates caused by some disturbance, a disturbance is also caused in a certain local area of the elastic medium, so that the medium particle A in the local area starts to move away from the equilibrium position. The movement of this particle A will push the adjacent medium particle B; then, the medium where the particle A locates compresses the adjacent medium where B locates, as shown in Fig. 1.

Since the medium has a certain elastic effect, when the adjacent medium B is compressed, a force against compression is generated, and this force also acts on A, which in turn returns A to the equilibrium position. In addition, since A obtains momentum and then moves forward due to inertia, it will continue to move forward, so that it will compress the adjacent medium on the other side, and the medium will also generate a force against compression. It also tends to balance the position movement. It can be seen that the particle A, which was initially disturbed, vibrates back and forth around the equilibrium position under the action of the elasticity and inertia of the medium. Similarly, the particle B pushed by A and even the particle C, D…farther away, they will vibrate back and forth around the equilibrium position. The propagation phenomenon of mechanical vibration like these medium particles from near to far is called the propagation of acoustic vibration or sound wave. Therefore, the sound wave is a kind of mechanical wave.

Fig. 1 Acoustic vibration
diagram

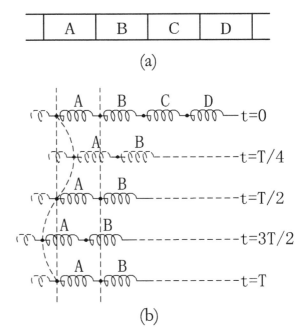

(a)

(b)

As can be seen from the above discussion, the propagation of sound waves is inseparable from the existence of elastic media. When the sound wave propagates in a fluid medium such as a gas or a liquid, it mostly appears as a longitudinal wave, that is, the propagation direction is consistent with the vibration direction of the particle, and when it propagates in a solid or other medium, it often performs as a transverse wave.

In addition, since the sound wave has a certain reversibility, its generation and reception process can be regarded as two opposite procedures. The vibration disturbance generates an acoustic wave, and the acoustic wave in return causes the vibration of the receiver to receive the acoustic signal.

2.1.1 Sound Pressure and Sound Pressure Level

The process of sound generation, propagation, and reception is discussed above, but it is not enough to describe the various properties of sound waves. Therefore, it is necessary to use a certain physical quantity to accurately describe the process of sound waves.

Assuming that the pressure in the particle changes from P_0 to P_1 after the particle is disturbed by the acoustic signal, the excess pressure generated by the acoustic perturbation signal, referred to as overpressure, can be expressed as:

$$p = P_1 - P_0 \tag{1}$$

This is the so-called sound pressure. In the process of sound propagation, the internal sound pressure is different under different mass particles at the same time, and as the same mass particle at different times, the sound pressure p changes with time. It can be seen that the sound pressure is generally a function of time and space, and it can be expressed as $p = p(x, y, z, t)$. Similarly, the amount of density change caused by this acoustic disturbance will also be a function of time and space, which can be expressed as $p' = p - p_0 = p'(x, y, z, t)$.

In addition, according to the previous discussion, the sound wave is the vibration propagation of the medium particles; then, the particles' vibrating velocity can also be used to describe the characteristic of the sound wave. But since the particle vibrating velocity is not easy to measure compared with the sound pressure, usually people measure the sound pressure to indirectly find other required physical quantities such as particle velocity. For the time being, sound pressure is also a physical quantity commonly used to describe the properties of the sound wave.

The space in which sound pressure presents is generally called as the sound field. The sound field in the free space of the outside world can be called the free field, while the sound field in a certain closed space is called the pressure field. The sound pressure somewhere in the sound field changes with time; then, the sound pressure value at a certain moment is called the instantaneous sound pressure, and the maximum instantaneous sound pressure in a certain time interval is called the peak sound pressure.

If the sound pressure changes with time according to the harmonic law, the peak sound pressure is the amplitude of the changed sound pressure. The effective sound pressure during this time can be expressed by the instantaneous sound pressure versus the root mean square.

$$p_{\text{eff}} = \sqrt{\frac{1}{T} \int_0^T p^2 dt} \tag{2}$$

where T represents this time interval. Generally, the sound pressure value measured by an electronic instrument such as a sound level meter is often an effective sound pressure. The sound pressure generally referred to is often the effective sound pressure indicated here.

The magnitude of sound pressure directly reflects the strength of the sound wave. The physical definition is the change amount of atmospheric pressure per unit area in the presence of the sound wave. The unit is generally Pa (Pa):

$$1\,\text{Pa} = 1\,\text{N/m}^2 \tag{3}$$

In addition, some other units are sometimes used to indicate sound pressure units, such as:

bar: 1 bar $= 100\,\text{kPa} = 103\,\text{mbar} = 106\,\mu\text{bar}$
Atm (a standard atmospheric pressure): 1 Atm $\approx 1 \times 105\,\text{Pa} = 1\,\text{bar}$
Psi: 145 Psi $= 1\,\text{MPa}$

It is well known that the frequency of sounds that can be heard by the human ear ranges from 20 Hz to 20 kHz (Wang et al. 2013). Then, the human ear has a tolerance range for the strength of the sound wave. For sounds at 1 kHz, the audible threshold of the human ear (which happens to be able to feel its presence) is about 20 μPa and the maximum sound pressure is about 20 Pa.

When measuring parameters such as sound pressure, sound intensity, and sound power in acoustics, it is often measured by a logarithmic scale. This is mainly based on the following two reasons: First, the strong sound power and the weaker sound power are usually different by more than a dozen orders of magnitude. Obviously, the logarithmic scale is more convenient than the absolute scale. Secondly, when the human ear receives the sound, especially the sound subjectively produced is generally not proportional to the absolute value of the intensity, but is approximately proportional to the logarithm of the intensity. Therefore, the logarithmic measure is generally accepted.

The logarithmic measure of sound pressure and sound intensity is called the sound pressure level and sound intensity level, and its unit is usually expressed in dB (decibel).

1. Sound pressure level (SPL)

$$SPL = 20 \log \frac{p}{p_0} = 10 \log \frac{p^2}{p_0^2} (\text{dB}) \tag{4}$$

In the above formula, p is the effective value of the sound pressure to be measured, and p_0 is the reference sound pressure, and the audible threshold sound pressure of the human ear is generally 20 μPa in the air.

2. Sound intensity level (SIL)

$$\text{SIL} = 10 \log \frac{I}{I_0} (\text{dB}) \tag{5}$$

Here, I is the sound intensity to be measured, and I_0 is the reference sound intensity which is generally 1 pW/m², i.e., 1×10^{-12} W/m² in the air. This is also the sound intensity corresponding to the sound pressure of 20 μPa, that is, the audible threshold sound intensity value of the 1 kHz sound.

The sound intensity indicates how many watts of sound energy pass through per unit area, and its numerical approximation can be regarded as the square of the sound pressure. From Eqs. (4) and (5), the SPL and the SIL are equal (when the characteristic impedance is 400 N s/m).

In addition, sound power is sometimes involved, which means the total sound energy radiated outward per unit time, within units of W or mW.

2.1.2   Acoustic Measurement Accuracy

In order to characterize the performance of an acoustic sensor, one usually uses a series of parameters to indicate its measurement accuracy, such as sound pressure sensitivity, minimum measurement accuracy, and equivalent noise sound pressure. The physical meaning of the definition of these parameters is described below.

1. Sound pressure sensitivity

Sound pressure sensitivity refers to the ratio of the open-circuit voltage outputted from the output of the receiving sensor to the sound pressure actually received on the receiving surface of the sensor, within volts per kPa (V/Pa).

$$\text{Sound pressure sensitivity} = \frac{\text{Output voltage}}{\text{Received sound pressure}} \tag{6}$$

Generally, for the convenience of calculation, the magnitude of sound pressure sensitivity is often expressed in decibels. The reference amount is usually 1 V/μPa. For example, the sound pressure sensitivity of 1 V/μPa is expressed in decibels as: -120 dB re 1 V/μPa.

2. Minimum measurement accuracy

In the field of sensors, in order to characterize the characteristics of the sensor, the minimum measurement accuracy is usually used. This parameter generally depends on two factors: instrument resolution and noise level. Therefore, there should be two ways to express it.

(1) $\frac{\text{Instrument minimum resolution}}{\text{Sensitivity}}$. Assuming an optical spectral analyzer (OSA) with a wavelength resolution of 20 pm and a typical grating temperature sensor with a sensitivity of 10 pm/°C, so the minimum measurement accuracy is 2 °C.

(2) $\frac{\text{Minimum noise pressure}}{\text{Sensitivity}}$. In addition to instrumental factors, interference from the external environment can also affect measurement accuracy. If the voltage generated by the acoustic sensor in the minimum noise environment is 0.001 mV and the sensitivity of the acoustic sensor is 10 mV/Pa, the sound pressure measurement accuracy should be 0.1 mPa.

In addition, when two values work simultaneously, the greater of the two should be used.

3. Equivalent noise sound pressure

In addition, the acoustic sensor is often equivalent to the sound pressure level to measure the effect of noise on the acoustic sensor. The smaller the equivalent noise sound pressure level, the better the noise immunity of the acoustic sensor. Its calculation method is as follows:

$$N_{eq} = \frac{\text{Sound pressure value}}{SNR(\text{Numerical value}) \times \sqrt{3} \text{ dB Bandwidth}} \tag{7}$$

For example, when the received sound pressure value is 400 mPa, the detected spectral signal-to-noise ratio (SNR) is 57.5 dB, and the 3 dB bandwidth of the spectral component is 50 Hz, then the equivalent noise sound pressure level obtained by this calculation is about 75 $\mu Pa/Hz^{1/2}$.

It should be pointed out that the sensor sensitivity and accuracy are the specific parameters to measure its index, and the equivalent noise sound pressure is a comprehensive performance of the sensor. In addition to the above parameters that characterize the measurement accuracy, the acoustic sensor often introduces parameters such as frequency range, sound pressure range, resonant frequency, and acoustic directivity, which are described in subsequent sections.

2.2 Design Guidelines and Fabrication

2.2.1 Design Guidelines of the Sensor Head

It is mentioned in the introduction section that in combination with the actual requirements for sensitivity and sensor element size, it is necessary to design an interference-type small-sized sensing structure. The FP structure is generally realized in the interference structure, and the infrasound acoustic coupling cavity (high-pressure microphone calibrator, B&K 4221) requires a 1/2-in. sensor head, while the FP structure inline is too small, so the external cavity FP structure is selected. The outer end of the fiber FP structure designed in this section is a reflecting surface, and the other one is composed of the surface of the thin-film transducer.

When the film structure is subjected to a low-frequency infrasound signal, it will vibrate to modulate the cavity length of the EFPI. When a stable single-wavelength laser is incident on the EFPI, the output power is synchronously changed due to the oscillation of FSR or wavelength. As can be seen from Fig. 2, the infrasound signal will be demodulated in the form of an output voltage signal. In addition, the wavelength measurement should be within the slope range of half the FSR, which is 1.8 nm in this figure.

It is known from the basic theory of acoustics that for a circular film fixed at the perimeter, the acoustic sensitivity is determined by the maximum deformation caused by the sound pressure change per unit amount ΔP (Ma et al. 2013), i.e.,

$$S = \frac{def}{\Delta P} = \frac{3(1 - \mu^2)r^4}{16Eh^3} \tag{8}$$

Here, r, h, μ, and E are radius, thickness, Poisson's ratio, and Young's modulus of the film, respectively. It can be seen from the above formula that in order to obtain a large acoustic sensitivity, a material having a large Young's modulus and a relatively small Poisson is required. The geometrical structure requires a small film thickness and a large radius. For this purpose, a polyethylene terephthalate (PET) film material was used in the experiment ($\mu = 0.39$, $E = 2.5$ GPa). And the geometrical

Fig. 2 Schematic diagram
of infrasound disturbance
into output signal

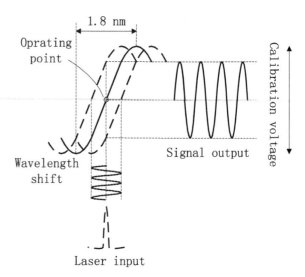

dimensions of the film are as follows: $h = 50\ \mu m$, $r = 9.6\ mm$. According to Formula (8), the sound pressure sensitivity can be calculated up to 4.32 $\mu m/Pa$, which is much larger than that reported in reference (Ma et al. 2013). In order to ensure a good linear relationship between the output power and the cavity length change, the cavity length variation should be less than 1/4 of the wavelength (Wang and Ma 2013), which is 375 nm. This indicates that the SPL of the low-frequency acoustic signal should be less than 73 dB.

For the circular film fixed at the perimeter, the fundamental resonance frequency can be expressed as the equation $f_n = \mu_n \frac{c}{2\pi a} = \frac{\mu_n}{2\pi a}\sqrt{\frac{T}{\sigma}}$. According to the formula, the basic resonant frequency of the sensing head is about 300 Hz (density is 1380 kg/m^3). Generally, near the resonant frequency, the deformation sensitivity of the film is high, a peak appears, and the sensitivity of the film deformation away from the resonant frequency is generally balanced. Therefore, in order to obtain a relatively flat frequency response, the maximum operating frequency of the sensor should be less than 1/3 of the fundamental resonant frequency (Xu 2005). That is, the maximum measurement frequency should be less than 100 Hz. In this chapter, the measured infrasound signal can only reach 20 Hz at most, which is obviously in line with the application conditions.

2.2.2 Manufacturing Process

The schematic diagram of the manufacturing process of the optical fiber EFPI infrasound sensor is shown in Fig. 3. Before starting to package the sensing head, it is necessary to design and fabricate various components such as a threaded cylinder,

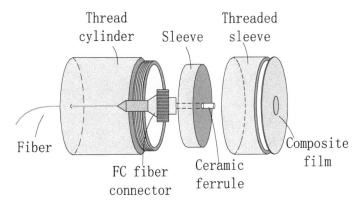

Fig. 3 Manufacturing process of optical fiber EFPI sensor head

a sleeve, and a threaded sleeve. The entire production process can be described as follows:

First, the pigtail of the FC fiber connector passes through a small round hole with a diameter of 1 mm at the bottom of the threaded sleeve (slightly larger than the pigtail diameter of 0.9 mm). Thus, the ceramic ferrule portion of the fiber joint can be fixed inside the threaded sleeve, but the axial direction is not fixed.

Then, use a small sleeve with a 3-mm-diameter hole at the bottom to allow a part of the ceramic ferrule (slightly smaller than 3 mm) to pass through the small hole, so that the ceramic ferrule can be fixed and the axial direction and the bottom of the threaded sleeve are vertical. In order to ensure sealing and stability, the two small hole portions are sealed with hot melt adhesive.

Next, a bottomless threaded sleeve with a thread matching the threaded cylinder is used to interface with the above structure, and they are connected by matching threads. In this way, the distance from the end face of the ceramic ferrule to the top plane of the threaded sleeve (prepared for placing the reflective surface of the film) can be flexibly adjusted by rotating the threaded sleeve.

Finally, in order to obtain a high-sensitivity low-frequency effect, a PET film is used as a transducing film to be flatly attached to the tip end of the threaded sleeve. In addition, a small piece of cold-rolled round aluminum foil (Goodfellow AL000320) is attached to the inner center of the PET film.

Thus, the fiber EFPI structure is completed. It should be mentioned that this small piece of aluminum foil has two functions: One is to improve the reflectivity of the film structure. The other is to ensure the smoothness and stability of the vibration plane. Because the deformation of the PET film at different positions during vibration is different, the composite film with aluminum foil is bonded, which forms a "bump" structure. And the "bump" is a relatively stable vibration reflecting surface when vibrating, which can ensure that the two reflecting surfaces of the ceramic ferrule end face and the "bump" are parallel when the film vibrates.

Figure 4 is a physical diagram of the completed optical fiber EFPI infrasound sensing head. The inner and outer diameters of the entire cylindrical structure are

Fig. 4 Physical map of the optical fiber EFPI infrasound sensor head

19.2/23.2 mm, the height is 36 mm, and the structural size is small. In the composite film structure at the front end of the sensor head, the PET circular film has a thickness of 50 μm, a radius of 9.6 mm, an aluminum foil thickness of 3 μm, and a radius of 3 mm. Compared with PET film, the quality and size of aluminum foil are very small, so the influence of aluminum foil on the detection sensitivity and resonance frequency of the sensing structure is almost negligible.

Using a broadband ASE source and an optical spectrum analyzer, the spectrum of the EFPI infrasound sensor head can be observed as shown in Fig. 5. Compared with a single PET film, the reflectivity of the composite film structure is significantly improved (around 0.87), and the resulting stripe contrast is about 15 dB, where the fringe free spectral range (FSR) can be expressed as

$$\text{FSR} = \frac{\lambda^2}{2nL} \tag{9}$$

It can be seen from the figure that the FSR is about 3.6 nm, which indicates that the cavity length is about 0.33 mm. This cavity length (i.e., the spacing between the ceramic ferrule end face and the end face of the composite film structure) can be adjusted by rotating the threaded sleeve as needed. In other words, the FSR can be flexibly adjusted.

2.3 Experimental Study of Infrasound Sensor

The schematic diagram of the optical fiber EFPI sensing system is shown in Fig. 6. The optical part of the sensing system consists of a 1570 nm DFB laser source (long-term stability: $\leq \pm 0.02$ dB), a fiber optic circulator, and an EFPI infrasound sensor head. The output optical signal will be converted into an electrical signal by the photodetector (PD, New Focus 1623), and then, the electrical signal carrying the infrasound disturbance information will be collected by the generator module (B&K LAN-XI 3160). The generator module here is a complete and independent analytical test system that combines data input and generator output. It is the ideal

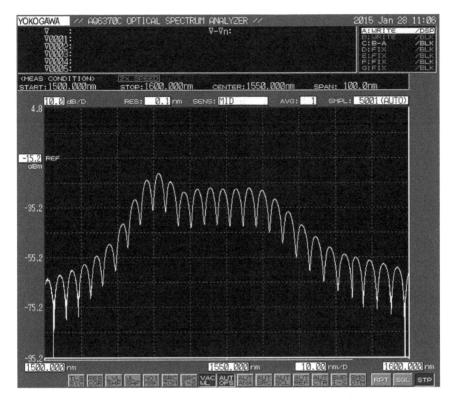

Fig. 5 Reflection spectrum of the optical fiber EFPI sensor

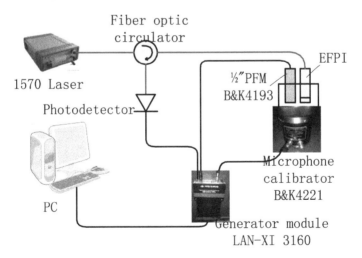

Fig. 6 Schematic diagram of the optical fiber EFPI sensing system

audio or electroacoustic test system device. It has two inputs and two outputs, and the other input is connected to a 1/2-in. standard pressure-field microphone (Pressure-field Microphone, B&K 4193) for calibration and comparison. Both the standard microphone and the manufactured EFPI infrasound sensor head are inserted into the high-pressure microphone calibrator (High Pressure Microphone Calibration, B&K4221; vocal range 0.01–1000 Hz; maximum sound pressure 164 dB). The high-voltage microphone calibrator is connected to an output of the generator module and driven by the generator module. The high-voltage microphone calibrator here is used as an infrasound source whose frequency and sound pressure amplitude are controllable. The data acquisition and signal output functions of the generator module are implemented by software operations in the computer.

In order to verify the accuracy of the sensor measurement, the high-voltage microphone calibrator is adjusted by software to maintain the acoustic signal at 71 dB sound pressure and the frequency is converted from 1 to 20 Hz (calibrated by the pressure-field microphone). Both the time-domain waveform and the corresponding Fourier transform spectrum can be recorded and displayed on the computer. Figures 7 and 8 show the time-domain waveforms of the detected 2 Hz and 3 Hz infrasound signals, respectively, from which one can clearly observe 20 and 30 times of amplitude vibrations within a time range of 10 s, showing time-domain waveform with approximate sine or cosine variation. That implies single-frequency infrasound disturbance signals at 2 Hz and 3 Hz, respectively, are demodulated qualitatively. And the sound pressure or sound pressure level (SPL) information can be indicated by the peak-to-peak values in the time-domain waveform.

In order to quantitatively determine the frequency information of the demodulated infrasound signals, a fast Fourier transform (FFT) processing is performed. The resulting frequency spectra are shown in Figs. 9 and 10, respectively. As can be seen from the figures, the SNR can reach more than 20 dB. This fully indicates that the fabricated fiber EFPI infrasound sensor can demodulate the infrasound disturbance signal well.

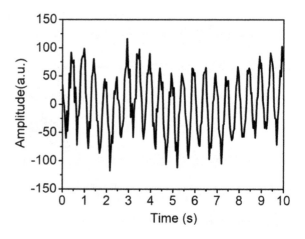

Fig. 7 2 Hz output time-domain waveform of the infrasound sensor

Fig. 8 3 Hz output time-domain waveform of the infrasound sensor

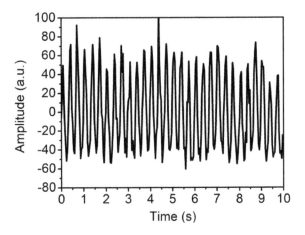

Fig. 9 2 Hz output frequency waveform of the infrasound sensor

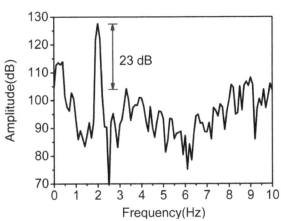

Fig. 10 2 Hz output frequency waveform of the infrasound sensor

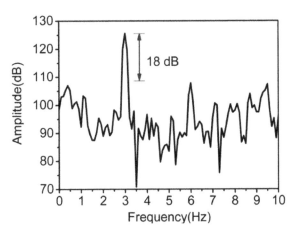

Fig. 11 1–20 Hz frequency response of the infrasound sensor

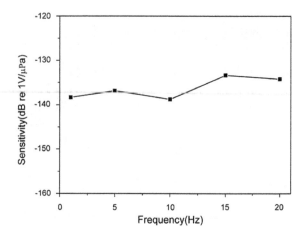

It is worth noting that the stability of the light source affects the signal demodulation of the intensity demodulation sensor. However, the 1570 nm DFB laser source used in this experiment has good stability, short-term 15-min stability: $\leq \pm 0.005$ dB; long-term 8-h stability: $\leq \pm 0.02$ dB. Therefore, the fluctuation of the light source has less influence on the EFPI infrasound sensor in this section.

Figure 11 shows the frequency response curve of the sensor head in the 1–20 Hz range in the infrasound band. It can be seen that the frequency response is relatively flat, and the sensitivity is concentrated in the range of -133 to -139 dB with fluctuations of ± 2.72 dB. The sensitivity at 1 Hz is as high as -138.3 dB re 1 V/μPa. Our original intention of an interference-type fiber infrasound sensor within high sensitivity and small size is well realized and experimentally demonstrated.

2.4 Discussion

The performance of the EFPI infrasound sensor in this chapter is compared with the recently reported schemes. The sensitivity value, frequency range, advantages, and disadvantages are listed in Table 1. It can be seen from the comparison that our fiber EFPI infrasound sensor has a great improvement in technical indicators and structural optimization.

3 Conclusion

In this chapter, a fiber optic EFPI infrasound sensor based on the composite film is developed, and the design and fabrication of the sensor head are performed by optimizing the parameters of structure, material, and size. And the high-sensitivity acoustic sensing application is realized in the infrasound frequency band

Table 1 Performance comparison of infrasound sensors reported in recent years

Sensing type	Sensitivity	Frequency range	Disadvantages	References
Capacitive type	3.5 mV/Pa @1 kHz	100 Hz–10 kHz	Susceptible to electromagnetic interference High cost	Je et al. (2013)
Piezoelectric type	2.0 mV/Pa	0.05–50 Hz	Same as above	Shields (2005)
Tubular MZI	Unknown	1–10 Hz	Huge size	Zumberge et al. (2003)
LD-SMI	0.25 Hz resolution	2–20 Hz	Complex structure	Li et al. (2013)
Our scheme	121 mV/Pa @1 Hz	1–20 Hz		

through experimental verification. Sensitive infrasound sensing applications. Firstly, combined with the requirements of infrasound detection and actual experimental conditions, the basic theory and design guidelines of the sensor are determined. Then, by selecting the relevant materials, dimensions, and structural design, a fiber EFPI infrasound sensor head with flexible adjustable cavity length is fabricated. Then, based on the experimental equipment and condition, the infrasound sensor is used for infrasound measurement, and the results show that high-sensitivity detection in the infrasound frequency band can be achieved. Finally, the comparison with our scheme and other recently reported schemes in terms of the advantages and disadvantages is analyzed and discussed.

References

G. Barruol, D. Reymond, F.R. Fontaine, O. Hyvernaud, V. Maurer, K. Maamaatuaiahutapu, Geophys. J. Int. **164**, 3 (2006)

A. Cusano, S. Campopiano, S. D'Addio, M. Balbi, S. Balzarini, M. Giordano, A. Cutolo, Optical fiber hydrophone using polymer-coated fiber bragg grating. Opt. Fiber Sens. (2006). https://doi.org/10.1364/OFS.2006.ThE85

G.L. D'spain, W.S. Hodgkiss, G.L. Edmonds, J. Acoust. Soc. Am. **89**, 3 (1991)

D.N. Green, J. Vergoz, R. Gibson, A.L. Pichon, L. Ceranna, Geophys. J. Int. **185**, 2 (2011)

P. Huke, R. Klattenhoff, C. Von Kopylow, R.B. Bergmann, J. Europ. Opt. Soc. Rap. Public. **8**, 13043 (2013)

C.H. Je, J. Lee, W.S. Yang, J. Kim, Y.H. Cho, J. Micromech. Microeng. **23**, 5 (2013)

A. Le Pichon, P. Mialle, J. Guilbert, J. Vergoz, Geophys. J. Int. **167**, 2 (2006)

C. Li, Z. Huang, X. Sun, Chin. Opt. Lett. **11**, 2 (2013)

J. Ma, H. Xuan, H.L. Ho, W. Jin, Y. Yang, S. Fan, I.E.E.E. Photonic, Tech. L. **10**, 25 (2013)

R.S. Matoza, A.L. Pichon, J. Vergoz, P. Herry, J.M. Lalande, H. Lee, I.Y. Che, A. Rybin, J. Volcanol. Geoth. Res. **200**, 1 (2011)

M. Morrissey, M. Garces, K. Ishihara, M. Iguchi, J. Volcanol. Geoth. Res. **175**, 3 (2008)

F.D. Shields, J. Acoust. Soc. Am. **117**, 6 (2005)

T. Sindelarova, D. Buresova, J. Chum, F. Hruska, Adv. Space Res. **43**, 11 (2009)

J.G.V. Teixeira, I.T. Leite, S. Silva, Photonic Sens. **4**, 3 (2014)

H. Tsuda, K. Kumakura, S. Ogihara, Sensors **10**, 12 (2010)

Q. Wang, Z. Ma, Opt. Laser Technol. **51** (2013)

S. Wang, P. Lu, H. Liao, L. Zhang, D. Liu, J. Zhang, J. Mod. Opt. **60**, 21 (2013)

G. Wild, S. Hinckley, S. Silva, IEEE Sens. J. **8**, 7 (2008)

J.P.F. Wooler, R.I. Crickmore, Appl. Opt. **46**, 13 (2007)

J. Xu, High temperature high bandwidth fiber optic pressure sensors. Diss. Virginia Tech. (2005). http://hdl.handle.net/10919/25988

M.A. Zumberge, J. Berger, M.A.H. Hedlin, E. Husmann, S. Nooner, R. Hilt, R. Widmer-Schnidrig, J. Acoust. Soc. Am. **113**, 5 (2003)

Specialty Fiber Grating-Based Acoustic Sensing

Wenjun Ni, Perry Ping Shum, Ping Lu, Xin Fu, Yiyang Luo, Ran Xia, and Lei Wei

Abstract Aiming at the high sensitivity, high precision, small size, and low cost acoustic detection requirements, a novel fiber sensor for curvature and acoustic wave measurement based on a thin core ultra-long period fiber grating (TC-ULPFG) has been proposed in this chapter. By tracking the power variation of different resonant wavelength caused by TC-ULPFG, high curvature sensitivity of 97.77 dB/m^{-1} is achieved, to best of our knowledge, which is highest than other structures at the same measurement range. Thus, the desired curvature property of the TC-ULPFG is used for acoustic measurement. The polyethylene terephthalate (PET) film is selected as a transducer, on which TC-ULPFG is tightly pasted. The acoustic pressure sensitivity of 1.89 V/Pa is two orders higher than other structures based on the diaphragm transducer, and the noise-limited minimum detectable pressure is 1.94 mPa/Hz$^{1/2}$ at 200 Hz. In addition, the frequency fluctuations are nearly ± 0.4 dB from 70 to 200 Hz and ± 0.2 dB from 1 to 3 kHz, respectively. Therefore, the proposed optical fiber acoustic sensor (OFAS) has a flat frequency response in relatively lower frequency.

W. Ni (✉) · P. P. Shum · Y. Luo · R. Xia · L. Wei
School of Electrical and Electronic Engineering, Nanyang Technological
University, Singapore, Singapore
e-mail: wenjun.ni@ntu.edu.sg

P. P. Shum
e-mail: epshum@ntu.edu.sg

Y. Luo
e-mail: luoyy@ntu.edu.sg

R. Xia
e-mail: N1906444L@e.ntu.edu.sg

L. Wei
e-mail: wei.lei@ntu.edu.sg

P. Lu · X. Fu
Huazhong University of Science and Technology, Wuhan, China
e-mail: pluriver@mail.hust.edu.cn

X. Fu
e-mail: fuxin@hust.edu.cn

© Springer Nature Singapore Pte Ltd. 2020
L. Wei (ed.), *Advanced Fiber Sensing Technologies*,
Progress in Optical Science and Photonics 9,
https://doi.org/10.1007/978-981-15-5507-7_6

The TC-ULPFG shows many advantages including high sensitivities of curvature, high acoustic pressure sensitivity, easy fabrication, simple structure, and low cost.

Keywords Thin core ultra-long period fiber grating (TC-ULPFG) · Optical fiber acoustic sensor (OFSA) · Flat frequency response

1   Introduction

In recent years, optical fiber acoustic sensor (OFAS) has been used in some important engineering applications, such as structural health monitoring (Hu et al. 2016), oil pipeline leakage detection (Wang and Liu 2013), and biomedical diagnosis (Tazawa et al. 2007). OFASs are in general immune to electromagnetic interference (EMI), resistant to corrosion, and highly sensitive. Therefore, OFASs are superior to electronic and mechanical acoustic sensors in harsh environment. In order to realize combination of lower cost, compact size, and high sensitivity, more and more optical fiber interferometric structures and optical fiber devices have been proposed and experimentally demonstrated. Some typical optical fiber interferometric structures, including Mach–Zehnder (MZ) (Pawar et al. 2016), Sagnac (Kang et al. 2014), Michelson (Liu et al. 2016a), and Fabry–Perot (FP) interferometer (Xu et al. 2014), are reported in the past decades, especially FPI based on thin films (Liu et al. 2016b; Ma et al. 2015). However, most of the structures mentioned above have relatively large size, lower conversion efficiency with identical transducers and easily influenced by external environment variation. Consequently, more attention has been paid to simple optical fiber device.

It has been reported in the last five years that some optical fiber devices, such as optical fiber taper (Xu et al. 2012; Li et al. 2011), fiber coupler (Li et al. 2015; Wang et al. 2014a), FP cavity based on pairs of fiber Bragg gratings (FBG-FP) (Lyu et al. 2016; Wang et al. 2014b), and long period fiber grating (LPFG) (Gaudron et al. 2012), are widely used as sensor heads in OFAS. The transducers of these optical fiber devices are always based on thin films, thin metal plate, composite materials and water, etc. FBG-FP is relatively difficult to fabricate; moreover, the repetitive rate of its fabrication is not high. Optical fiber taper needs to be inserted in other interferometric structure to introduced modal interference. Optical fiber coupler can measure the wideband response. However, the frequency response is always not flat. Compared with the above-mentioned optical fiber devices, ultra-long period fiber grating (ULPFG) has significant superiority in the field of acoustic measurement, such as a single optical device without other cascaded interferometric structure and easy to fabricate. The most important advantage is that ULPFG possesses high curvature sensitivity. Since acoustic measurement based on ULPFG is dynamic curvature detection in fact, ULPFG is a novel optical device and good candidate for dynamic parameters sensing. Recently, LPFG has been written in various of special fiber, including thin core fiber (TCF) (Fu et al. 2015), photonic crystal fiber (PCF) (Tian et al. 2013), and few mode fiber (FMF) (Wang et al. 2015). But to best of our knowledge, no one has used TCF to inscribe ULPFG.

In this chapter, a novel optical device TC-ULPFG with high curvature sensitivity is proposed and demonstrated. By tracking the power variation of different resonant wavelength, curvature sensitivity of 97.77 dB/m^{-1} is achieved, corresponding to the resonant wavelength of 1570 nm. Thus, TC-ULPFG is used for acoustic measurement, which has flat frequency response in two bands, 70–200 Hz and 1–3 kHz, respectively. Additionally, the minimum detectable pressure level is 1.94 $mPa/Hz^{1/2}$ and the sensitivity of acoustic pressure is 1.89 V/Pa. The input acoustic signal is recovered in the method of intensity demodulation. The proposed sensor head shows many advantages including multi-parameters measurement, high sensitivity of curvature, easy fabrication, simple structure, and low cost.

2 Main Body Text

2.1 Static Properties of the Sensor Head

The TC-ULPFG is inscribed by the focused CO_2 laser pulse (SYNRAD, 48-series) on one side of TCF using point-by-point method, and the experiment setup of fabrication process has been published recently (Ni et al. 2016). It is obvious that the differences between TC-ULPFG and thin core long period fiber grating (TC-LPFG) are number of periods and grating pitch. TC-ULPFG has an ultra-long grating pitch, which is twice the published TC-LPFG. The number of periods, grating pitch, and the diameter of TCF (YOEC, 15-80-U16, 80 μm) are 30, 1 mm, and 80 μm, respectively. The transmission spectrum of TC-ULPFG is shown in Fig. 1, and the inset is schematic diagram of the sensor head. The proposed structure is composed of five sections named Input-SMF (ISMF), Input-TCF (ITCF), TC-ULPFG, Output-TCF (OTCF), and Output-SMF (OSMF), which can be fabricated only by the fusion splicer (Fujikura, FSM-60S). The ITCF and OTCF are cut off by 15 cm with fiber cleaver (Fujikura, CT-38) to avoid the influence of modal interference (Fu et al. 2015; Ni et al. 2016). Because the length of TCF is relatively long, nearly all of the cladding modes excited by the mode field mismatch or TC-ULPFG will undergo entirely loss in the TCF. The transmission of all the fiber modes will experience with the following two process. Firstly, cladding modes will be excited in ITCF at the first splicing point due to the mode field mismatch between ISMF and ITCF, and will undergo completely loss in ITCF because of the long length of ITCF (Fu et al. 2015). Secondly, cladding modes will be excited again at the grating region due to the mode coupling between core and cladding modes in TC-ULPFG (Dianov et al. 1996), and will disappear in OTCF for the same reason. Since the coupling between the core and cladding modes is wavelength dependent, several resonant dips will be introduced in the transmission spectrum. It can be seen from Fig. 1 that there are four dominant dips generated by TC-ULPFG which are named as dip1, dip2, dip3, and

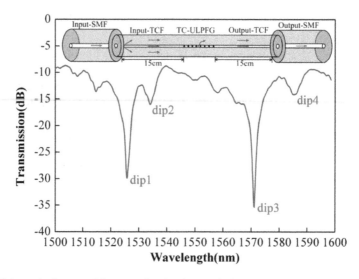

Fig. 1 Schematic diagram of the sensor head and transmission spectrum

dip4 from left to right, respectively. The location of dip3 is close to 1570 nm, and the extinction ratio is nearly 25 dB. Moreover, the 3 dB bandwidth of dip3 is 3.5 nm, which is narrower than any reported common LPFG. When the number of scanning circles is increasing, the 3 dB bandwidth is generally decreasing and the extinction ratio is generally increasing. The 3 dB measurement can be detected by the OSA (optical spectrum analyzer) or the software of OriginLab. Taking these advantages of TC-ULPFG into consideration, the resonant wavelength of TC-ULPFG can be well used for the intensity demodulation.

Curvature is a static parameter, so the measurement process is relatively simple. Only an amplified spontaneous emission (ASE) light source, optical spectrum analyzer (OSA, Yokogawa AQ6370c) and two-dimensional (2D) translation stages (Zolix) are needed. It is worth noting that the sensor head should keep straight line without external axial tension. The curvature is increased by reducing the distance between two translation stages. As shown in Fig. 2a, the intensity of dip1 and dip3 is increasing with the curvature increasing from 0.4243 to 0.5475 m^{-1}. Furthermore, another two dips are gradually disappearing. Thus, dip1 and dip3 are selected to detect the curvature variation. The extinction ratio of dip1 and dip3 decreases with the increased curvature due to part of the cladding modes leaking out. As shown in Fig. 2a, the power of the dip2 and dip4 is generally decreasing. As we all know, there are sidelobes in TC-ULPFG. Thus, the two different dips corresponding to the two cladding modes have mode competition. If the power of the main dip is decreasing, the power of the sidelobes of the TC-ULPFG will increase. The following figure shows that dip1 and dip3 are the dominant resonant wavelength. Therefore, the trend of dip1 and dip3 is different from that of dip2 and dip4. When curvature increases from 0.4243 to 0.5475 m^{-1}, linear intensity variations of dip1 and dip3 with the

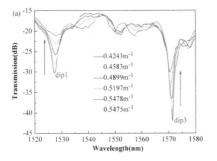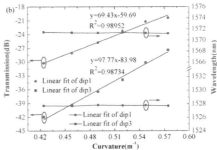

Fig. 2 a Transmission spectrum with curvature increasing from 0.4243 to 0.5475 m^{-1}. **b** Linear fit of dip1 and dip3 with sensitivities of 69.43 dB/m^{-1} and 97.77 dB/m^{-1}, respectively

sensitivities of 69.43 and 97.77 dB/m^{-1} are achieved, as shown in Fig. 2b. The blue line of the circle point is the linear fit of dip1, and the red line of the square point is the linear fit of dip3. Different sensitivities of dip1 and dip3 are caused by the different order of cladding modes and different diffraction order. From Fig. 2b, it can be seen that there is nearly no wavelength shift. Only intensity variation of dip1 and dip3 can be observed. Therefore, high curvature sensitivities based on intensity demodulation can be obtained by the proposed TC-ULPFG. As is known to all that the acoustic signal detection can be converted to the dynamic curvature measurement. Given the above discussion, an OFAS can be realized based on TC-ULPFG because of its high curvature sensitivities.

2.2 Theoretical Analysis of Acoustic Signal Demodulation

For acoustic sensing, good transducer is a vital device in acoustic signal measurement processing. Membrane is a very common transducer mentioned by the last report. The PET film is selected as the following acoustic sensing, as shown in the inset of Fig. 3. The acoustic conversion setup is based on a circular PET film where a TC-ULPFG is adhered on the central axis. It should be noticed that the TC-ULPFG has a tiny pre-bending to avoid frequency doubling phenomenon (Gaudron et al. 2012). When the acoustic wave is applied on the membrane, the acoustic signal is converted to the vibration. Consequently, the membrane will deform due to its vibration. The film deformation equation can be described as (Wang et al. 2016):

$$S_{\mathrm{def}} = \frac{\Delta L}{\Delta P} = \frac{3(1 - \mu^2)r^4}{16Eh^3} \tag{1}$$

In order to study the characteristics of the frequency response, the PET film can be regarded as a thin plate. Similarly, the resonant frequency f_{res} of the diaphragm can be expressed as (Ma et al. 2013):

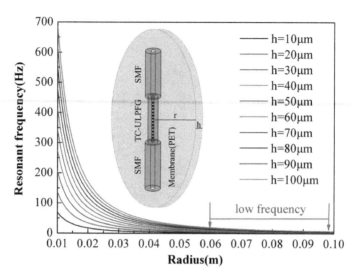

Fig. 3 Relationship between resonant frequency and radius with different thickness of PET film

$$f_{res} = \frac{10.21h}{2\pi r^2} \sqrt{\frac{E}{12\rho(1-\mu^2)}} \qquad (2)$$

In the above equations, r, h, μ, ρ, and E are the radius, thickness, Poisson ratio, mass density, and Young's modulus of the PET film, respectively. The parameters of PET film are as follows: $\mu = 0.39$, $\rho = 1.38 \times 10^3 \, kg/m^2$, and $E = 2.5 \, GPa$. As shown in Fig. 3, the resonant frequency is in negative correlation with the film thickness, and the lower resonant frequency is mainly generated with the radius ranging from 0.06 m to 0.1 m. In Sect. 2, the length of the TC-ULPFG is 3 cm. The TC-ULPFG must be pasted on the PET film. In our work, the adhesive tape is applied to fix the TC-ULPFG on the middle of the membrane. Therefore, the diameter of the membrane should have an enough length to fix the device. If the diameter of the membrane is too short, the grating region is easy to be broken. Thus, the diameter of the PET film is about three times longer than the TC-ULPFG. When the diameter of the membrane is a certain value, the resonant frequency is increasing with the thickness of the membrane increasing. Therefore, the film thickness is positive correlation of the resonant frequency. According to Eq. (2), it also can be seen that the relationship between resonant frequency and the film thickness is positive correlation. In order to detect lower frequency, radius is better to be selected as the average value which is 0.08 m. It can be seen from Fig. 3 that the lower resonant frequency bands are close to a straight line no matter how to change the thickness of PET film. Thus, the thickness of PET film is also selected as the mid-value which is 50 μm. According to Eq. (1), the static pressure sensitivity induced by the membrane deformation is 208 cm/Pa. In addition, the resonant frequency of the PET film is calculated to be

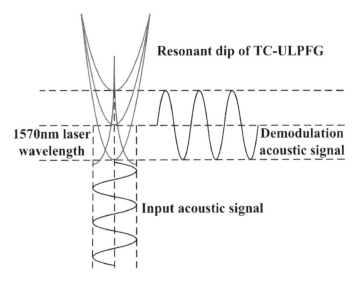

Fig. 4 Principle of the intensity demodulation

5.36 Hz from Eq. (2). The two parameters obtained by the above analysis are very important in acoustic measurement fields.

The acoustic signal measurement based on TC-ULPFG is dynamic curvature detection in fact. As shown in Fig. 2(b), there is nearly no wavelength shift of dip1 and dip3, but only intensity variation while measuring the static curvature (Gong et al. 2011). The 1570 nm distributed feedback laser (DFB, long-term stability: $\leq \pm 0.02$ dB) laser is used as the laser source since the resonant wavelength of dip3 is 1570 nm. The schematic diagram of the intensity demodulation is exhibited in Fig. 4. When the input acoustic signal is applied on the PET film, the laser intensity will vary periodically. Then, the output signal will be a time domain waveform after photovoltaic conversion. Therefore, the acoustic signal can be real time demodulated by detecting the intensity of output optical signal.

2.3 Experimental Results and Discussions

The experimental setup of acoustic measurement is shown in Fig. 5, including 1570 nm distributed feedback (DFB) laser, acoustic source, membrane, TC-ULPFG, photoelectric detector (PD), and oscilloscope. The membrane is closely pasted on the hollow cylinder, which the bottom is closed and the top is open. The placement of the membrane is shown in the following schematic diagram. Additionally, the membrane is parallel to the bottom and the surface of the membrane must keep flat. TC-ULPFG fixed by adhesive is pasted in the middle of the membrane, and the acoustic source is placed in the above of the membrane. The distance between

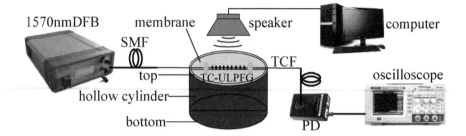

Fig. 5 Schematic diagram of the experimental setup

acoustic source and membrane is controlled to nearly 10 cm to sustain a consistent response of the membrane. The acoustic source is controlled by the computer to regulate acoustic frequency. The light intensity variation is detected by the PD, and displayed on the oscilloscope.

In view of dynamic parameter sensing, TC-ULPFG has a dominant advantage of high curvature sensitivity. A single frequency acoustic signal is given by the computer, and it is recovered by the demodulation method mentioned before. The inset of Fig. 6 shows the time domain waveform at the frequency of 200 Hz. The frequency spectrum response of 200 Hz is obtained by taking fast Fourier transform (FFT), which is shown in Fig. 6. It is observed that there are odd harmonics and even harmonics in the frequency spectrum simultaneously. The existence of the harmonics is caused by the indoor reflection of the acoustic wave. The signal-to-noise ratio (SNR) at 200 Hz is 35 dB. The power frequency of 50 Hz can be nearly neglected because it is close to the noise level. The 3 dB bandwidth at 200 Hz is 2.12 Hz, and the acoustic pressure applied on the PET film is 78 dB, corresponding

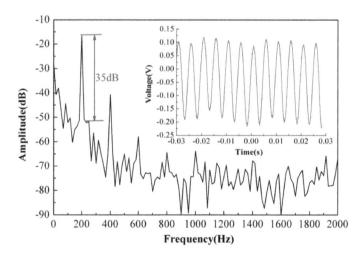

Fig. 6 Frequency spectrum response and time domain waveform at 200 Hz

Table 1 Comparison result of different structure for curvature and acoustic measurement

Structure	Curvature sensitivity	Acoustic pressure sensitivity range	Minimum detectable pressure	Reference
Inline MZ interferometer	70.03 dB/m^{-1}			Gong et al. (2011)
Diaphragm-based LPFG	13.5 nm/m^{-1}	60.52 mV/Pa	1.42 mPa	Gaudron et al. (2012)
Diaphragm-based FP		13.15 mV/Pa	75μPa/Hz$^{1/2}$	Ma et al. (2013)
Our scheme	97.77 dB/m^{-1}	1.89 V/Pa	1.94 mPa/Hz$^{1/2}$	

to a noise-limited minimum detectable pressure level of 1.94 mPa/Hz$^{1/2}$. Besides, the sensitivity of acoustic pressure is 1.89 V/Pa, which is two orders higher than the previously reported sensor of diaphragm-based FP cavity, etc. The comparison result is listed in Table 1.

The proposed TC-ULPFG has a very high acoustic pressure sensitivity which is mentioned above. Furthermore, in order to investigate the characteristics of the frequency response, a wide frequency test is carried out in a continuous frequency spectrum. In the acoustic wave experiment, the frequency from 70 Hz to 3 kHz can be measured, as shown in Fig. 7a. The experiment result exhibits that the sensor head or the transducer is sensitive to the relatively lower frequency, which is consistent with the simulation result in Fig. 3. The proposed structure can measure infrasound and low frequency acoustic signal according to the theoretical analysis in Sect. 3. Therefore, the strong response appears when the frequency is lower than 200 Hz. However, the response to other frequencies is generally decreasing with the acoustic frequency increasing since the resonant frequency of the selected PET film is 5.36 Hz. When the acoustic frequency is higher than 1 kHz, the frequency response approximately trends to be stable because relatively higher frequency is far away from the resonant

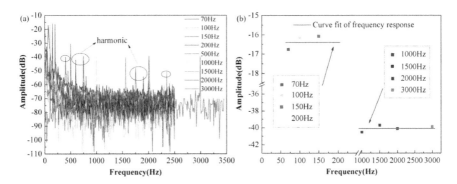

Fig. 7 a Frequency response ranges from 70 Hz to 3 kHz. **b** Frequency fluctuation from 70 to 200 Hz and from 1 to 3 kHz

frequency. In Fig. 7b, the frequency fluctuations are nearly ±0.4 dB from 70 to 200 Hz and ±0.2 dB from 1 to 3 kHz, respectively. The red line of the square point is the curve fit of the frequency response from 70 Hz to 3 kHz. As a consequence, the proposed TC-ULPFG has a flat frequency response when the measurand is relatively close or far away from the resonant frequency.

3 Conclusion

In summary, we propose a novel fiber sensor based on TC-ULPFG for curvature and acoustic measurement. TC-ULPFG is fabricated by the focused CO_2 laser pulse on one side of TCF by the method of point-by-point. The high curvature sensitivities of 69.43 and 97.77 dB/m^{-1} are achieved by tracking the power variation of dip1 and dip3, respectively. TC-ULPFG has a better curvature property, which is highest than other structures at the same measurement range. Additionally, this special optic device is used for acoustic measurement. The PET film is selected as the transducer, the thickness and radius of which are 50 μm and 8 cm, respectively. According to the theoretical analysis, membrane deformation and the resonant frequency are 208 cm/Pa and 5.36 Hz. In acoustic experiment, the acoustic pressure sensitivity is 1.89 V/Pa which is two orders higher than other structures of diaphragm-based transducer. The noise-limited minimum detectable pressure level is 1.94 $mPa/Hz^{1/2}$ at 200 Hz. Moreover, the frequency fluctuations are nearly ±0.4 dB from 70 to 200 Hz and ±0.2 dB from 1 to 3 kHz, respectively. As a consequence, the proposed OFAS has a flat frequency response in relatively lower frequency. The TC-ULPFG shows many advantages including high sensitivities of curvature, high acoustic pressure sensitivity, easy fabrication, simple structure, and low cost. Therefore, the proposed sensor may have a great potential in the fields of engineering system.

References

E.M. Dianov, S.A. Vasiliev, A.S. Kurkov, O.I. Medvedkov, V.N. Protopopov, In-fiber Mach-Zehnder interferometer based on a pair of long-period gratings, in *22nd European Conference on Optical Communication(ECOC'96)*. (Academic, 1996), pp. 65–68

C. Fu, X. Zhong, C. Liao, Y. Wang, Y. Wang, J. Tang, S. Liu, Q. Wang, Thin-core-fiber-based long-period fiber grating for high-sensitivity refractive index measurement. IEEE Photonics J. **7**(6), 1–8 (2015)

J.O. Gaudron, F. Surre, T. Sun, K.T.V. Grattan, LPG-based optical fibre sensor for acoustic wave detection. Sens. Actuators A Phys. **173**(1), 97–101 (2012)

Y. Gong, T. Zhao, Y. Rao, Y. Wu, All-fiber curvature sensor based on multimode interference. IEEE Photon. Technol. Lett. **23**(11), 679–681 (2011)

C. Hu, Z. Yu, A. Wang, An all fiber-optic multi-parameter structure health monitoring system. Opt. Express **24**(18), 20287–20296 (2016)

J. Kang, X. Dong, Y. Zhu, S. Jin, S. Zhuang, A fiber strain and vibration sensor based on high birefringence polarization maintaining fibers. Opt. Commun. **322**, 105–108 (2014)

Y. Li, X. Wang, X. Bao, Sensitive acoustic vibration sensor using single-mode fiber tapers. Appl. Opt. **50**(13), 1873–1878 (2011)

F. Li, Y. Liu, L. Wang, Z. Zhao, Investigation on the response of fused taper couplers to ultrasonic wave. Appl. Opt. **54**(23), 6986–6993 (2015)

L. Liu, P. Lu, H. Liao, S. Wang, W. Yang, D. Liu, J. Zhang, Fiber-optic michelson interferometric acoustic sensor based on a PP/PET diaphragm. IEEE Sens. J. **16**(9), 3054–3058 (2016a)

L. Liu, P. Lu, S. Wang, X. Fu, Y. Sun, D. Liu, J. Zhang, H. Xu, Q. Yao, UV adhesive diaphragm-based FPI Sensor for very-low-frequency acoustic sensing. IEEE Photon. J. **8**(1), 1–9 (2016b)

C. Lyu, Y. Liu, C. Wu, Wide bandwidth dual-frequency ultrasound measurements based on fiber laser sensing technology. Appl. Opt. **55**(19), 5057–5062 (2016)

J. Ma, H. Xuan, H.L. Ho, W. Jin, Y. Yang, S. Fan, Fiber-optic Fabry-Perot acoustic sensor with multilayer graphene diaphragm. IEEE Photon. Technol. Lett. **10**(25), 932–935 (2013)

J. Ma, Y. Yu, W. Jin, Demodulation of diaphragm based acoustic sensor using Sagnac interferometer with stable phase bias. Opt. Express **23**(22), 29268–29278 (2015)

W. Ni, P. Lu, C. Luo, X. Fu, L. Liu, H. Liao, X. Jiang, D. Liu, J. Zhang, Bending direction detective fiber sensor for dual-parameter sensing based on an asymmetrical thin-core long-period fiber grating. IEEE Photonics J. **8**(4), 1–11 (2016)

D. Pawar, C.N. Rao, R.K. Choubey, S.N. Kale, Mach-Zehnder interferometric photonic crystal fiber for low acoustic frequency detections. Appl. Phys. Lett. **108**(4), 041912 (2016)

H. Tazawa, T. Kanie, M. Katayama, Fiber-optic coupler based refractive index sensor and its application to biosensing. Appl. Phys. Lett. **91**(11), 113901 (2007)

F. Tian, J. Kanka, B. Zou, K.S. Chiang, H. Du, Long-period gratings inscribed in photonic crystal fiber by symmetric CO_2 laser irradiation. Opt. Express **21**(11), 13208–13218 (2013)

Z. Wang, Z. Liu, Design and experiment of data acquisition system of submerged buoy. Ocean Technol. **32**(4), 6–10 (2013)

S. Wang, P. Lu, L. Zhang, D. Liu, J. Zhang, Optical fiber acoustic sensor based on nonstandard fused coupler and aluminum foil. IEEE Sens. J. **14**(7), 2293–2298 (2014a)

X. Wang, L. Jin, J. Li, Y. Ran, B. Guan, Microfiber interferometric acoustic transducers. Opt. Express **22**(7), 8126–8135 (2014b)

B. Wang, W. Zhang, Z. Bai, L. Wang, L. Zhang, Q. Zhou, L. Chen, T. Yan, CO_2-laser-induced long period fiber gratings in few mode fibers. IEEE Photon. Technol. Lett. **27**(2), 145–148 (2015)

S. Wang, P. Lu, L. Liu, H. Liao, Y. Sun, W. Ni, X. Fu, X. Jiang, D. Liu, J. Zhang, H. Xu, Q. Yao, Y. Chen, An infrasound sensor based on extrinsic fiber-optic Fabry-Perot interferometer structure. IEEE Photon. Technol. Lett. **28**(11), 1264–1267 (2016)

B. Xu, Y. Li, M. Sun, Z. Zhang, X. Dong, Z. Zhang, S. Jin, Acoustic vibration sensor based on nonadiabatic tapered fibers. Opt. Lett. **37**(22), 4768–4770 (2012)

F. Xu, J. Shi, K. Gong, H. Li, R. Hui, B. Yu, Fiber-optic acoustic pressure sensor based on large-area nanolayer silver diaghragm. Opt. Lett. **39**(10), 2838–2840 (2014)

Electrospinning Nanofibers

Shaoyang Ma and Tao Ye

Abstract With the rapid development of human technology, we are continuously facing new problems and challenges. On the one hand are the limited natural resources, and on the other hand are the increasing demands for convenient lifestyle. It is not a good idea to sacrifice one aspect to satisfy the other, but we should seek a balance between sustainable development and the comfortable living concept. Thus, it becomes a very crucial task to explore the low-cost, large-scale, and environmentally friendly fabrication methods for wide applications. Electrospinning is a top-down method in which polymeric or melt components are drawn out from a solution system onto a collector by electrostatic force. In comparison with other methods, including drawing, template synthesis, chemical vapor deposition, and so on, electrospinning offers some attractive features. One of the most important advantages is its industrial scalability, which makes it possible to directly transfer the results from laboratory research to the industry. In general, the unique advantages of electrospun nanostructures, including high spatial interconnectivity, high porosity, and large surface-to-volume ratio, make it a promising fabrication method in a wide range of applications. This chapter covers aspects of information relating to electrospinning nanofibers, including the materials for nanofiber fabrication, processing mechanism and parameters of electrospinning techniques, special electrospinning techniques, and potential applications of electrospun nanofibers.

Keywords Electrospinning techniques · Electrospun nanofibers · Materials parameters · Applications

S. Ma (✉)
Key Laboratory of All Optical Network and Advanced Telecommunication Network of Ministry of Education, Institute of Lightwave Technology, Beijing Jiaotong University, 100044 Beijing, China
e-mail: mashaoyang@bjtu.edu.cn

T. Ye
Materials Research Institute, Pennsylvania State University, University Park, PA 16802, USA
e-mail: tvy5161@psu.edu

© Springer Nature Singapore Pte Ltd. 2020
L. Wei (ed.), *Advanced Fiber Sensing Technologies*,
Progress in Optical Science and Photonics 9,
https://doi.org/10.1007/978-981-15-5507-7_7

1   Introduction

Recently, fibers and textiles with multi-materials add-on or built-in multi-functional units have become a hot research area for researchers. Fiber- and textile-based flexible electronics have shown great possibilities in numerous applications including flexible circuits, skin-like pressure sensors, transistors, memory devices, and displays. But similar like other mainstream issues come across during the human society development and need to be addressed, how to guarantee low-cost and production scale simultaneously is the problem that constrains the application of fibers and textiles.

Electrospinning is an excellent solution to address these problems. Compared to other material processing methods, electrospinning is more suitable for large-scale commercial applications to produce long-lasting continuous polymer micro- and nanofibers (Huang et al. 2003), due to all the merits including easy accessibility, simple synthesis, well-controlled parameters, and low cost.

First raised in early twentieth century, electrospinning has been utilized to process materials for over a hundred years, but not until recent decades, has it been widely applied to various researching fields. With the well-developed techniques and versatile functionalities, electrospinning method is widely adopted in various applications, including energy conversion and storage, wearable smart devices, environment-related applications, and biomedical applications.

2   Main Body Text

Ultrathin one-dimensional structures like micro- and nanofibers, compared to three-dimensional bulk and two-dimensional film structures, possess numerous advantages, such as large surface-to-volume ratio, tensile strength, and good flexibility. And these amazing characteristics make fibers a competitive candidate for large quantity of applications, including energy harvesting, multi-functional sensing, bio-tissue growing, and many other fields.

There are many ways to fabricate one-dimensional structures, including fiber drawing, template growing, vapor growing, self-assembly, and electrospinning. Compared to other synthesis methods, electrospinning is more suitable for large-scale commercial applications to produce long-lasting continuous polymer micro- and nanofibers (Huang et al. 2003), due to all the merits mentioned in previous section.

The concept of electrospinning could retrospect to very early of the last century, first patented by Formhals (1943), and was developing slowly during the following decades. It was not until the mid-1990s that electrospinning started to regain the attention of an increasing number of researchers as a promising method to fabricate continuous micro- and nanofibers, and developed rapidly since the entry of this century as shown by Fig. 1. For the past two decades, this versatile synthesis technique has been applied to a wide range of researching fields, including energy conversion

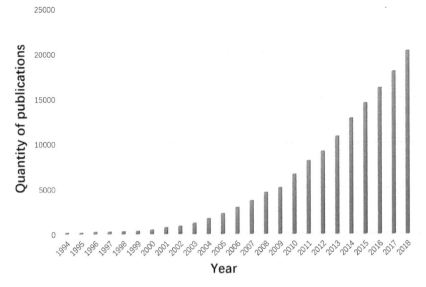

Fig. 1 Annual number of publications related to electrospinning

and storage, environment-related applications, and biomedical application and will be discussed in the following sections.

The apparatus for electrospinning is showed by Fig. 2. There are several necessary parts for a standard electrospinning setup, including a spinneret (usually a needle with a syringe), a pump that can maintain the feeding, a high-voltage power source (usually several tens of kilovolts), and a conductive collector to gather the resultant.

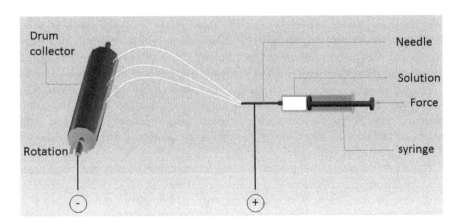

Fig. 2 An electrospinning apparatus with a rotating collector

There are several features that electrospinning process must fulfill, including (1) the solvent need to be selected properly to dissolve the polymer material; (2) suitable vapor pressure should be designed and maintained for the fiber integrity while avoiding over-harden; (3) the viscosity and surface tension of the solvent should be carefully balanced; (4) the power source must offer a voltage large enough to overcome the surface tension; and (5) the distance between spinneret and collector also need to be appropriate for full evaporation of the solvent. More specifics will be discussed in the coming sections.

2.1 Materials of Electrospinning

To understand electrospinning, the mechanism behind the process need to be investigated and learned. Different from the traditional fibers with relevant large diameter, which is fabricated by drawing method using a die involving an external mechanical force to form individual strand of fiber, electrospinning process adopt high voltage to generate a jet from polymer solution droplet. One great advantage of electrospinning is that it can process not only molten polymers but nearly all soluble polymers and materials that can be made into a suspension. Thus, there are many different types of polymers and precursors that can be electrospun for the formation of fibers. According to the application requirements, materials should be selected carefully. Some polymer and polymer composites nanofibers can be obtained directly while some ceramics and carbon nanotubes need further processing after electrospinning operation.

For each electrospinning application, proper materials must be selected and modified to obtain certain morphologies and properties to match the desired functions of the expected applications. The electrospinning technology can generate ultrathin fibers by using a wide variety of materials that contains, polymers such as polyacrylonitrile(PAN), polyvinylalcohol (PVA), polyvinylpyrrolidone (PVP), polyethylene glycol (PEG), polyvinylidene fluoride (PVDF), and polystyrene(PS); metal oxides/ceramics such as Fe_2O_3, CuO, NiO, and TiO_2; mixed metal oxides such as $TiNb_2O_7$, $NiFe_2O_4$, and $LiMn_2O_4$; composites such as carbon/SnO_2, PVA/TiO_2, Nylon-6/gelatin, graphene/TiO_2, and collagen/hydroxyapatite; and carbon-based materials such as carbon nanotubes, graphite, and graphene. Combined with controlled calcinations treatment, 1D hierarchical nanofibers can be synthesized easily by electrospinning.

The most commonly used materials for electrospinning are polymers. One great advantage of commercialization of polymers is that the cost of polymers is generally cheap because they are easy to synthesize. Polymers consist of repeating units of monomers bonded to one another, which form long molecular chains. For example, the polymer named polyethylene is made up of the repeating units of $[-CH_2CH_2-]_n$. One important value to describe a polymer is its molecular weight. As polymer chain consists of many repeating units, the molecular weight of the polymer is the sum of all the units. Higher molecular weight will make it harder to dissolve in solvents and cause a higher viscosity of the solution.

Electrospun polymer fibers are usually used as they are fabricated due to their processability and flexibility, but ceramic fibers that are hard and brittle must be electrospun from precursors followed by sintering process. Ceramics are a group of compounds composed of both metallic and non-metallic elements, usually oxides. Another common material for electrospun fiber is carbon. Carbon is the most fundamental element for organic compounds. Thus, many carbonaceous precursors can be made into pure carbon. Carbon nanofibers are known to be good conductor for electricity. And carbon nanotubes have also been applied in many electronics applications (McEuen 1998).

2.2   Electrospinning Process

After high electric potential difference is applied between the spinneret and the collector, the droplet at the needle tip will be charged and stretched under the combined forces of static electric force and liquid surface tension. With the increase of the electric potential difference, the static electric force will overcome the surface tension and the solution will form a Taylor cone (shown as Fig. 3), and a stream of

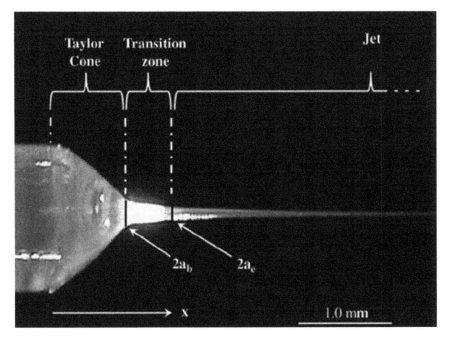

Fig. 3 Photograph of a Taylor cone in the process of electrospinning (Reneker 2008)

liquid will erupt from the surface. The liquid jet will stay in a single stream status if molecular cohesion overcomes electrostatic repulsion or break up into multiple streams otherwise. And with the evaporation of the solution solvent, the jet dries in flight and is then elongated with small bending due to the electrostatic repulsion, and finally reaches the collector to form micro- and nanoscale uniform fibers.

During the electrospinning process, there are a number of factors that can determine the final morphology and property of the resultant fibers, including the polymer solution parameter such as surface tension and viscosity, processing conditions including the voltage and distance, and the ambient condition like humidity. In the following sections, the factors that can affect electrospinning process will be introduced in detail.

2.2.1 Polymer Solution Parameters

Materials for electrospinning process must be maintained in a liquid form, molten or more commonly, dissolved in a solution. Thus, the properties of polymer solution play a crucial role in the electrospinning process and have a significant effect on the morphology of the products. When electrospinning process starts, the solution will be charged and drawn from the droplet at the spinneret under the force of electric field. The electrical property of the polymer solution, together with its viscosity and surface tension, will determine how the solution will be stretched. Meanwhile, the solution viscosity is also affected by the evaporation of the solvent and the solubility of the polymer in the solvent. The key factors of the solution include the surface tension, the polymer solubility, the viscosity, and the evaporation property and the conductivity of the solution.

Surface tension is the elastic tendency of a fluid surface, which makes it acquire the least surface area possible. At liquid–air interfaces, surface tension results from the greater attraction of liquid molecules to each other (due to cohesion) than to the molecules in the air (due to adhesion). In electrospinning process, after high electric potential difference is applied between the spinneret and the collector, the droplet at the needle tip will be charged and stretched under the combined forces of static electric force and liquid surface tension. With the increase of the electric potential difference, the static electric force will overcome the surface tension and the solution will form a Taylor cone (shown in Fig. 3), and a stream of liquid will erupt from the surface. The liquid jet will stay in a single stream status if molecular cohesion overcomes electrostatic repulsion or break up otherwise. If the resultant is collected as droplets, then the process should be called electrospraying instead of electrospinning where the liquid jet breaks up into multiple streams. And with the evaporation of the solution solvent, the jet dries in flight and is then elongated with small bending due to the electrostatic repulsion, and finally reaches the collector to form micro- and nanoscale uniform fibers (Li and Xia 2004a). For a pure liquid system, the surface tension of the solution is negatively correlated with temperature. But for an organic–aqueous mixture, the surface tension is more complicated as the small amount of organic molecules will significantly affect the surface tension due to hydrophobic behavior.

Another factor is the polymer solubility, since although electrospinning could use molten polymers, but for most cases, researches choose polymer solutions. As different solvent possesses different level of electro-spinnability (Jarusuwannapoom et al 2005), selecting the proper solvent is of great significance to dissolve the polymer and conduct the electrospinning process. Thus, the polymer solubility in a particular solvent is expected to have influence on the morphology of obtained electrospun fiber (Wannatong et al. 2004). As the molecular weight of polymer is far larger than low-molecular weight compounds, the size difference between polymer and solvent molecules is remarkable, which makes the solubility more complicated considering the viscosity and structure effects together. When the polymer is infiltrated in solvent, the solvent molecules will diffuse into polymer bulk slowly and generate a swollen gel. After the inter-bonds within the polymer molecules are broken, there will be a homogeneous polymer solution.

Another factor that has profound effect on electrospinning process and resultant morphology is solution's viscosity, which is usually related to the extent of molecule chains entanglement within the polymer solution. Too low viscosity will lead to electrospraying, and polymer particles will be obtained on the collector rather than fibers. Relatively lower viscosity caused by lower polymer chain entanglements can generate fibers probably accompanied by beads. Only solution with appropriate viscosity can produce desired smooth fibers. Therefore, the factors affecting solution's viscosity will have influence on electrospinning process and the resulting fibers, too. The intrinsic viscosity relates to the average molecular weight of polymers. The choosing of solvent is also important as in proper solvent, and the long-chain molecules of polymer are enclosure by solvent molecules, which can help reduce intercontact between polymer molecules, resulting in favoring uncurled configurations and decreasing the solution viscosity. Besides, the increasing of temperature also causes decrease in solution viscosity due to higher polymer chain mobility.

The effect of evaporation property on electrospinning resultant is obvious. The solvent of the jet will keep evaporating during the flight toward the collector. If the solvent is evaporated completely, the resultant should be well-formed fibers while incomplete evaporation will cause solvent residue and a thin film is likely to be obtained. Furthermore, solution conductivity is also important as the stretching of the polymer solution is caused by electrostatic repulsion. So, by adjusting the solution conductivity with extra ions, the carried charge could be increased, which can help in the stretching of polymer chains and avoiding beaded fibers.

2.2.2   Processing Conditions

During the electrospinning process, various of external parameters exerting on solution jet, including applied voltage, feed rate, distance between the spinneret and collector, and the type of collector, together determine how the morphology of the resultant fiber is.

First crucial element in electrospinning process is the high voltage applied between the spinneret and the collector. When the polymer solution is connected with a high voltage, charges will be induced. When the electrostatic force grows enough to overcome the surface tension, together with the electric field generated by the potential difference between the spinneret and the collector, electrospinning process starts to work. The voltage should be selected to match with the feed rate in order to form a stable Taylor Cone. Because higher voltage may induce more charges, accelerating the jet and draw larger amount of solution from the spinneret, which will cause smaller and unstable Taylor Cone.

Feed rate is the volume of solution being processed per unit time during electrospinning. For a certain voltage, there is a corresponding feed rate that can guarantee a stable Taylor Cone. With the increasing of feed rate, the diameter or beads size also increases since there is a greater volume drawn away from the spinneret. But there is a limitation of the upper feed rate, because increased volume brought by larger feed rate needs more charges for the stretching of solution, and more charges will accelerate the solution and decrease the flight time. However, greater volume requires more time for the evaporation of solvent, which contradicts the previous situation. Thus, a lower feed rate is more desirable for the electrospinning process as there will be long time for the completed evaporation of solvent.

The effect of temperature is simple and obvious. Higher temperature will increase the evaporation rate while reducing the viscosity of the solution. Meanwhile, the polymer solubility is also enhanced. Thus, the produced electrospun fibers tend to possess a smaller but more uniform diameter (Mit-uppatham et al. 2004).

To establish the electrospinning process, there must be an electrical field between the spinneret and the collector. So, the most commonly used collectors are chosen as conductive material like aluminum foil properly electrically grounded or connected to negative voltage. On a conductive collector, the distributed electrospun fibers are likely to be denser than that on an insulating collector, where the charges will accumulate and repulsion forces tend to drive fibers to form a loose structure. But this mechanism can be utilized to fabricate 3D fiber structures.

The final factor needed to be mentioned is the distance between the spinneret and the collector, as by changing the distance can vary the electrical field and flight time directly. If the distance is too small, the flight time will be possibly not enough for the completed solvent evaporation, especially when shorter distance will lead to larger electrical field, which will accelerate the jet meanwhile. In this case, extra residual solvent will exist when the resultant reaches the collector and the fibers may merge to form polymer nets or thin films instead of individually fibers. Decreasing the distance can be regarded as increasing the voltage supply.

2.2.3   Ambient Parameters

The effect of environmental on electrospinning process is a very complicated issue. The interaction between the surroundings and the polymer solution may be reflected on the morphology of the electrospun fibers. For example, high humidity level will possibly lead to the formation of pores on fiber surface. Because at normal atmosphere, water tends to condense on the fiber surface when humidity is high. This happens especially when the solvent is volatile. This has been verified by experiments using Polysulfone (PS) dissolved in Tetrahydrofuran (THF). When the humidity is over 50%, there will appear circular pores with the size and depth increasing with humidity until saturation (Casper et al. 2004). The environmental humidity also affects the evaporation rate of the solvent in the solution. When the humidity is very low, a volatile solvent is able to dry very soon, even faster than the solution removal from the spinneret. In this case, there may be only several minutes before the electrospinning process is terminated because the spinneret tip is clogged (Baumgarten 1971).

Besides the humidity, the atmosphere composition of the electrospinning environment also has effect on the processing. Under high electrostatic field, different behavior comes from different gases. Take helium as an example, under high electrostatic field, helium will break down that makes electrospinning not going to happen.

When the electrospinning process is carried within an enclosed environment, the pressure may become another factor that has influence. Generally speaking, reducing the atmosphere pressure does not good to electrospinning process. When the pressure is low, the polymer solution tends to flow out the spinneret automatically. Thus, the jet initiation becomes unstable. If the pressure keeps dropping, bubbles will occur at the tip of spinneret. At some extreme cases, electrospinning process is not able to establish due to the direst discharge of the electrical charges.

2.3 Special Electrospinning Techniques

Besides the conventional electrospinning method introduced above, there are many special electrospinning techniques, which can serve to fabricate specialty fibers. This greatly enriches the production of electrospun fibers and promotes the range of electrospun fiber application.

2.3.1 Highly Aligned Electrospun Fibers and Single Electrospun Fibers

First is the continuous electrospinning of aligned nanofibers. The approaches to gain highly aligned electrospun nanofibers are mainly by customizing collectors as Fig. 4 shows. One most commonly adopted way is to use high speed rotating drum as the collector, and the rotating speed range is from 2000 to 4000 rpm (Baniasad et al. 2015). Another approach is to replace the highly rotating drum with a rotating disk with an extremely sharp edge. Furthermore, the interdigital electrode is also an ideal substitute as the collector. By varying the configurations of the interdigital electrode, researchers can get parallel electrospun nanofibers with different length (Ke et al. 2017). Besides the frequently used methods above, the collector can also

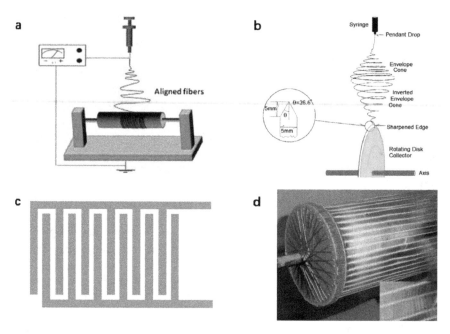

Fig. 4 Different approaches to acquire aligned electrospun nanofibers by **a** highly speed rotating drum (Prabhakaran et al. 2013), **b** sharp-edged rotating disk (Xu et al. 2004), and **c, d** parallel distributed electrodes (Ke et al. 2017; Katta et al. 2004)

be replaced by evenly distributed metal wire, parallel auxiliary electrodes, or U-shape collector. Similar methods of modifying collector are also used to get single electrospun fibers, including using cardboard frame, a substrate with micro-channels, and parallel electrodes.

2.3.2 Wet-Electrospinning

Sometimes, the electrospun fibers are preferable with 3-dimensional structure, so the wet-electrospinning system is established. If the collector of electrospinning process is immersed into solvent, or replaced by solvent container directly as Fig. 5 shows, this process is called wet-electrospinning.

In a previous study, pullulan nanofiber was fabricated by wet-electrospinning to explore how the solvent concentration can affect the fiber diameter. In this study, ethanol coagulation bath was used to collect the electrospun pullulan nanofibers (Kong and Ziegier 2014).

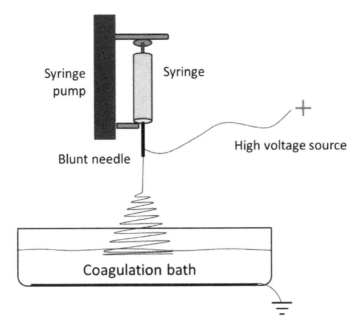

Fig. 5 Schematic drawing of the wet-electrospinning setup (Kong and Ziegier 2014)

2.3.3 Core–Shell Electrospinning

Except customizing the collector, the spinneret is another component that can be remolded to expand the morphology and functions of electrospinning products. For example, by using a special spinneret, as Fig. 6 shows, core–shell electrospun nanofiber is obtainable (Li and Xia 2004b). The special spinneret is made of two coaxial capillaries, through which two different, usually incompatible liquids are ejected to form a continuous coaxial jet simultaneously.

This core–shell electrospinning method is often used to process precursors of inorganic material. By calcining the electrospun core–shell nanofiber under high temperature condition, the as-spun nanofiber will convert into desired inorganic components (Zhang et al. 2012). And by adding functional particles into core liquid, which will be removed later, hollow nanofibers with decorated surfaces are obtained as Fig. 7 illustrates.

2.3.4 Melt Electrospinning

In some special cases like biomedical applications, the residual solvent is unwanted due to the potential toxicity to cells and tissues, or the residual repulsive charge may compromise the consistent density and thickness of electrospun fabrics. Then, melt electrospinning is a way to address these problems. The melt electrospinning setup

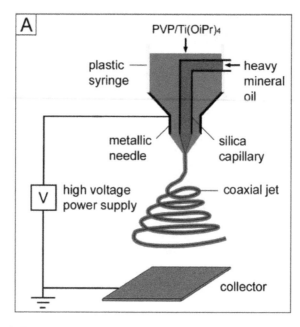

Fig. 6 Schematic illustration of the setup for core–shell electrospinning (Li and Xia 2004b)

Fig. 7 TEM images of coaxial nanofibers when **a** as-spun and **b** calcined at 450° and **c** corresponding SEM image (Zhang et al. 2012)

is illustrated in Fig. 8 in which the key part is a heating module that can melt the polymer material.

Nowadays, melt electrospinning has been widely used in industry for many applications including filtration, reverse osmosis membranes, fuel cell catalyst technologies textiles, separator, and biomedical uses. Melt electrospinning can be further combined with core–shell electrospinning technology to explore its joint application.

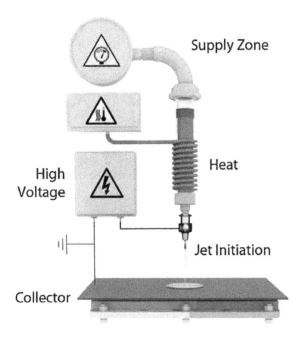

Fig. 8 Schematic of melt electrospinning onto a static collector (Brown et al. 2016)

2.3.5   Electrospray Technique

Electrospray, sharing the same mechanism and apparatus with electrospinning, also works on the principle of applied electric fields. Compared to electrospinning, the solution for electrospray is with less viscosity or smaller molecular weight, so that when applied with a high-voltage difference, the electrostatic repulsion overcomes the molecular cohesion and continuous liquid jet will break up into highly charged liquid droplets and radially disperses. By regulating the feed rate and applied voltage, the size of the electrosprayed droplets can be controlled precisely, from micro- to nanolevel (Fig. 9).

Due to its versatility and feasibility, electrospray has been involved in many nanoscience and nanotechnology area, especially in micro/nanoparticles fabrication for biomedical applications such as bone tissue engineering and drug delivery.

2.4   Application Examples of Electrospinning Nanofibers

As discussed in above sections, the electrospun nanofibers have been widely investigated in terms of the material science and engineering viewpoint. With the unique material properties and facile control of the fiber morphology/structure, various application fields of the electrospun nanofibers have been proposed by now, such as

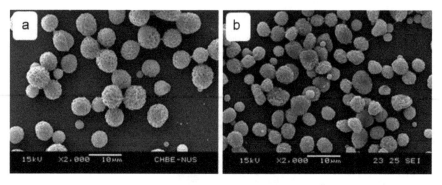

Fig. 9 Electrosprayed particles from **a** 5% w/v HMw PLGA and **b** 6% w/v LMw PLGA in pure DCM (5% w/v of BSA served as the core at a shell:core flow rate of 1.0:0.1 ml/h) (Zamani et al. 2014)

energy and electronics, bioengineering and biotechnology, defense and security, environmental engineering. Among these fields, the needing of electrospun nanofibers with enhanced mechanical, electrical, chemical, and physical properties have been inspired and they are believed to make the new era of the material research.

2.4.1 Energy Applications

Nowadays, the search for clean and sustainable energy sources is attracting increasing attention due to severe environmental issues raised from the massive usage of the traditional energy resources, such as coal, fossil fuels, and natural gas. Thus, human society have invested much time and effort to develop new energy techniques, such as batteries, catalysts, supercapacitors, fuel cells, and solar cells. Electrospun porous nanofiber mats made of functional material showed excellent energy storage and transfer behaviors in these research fields. The electrospun nanofibers have shown great potential in enhancing the performance of various batteries, increasing energy density of capacitors and fuel cells, and power conversion efficiency (PCE) of solar cells.

Batteries

According to a large amount of recent reports, the electrospun nanofiber mats can be used within various batteries to enhance their performance by introducing higher specific surface area, internal surface area, large pore volume, higher conductivity, and better 1D charge carrier transport. First, battery membranes made from electrospun fiber mats with tunable porosities have shown much enlarged specific areas to enhance the separation performance. For example, the electrospun sulfonated styrene nanofiber membranes expressed excellent affinity to liquid electrolytes and high ionic conductivity when acted as separators. Second, various morphologies (particles or nanoparticles) of transition metal oxide coated on the surface of carbon nanofibers (CNF) generate more access channels for Li^+ transport into the inner active sites of anodes and remarkably increase the electron transport rate. Furthermore, the coaxial nanofiber composite ($LiCoO_2$ core with MgO shell) electrodes fabricated by electrospinning, which exhibits excellent reversibility, small impedance growth, and better cyclability since the $LiCoO_2$ cannot directly contact electrolyte (Gu et al. 2007).

Catalysts

As far as we known, the catalysts' performance is determined by the number of active sites exposed to the precursors. The electrospinning method exbibits the facile production of the porous nanofiber mats with high specific area, which has shown great potential in novel catalysts fabrication. Electrospun TiO_2 nanofiber mats modified with Pt, Pd, and Ru nanoparticles have been successfully fabricated, their performance toward Suzuki coupling reactions has been tested, and the results showed the novel metal coated TiO_2 catalysts can shorten the reaction duration (Formo et al. 2009). The performance enhancement is attributed to the quasi-one-dimensional transport of the electron within the nanofibers as well as the dramatically reduction of the grain boundaries between the catalytic nanoparticles.

Supercapacitors

Supercapacitors are intensively investigated as an energy storage device due to their rapid charging/discharging capacity, long cycle lifetime, and high power density. They are widely used in large-range applications, from handheld devices to electric vehicles even to power station. For a supercapacitor, higher operating voltage and capacitance value are key points to increase the overall energy density of supercapacitors. Increase the surface contact area of the electrodes can be a simple method to obtain higher capacitance performance, and the electrospun nanofiber mats possess this advantage. High temperature activated electrospun CNFs have shown great potential to make high performance supercapacitor electrodes due to the porous surficial structures and 1D morphology. The commonly used precursors for the CNFs synthesis are some polymers (Kim et al. 2004), such as polyacrylonitrile (PAN), isotropic pitch precursor (IPP), poly(imide) (PI), (Kim 2005). polyamic acid (PAA), and polybenzimidazol (PBI). These polymer solutions are first loaded into a typical electrospinning setup to obtain the nanofibers and then carbonized under high temperature in inert atmospheres (N_2 or Ar) to obtain the CNFs. Ruthenium nanoparticles doped CNFs have been synthesized by electrospinning the PAN/Ruthenium acetylacetonate solution and the specific capacitance of the nanofiber-based device jumped from just 140 to 391 F/g. (Ju et al. 2007).

Fuel Cells

Platinum is extensively used in fuel cells for oxidation of fuels (such as hydrogen and methanol) due to its high activity and chemical stability, but the Pt in earth is very rare so the price of Pt-based materials is very high. To reduce the usage amount of Pt in fuel cells, the Pt nanoparticles can be coated on the surfaces of the electrospun carbon nanofibers to form a composite electrocatalyst, which can be used for the oxidation of methanol (Li et al. 2008). With enhanced specific surface area and porosity, the electrospun Pt and PtRh nanowires show better catalytic performance than the traditional Pt nanoparticle catalysts (Kim et al. 2009).

Thin Film Solar Cells

Thin film solar cells are fabricated by depositing several function layers, such as transparent metal oxides electrodes (fluorine-doped tin oxide and indium-doped tin oxide), light absorbers (perovskite and dyes), charge carrier transport layers, and metal electrodes, which can be used for converting sunlight into electricity. In order to fabricate high efficiency device, the efficiently transport/extraction of the photo-generated charge carriers (especially for electrons) should be achieved. Thus, high surface area, reduced crystal defects, and 1D pathway are urgent needed for the charge carriers transport layer synthesis. TiO_2 nanofiber mats are proposed for the high effi-ciency solar cell fabrication, and they are obtained by electrospinning polymers [PVP and polyvinyl acetate (PVAc)] and TiO_2 precursors (titanium isopropoxide) mixture in DMF. After high temperature annealing, the anatase TiO_2 nanofibers are grinded and fabricated into thin film electrodes to transport photogenerated elec-trons, the enlarged surface area, reduced crystal defects, and 1D electron pathway along the nanofiber will lead to the enhancement of light absorbers loading and sunlight harvesting efficiency (Chuangchote et al. 2008). Hollow rice grain-shaped TiO_2 nanostructures with abundant nanopores, reduced grain boundaries, direct elec-tron transportation, and large surface area were prepared by simple electrospinning of PVAc (Mw $=$ 500,000) and titanium isopropoxide mixture followed with high temperature calcination. Then, they were used as electron transport material in a perovskite solar cell and the champion PCE of 14.2% is achieved for 1 cm^2 device, which was 47.9% higher than its planar counterpart (with an efficiency of just 9.6%, Fig. 10) (Ma et al. 2019).

2.4.2   Electronic Applications

Nanofibers made of sensing materials have shown the high sensing behavior with fast response sensitivity. It is widely accepted that the electrospun nanofibers with good mechanical durability, high surface area, and aligned charge carrier transport can supply better structure reinforcement and achieve high sensing performance.

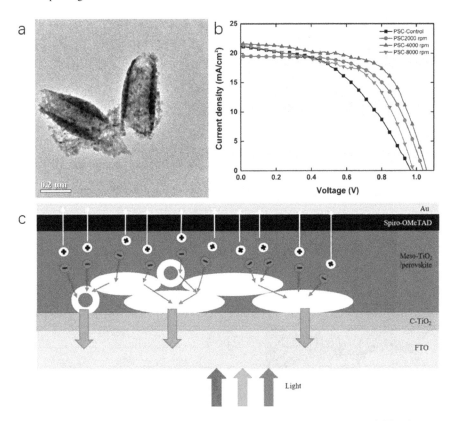

Fig. 10 **a** The TEM image of the fabricated rice grain-shaped TiO$_2$ nanostructures. **b** The photocurrent–voltage curve of the champion device with an active area of 1 cm^2. **c** The proposed mechanism for the enhanced performance of the rice grain-shaped TiO$_2$ nanostructure-based PSCs (Ma et al. 2019)

Motion Sensors

Piezoelectric materials, which can convert mechanical deformation into electric signals, are widely used to fabricate motion sensors, which are demanded for the wearable devices for power generation and body movement sensing. The electrospun fibers exhibit high mechanical endurance and enhanced piezoelectric property because of their unique morphologies. By applying the stretching-induced alignment method upon just-prepared poly[(vinylidenefluorid-co-trifluoroethylene)] P(VDF-TrFE) fibers, the highly oriented P(VDF-TrFE) fiber mat can be obtained. The 80% aligned electrospun P(VDF-TrFE) fibers show high output voltage of 84.96 mV (Fig. 11). The aligned electrospun P(VDF-TrFE) fiber bundle was integrated in daily cloth to monitor real-time body gestures (Ma et al. 2018).

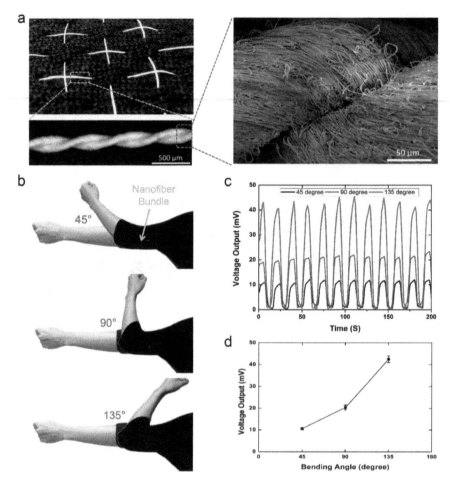

Fig. 11 a Twisted electrospun P(VDF-TrFE) fiber bundle integrated into cloth. **b** Monitoring different body movements (45°, 90°, and 135°). **c** Corresponding voltage outputs and **d** their average values for different body movements (Ma et al. 2018)

Chemical Sensors

When chemicals (usually liquid or gas) contacts with some nanostructured semi-conductors, the charge transfer between the chemicals and the semiconductors will lead to resistance change of the semiconductors. One can get the information about the chemicals by fabricating chemical sensors based on these semiconductors by tracking the electric resistance change. It is commonly known that the size of the nanomaterials will affect the internal free charge carriers transport, so the chemical sensors fabricated with electrospun nanomaterials may exhibit faster and more sensi-tive response when compared to the conventional powder samples. For example, the electrospun SnO_2 and MoO_3 nanowires have been observed with higher sensitivity and faster response time in detecting H_2S and NH_3 than their thin film counterparts, respectively (Sawicka et al. 2005). Also, the detection threshold of the chemical sensors based on the electrospun nanomaterials is very low, such as electrospun ZnO nanowires can detect ethanol vapor with a concentration of just 10 ppm (Wu et al. 2009).

Actuators

Contrary to motion sensors, the actuators are a kind of devices designed to convert electrical energy into various mechanical motions. The electrospun flexible nanofiber mats can be used to enhance the mechanical properties of the device by increasing the strain since the electrolyte can better penetrate into the porous structure of the electrospun nanofiber membranes. The response rate of the electrospun nanofiber mat-based actuator can be increased since high ion mobility can be guaranteed. Some thermal sensitive nanofiber mats have been fabricated for the construction of unique thermal response actuators.

Optoelectronics

With regard to the optoelectronics, the 1D electrospun nanofibers with high conductivity can efficiently transport the photogenerated charge carriers. When compared to the traditional thin film, an electrospun Al doped ZnO nanowire shows higher sensitive photoresponse under below bandgap light illumination. Intrinsic p-type quasi-1D CuO nanofibers fabricated with rotating electrospinning method have been used to construct low-cost and photo-sensitive field-effect transistors (Wu et al. 2006).

2.4.3   Bioengineering and Biotechnology

The nanofibers made of biocompatible materials are widely investigated for implanting into patient's body for the purpose of repairing the diseased tissues (Fig. 12, Yang et al. 2012). The well-designed bio-friendly nanofiber mats have shown the capacity to separate the bio-interfaces by using different chemical ligands to functionalize the nanofiber surfaces. The separation efficiency can be remarkably enhanced due to the high surface area of the electrospun nanofibers. The electrospun nanofibers are also regarded as drug carrier with controllable release function. That is, the electrospun nanofiber mats loaded with specific drug can be used for surgery wound recovery and controllable deliver the required amount of drug through patient's digestive system by encapsulating into capsules.

Fb2	Fa2 + Polyplex	Fa2	Control	

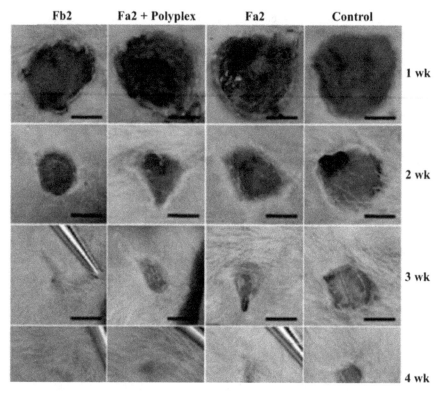

Fig. 12 Diabetic skin wound using a rat model for comparison of control and those subjected to delivery of pbFGF polyplexes from electrospun poly(ethylene imine)/PEG (2 kDa) core/shell fibers (Fa2: blank fibers and Fb2: fibers with pbFGF polyplexes in the core) (Yang et al. 2012)

2.4.4 Defense and Security

The electrospun nanofiber mats can be used as protective clothing for the soldiers' safety. By incorporating the nanofiber mats, the protective cloth can kill/degrade the biological warfare agents and harmful chemicals. Work together with the chemical sensors, the protective clothing can perception the danger in advance to avoid body injure from chemical weapons.

2.4.5 Environmental Engineering

In the aspect of environmental engineering, the electrospun nanofiber mats can be used to filter dusts and hazardous substances. Polluted air or solutions can be purified by these nanoscale porous membranes. Also, the bacteria or organic groups within the air or solutions can be killed or degraded by the functionalized nanofibers by surficial decorated with some functional chemical groups, and this technique has

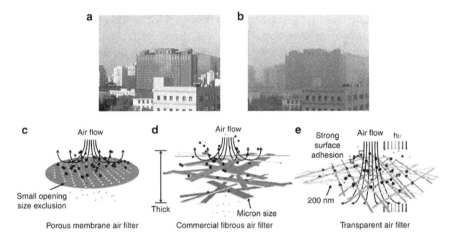

Fig. 13 a, b Photographs of a place in Beijing during sunny and hazy day with high PM2.5 level. **c–e** Schematics of different air filters that capture PM particles (Liu et al. 2015)

shown great potential to protect people outside/inside a room (Fig. 13, Liu et al. 2015).

3 Conclusion

In this chapter, information about electrospinning nanofibers is comprehensively introduced, including the materials for nanofiber fabrication, processing mechanism and parameters of electrospinning techniques, special electrospinning techniques, and potential applications of electrospun nanofibers.

The electrospinning technology can generate ultrathin fibers by using a wide variety of materials that contain the most commonly used polymers, metal oxides/ceramics, and composites. There are several necessary parts for a standard electrospinning setup, including a spinneret (usually a needle with a syringe), a pump that can maintain the feeding, a high-voltage power source (usually several tens of kilovolts), and a conductive collector to gather the resultant. During the electrospinning process, there are a number of factors that can determine the final morphology and property of the resultant fibers, including the polymer solution parameter such as surface tension and viscosity, processing conditions including the voltage and distance, and the ambient condition like humidity. Besides the conventional polymer fibers, nanofibers with special morphology and property can be customized with special electrospinning techniques such as wet- electrospinning, melt electrospinning, core–shell electrospinning, and electrospray technique. The resulted electrospun nanofibers can be applied in various situations like energy harvesting and storage, all kinds of sensing applications, biomedical applications, and environmental protection.

Electrospinning is an old technique that has existed for nearly one century, but it still has so many possibilities in potential innovation and can facilitate the large-scale fabrication of nanofibers. Electrospun nanofibers have attracted extensive interest as their diverse applications. And with the problems being continuously addressed and new modules being developed, this technique is expected to be applied on a broader platform and contributed to the improvement of daily life in the foreseeable future.

References

M. Baniasad, J. Huang, Z. Xu, S. Moreno, X. Yang, J. Chang, M.A. Quevedo-Lopez, M. Naraghi, M. Minary-Jolandan, A.C.S. Appl, Mater. Interfaces **7**, 5358 (2015)

P.K. Baumgarten, J. Colloid Interf. Sci. **36**, 75 (1971)

T.D. Brown, P.D. Dalton, D.W. Hutmacher, Prog. Polym. Sci. **56**, 116 (2016)

C.L. Casper, J.S. Stephens, N.G. Tassi, D.B. Chase, J.F. Rabolt, Macromolecules **37**, 573 (2004)

S. Chuangchote, T. Sagawa, S. Yoshikawa, Appl. Phys. Lett. **93**, 033310 (2008)

A. Formhals, US Patent 1975504 (1943)

E. Formo, M.S. Yavuz, E.P. Lee, L. Lane, Y.N. Xia, J. Mater. Chem. **19**, 3878 (2009)

Y.X. Gu, D.R. Chen, X.L. Jiao, F.F. Liu, J. Mater. Chem. **17**, 1769 (2007)

Z.M. Huang, Y.Z. Zhang, M. Kotaki, S. Ramakrishna, Compos. Sci. Technol. **63**, 2223 (2003)

T. Jarusuwannapoom, W. Hongrojjanawiwat, S. Jitjaicham, L. Wannatong, M. Nithitanakul, C. Pattamaprom, P. Koombhongse, R. Rangkupan, P. Supaphol, Eur. Polym. J. **41**, 409 (2005)

Y.W. Ju, G.R. Choi, H.R. Jung, C. Kim, K.S. Yang, W.J. Lee, J. Electrochem. Soc. **154**, A192 (2007)

P. Katta, M. Alessandro, R.D. Ramsier, G.G. Chase, Nano Lett. **4**, 2215 (2004)

J.Y. Ke, H.J. Chu, Y.H. Hsu, C.K. Lee, Proc. SPIE 10164 (2017)

C. Kim, J. Power Sources **142**, 382 (2005)

C. Kim, Y.O. Choi, W.J. Lee, K.S. Yang, Electrochim. Acta **50**, 883 (2004)

H.J. Kim, Y.S. Kim, M.H. Seo, S.M. Choi, W.B. Kim, Electrochem. Commun. **11**, 446 (2009)

L. Kong, G.R. Ziegler, Food Hydrocoll. **38**, 220 (2014)

D. Li, Y. Xia, Adv. Mater. **16**, 1151 (2004a)

D. Li, Y. Xia, Nano Lett. **4**, 933 (2004b)

M.Y. Li, G.Y. Han, B.S. Yang, Electrochem. Commun. **10**, 880 (2008)

C. Liu, P.C. Hsu, H.W. Lee, M. Ye, G. Zheng, N. Liu, W. Li, Y. Cui, Nat. Comm. **6**, 6205 (2015)

S. Ma, T. Ye, T. Zhang, Z. Wang, K. Li, M. Chen, J. Zhang, Z. Wang, S. Ramakrishna, L. Wei, Adv. Mater. Technol. **3**, 1800033 (2018)

S. Ma, T. Ye, T. Wu, Z. Wang, Z. Wang, S. Ramakrishna, C. Vijila, L. Wei, Sol. Energy Mater. Sol. Cells **191**, 389 (2019)

P.L. McEuen, Nature **393**, 6680 (1998)

C. Mit-uppatham, M. Nithitanakul, P. Supaphol, Macromol. Chem. Phys. **205**, 2327 (2004)

M.P. Prabhakaran, E. Vatankhah, S. Ramakrishna, Biotechnol. Bioeng. **110**, 2775 (2013)

D.H. Reneker, A.L. Yarin, Polymer **49**, 2387 (2008)

K.M. Sawicka, A.K. Prasad, P.I. Gouma, Sens. Lett. **3**, 31 (2005)

L. Wannatong, A. Sirivat, P. Supaphol, Polym. Int. **53**, 1851 (2004)

H. Wu, D.D. Lin, W. Pan, Appl. Phys. Lett. **89**, 133125 (2006)

W.Y. Wu, J.M. Ting, P.J. Huang, Nanoscale Res. Lett. **4**, 513 (2009)

C.Y. Xu, R. Inai, M. Kotaki, S. Ramakrishna, Biomaterials **25**, 877 (2004)

Y. Yang, T. Xia, F. Chen, W. Wei, C. Liu, S. He, X. Li, Mol. Pharm. **9**, 48 (2012)

M. Zamani, M.P. Prabhakaran, E.S. Thian, S. Ramakrishna, Int. J. Pharm. **473**, 134 (2014)

X. Zhang, V. Thavasi, S.G. Mhaisakar, S. Ramakrishna, Nanoscale **4**, 1707 (2012)

Nanofibers for Gas Sensing

Wei Liu and Ling Zhu

Abstract This chapter reports a comprehensive review of the nanofiber gas sensor for enabling fast, relatively inexpensive, and minimal monitoring of the target gas concentration. The front part provides detail information on the sensing mechanism, evaluation criteria and application fields of the nanofibers-based gas sensor. The sensing mechanism mainly divided into the ionosorption model and oxygen-vacancy model, which can determine the fundamental factor for the evaluation of sensing performance. The main application fields focus on public safety, food processing, environmental monitoring, and disease diagnosis so far. The rear section then discusses the effect of the composition and morphology of nanofibers for the sensing performance in detail. Construction of multiple heterojunction components and various dopants, including the noble metal and the rare-earth, can effectively improve the sensing performance of the nanofiber-based gas sensor, which can provide more oxygen-vacancy and active sites. Meanwhile, the control of morphology provides a larger specific surface area and more gas diffusion channel. This chapter might bring further development and evolution of sensors based on nanofibers for the detection of various analytes.

Keywords Nanofibers · Metal oxide · Gas sensor · Sensing materials · Sensing performance

W. Liu (✉)
College of Electronics Science and Technology,
Shenzhen University, 518060 Shenzhen, People's Republic of China
e-mail: weil@szu.edu.cn

L. Zhu
Key Laboratory of Optoelectronic Device and Systems of Ministry of Education and Guangdong Province, College of Optoelectronic Engineering, Shenzhen University, 518060 Shenzhen, People's Republic of China
e-mail: zhuling@szu.edu.cn

© Springer Nature Singapore Pte Ltd. 2020
L. Wei (ed.), *Advanced Fiber Sensing Technologies*,
Progress in Optical Science and Photonics 9,
https://doi.org/10.1007/978-981-15-5507-7_8

1 Introduction

With the development of the Internet of Things, big sensor data will leverage capabilities such as cloud infrastructures for data storage, processing, and visualization.
Access to this data will become pervasive, especially through mobile devices (Kim
et al. 2015). Gas sensors are expected to be one of the largest generators of data in our
everyday life, which can be converted into visual signals through the sensor output.
It brings a huge hope for the rapid in situ detection of a series of volatile organic
compounds (VOCs) and toxic gases derived from industrial sources, automobile
engines, and biomarkers in human exhalation (Lu et al. 2016). It is well-known that
gas sensor performance features such as sensitivity, selectivity, response and recovery
times, stability, durability, reproducibility, and reversibility are largely influenced by
the properties of the sensing materials. Many various sensing-active materials, such
as inorganic materials, conducting polymers, carbon materials, as well as multiple
composites, have been used as sensing materials to detect the targeted gases based
on various sensing techniques and principles (Ramgir et al. 2010). There are two
important functions in gas sensing processes, receptor and transducer functions. The
receptor function explains the importance of the surface reactions dominated by the
chemisorbed oxygen species on the surface of the sensing materials. On the other
hand, the transducer function is related to the transduction of the sensing signal
induced by the chemical reactions into an output signal such as resistance (Kim
et al. 2017a, b). Besides, it is worth noting that the performance of gas sensors is also
strongly affected by the specific surface area and the number of active sites of sensing
materials from the fundamental understanding of basic gas sensing mechanisms and
structural design principles. For this reason, many efforts have been focused on
different morphology and composite materials for high-performance sensors with
enhanced gas sensing properties.

   Among different morphology and composite nanomaterials, nanofibers benefiting
from the advantages of large surface-area-to-volume ratio, tailored pore structures,
large stacking density, and ease of surface modification which can achieve high
sensitivity, selectivity, fast response and recovery, and good reversibility (Zhou et al.
2019a, b). Generally, the nanofibers are deposited on a fixed collector in a three-
dimensional nonwoven membrane structure with a wide range of nanofiber diameter
distribution, which can further provide interconnected porous structure and more
gas diffusion channels (Liu et al. 2019a, b, c, d). For this reason, nanofibers are
expected to be an ideal candidate as the structure of sensing materials. However,
single-component nanofibers have been difficult to meet all the demand for ideal
sensors. To solve this problem, many previous studies have shown that the gas
sensing performance of sensors can be remarkably improved through the integration of multiple composites and dopants, including the noble metals and the rare
earth elements. The multiple composites can construct the heterogeneous structure,
and the intimate electrical contact on the surface of different nanofiber materials

leads to charge transfer and the formation of an electron depletion layer, which can improve gas sensing performance. The doped elements can enter the nanofiber lattice, changing the crystal structure or electronic structure of the host material, and change the morphology of the material during synthesizing processes (Wang et al. 2010).

This chapter focuses on nanofiber-based gas sensors with high surface area, tenable morphology, and chemically nanostructure effect as well as the relevant basic theories and concepts. Various factors that have an influence on the sensing performances of nanofiber-based gas sensors, such as multiple compositions, dopants, morphology, and surface-modified properties of nanofiber materials, along with the in-depth elucidation on the gas sensing mechanism, are thoroughly discussed and analyzed. And also, this chapter can serve as valuable guidance for new researchers in the gas sensing area, including the basics, the sensing mechanism, and the advanced application of nanofiber-based gas sensors.

2   Main Body Text

2.1   Nanofiber Gas Sensors: Sensing Mechanism, Evaluation Criteria, and Application Fields

2.1.1   Sensing Mechanism

Extensive research efforts are directed towards the development of highly sensitive gas sensors using nanofiber materials, which is a typical resistive gas-sensor. Understanding the sensing mechanism is necessary to evaluate the merit and demerit of nanofiber materials with different nano-sized crystals and grains. In terms of the material properties, the gas-sensor based on the nanofiber materials is subject to the semiconductor gas sensor, which can be distinguished as n-type and p-type semiconductors. The resistance of n-type semiconductors is decreased upon contact with a reducing analyte and increased with an oxidizing one. In contrast, p-type semiconductors have the opposite trend. In detail, two main sensing mechanism is currently approbatory as Gurlo and Riedel mention, the "Ionosorption" and "Oxygen-Vacancy" models (Gurlo and Riedel 2007), since they have been repeatedly discussed in mechanism studies. It must also be kept in mind that the mechanism may be influenced significantly by the materials and other testing conditions.

2.1.2   Ionosorption Model

In the ionosorption model, the detection of gases can be subdivided between reception and transduction function. The receptor function concerns the ability of the oxide surface to interact with the target gas. As shown in Fig. 1a, the adsorbed oxygen acts as a receptor, which can react with the target gas to produce H_2O and CO_2. More

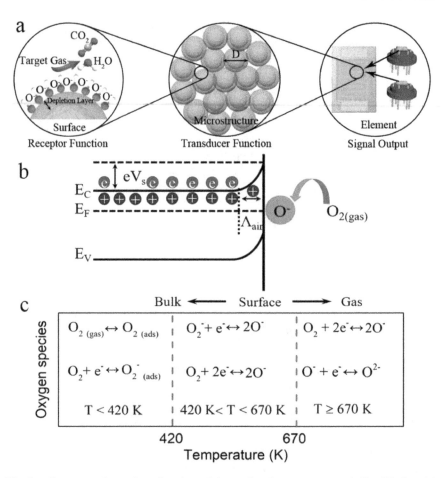

Fig. 1 a Receptor and transducer functions of the semiconductor gas sensor. **b** Simplified model illustrating band bending after the ionosorption of oxygen on the surface sites of semiconductor (E_C, E_V, and E_F represent the energy of the conduction band, valence band, and the Fermi level, respectively, while Λ_{air} and eV_s represent the thickness of the electron depletion layer and the potential barrier height, respectively. The e^- and $+$ represent conducting electrons and the donor sites, respectively). **c** The oxygen species at different temperatures on the surface of the semiconductor. Reproduced from ref. (Franke et al. 2006). Copyright 2006, Wiley-VCH

specifically, oxygen is adsorbed on the grains as the negative charge ions when the semiconductor oxide exposure to the air, and inducing the formation of the electron depletion layer. Upon exposure to the target gas, the adsorbed oxygen is consumed and decreased down to a steady-state level, resulting in the change of electronic on the oxide surface. After the reception of the target gas by a change in ionosorbed oxygen concentration, the conductivity of oxide is quantitatively varied in compliance with its transduction function. This function has been explained by assuming a formation of double Schottky barriers for the transport of electrons through grain boundaries. The ionosorbed oxygen scatters electrons within the Debye length (δ) of oxide reducing its electron mobility. The resistance and gas sensitivity of the device hardly depend on the grain size (diameter, D). For large grains ($D \gg \delta$) the sensing mechanism is controlled by the grain boundaries. For ultra-fine nanoparticles, there are two possible mechanisms as a function of the grain size. If the grain size is bigger than twice the Debye length of oxide, a conduction channel (L) with bulk mobility exists within a diameter ($L = D\text{-}2\delta$) from the grain center. In contrast, if the grain size is smaller or equal to twice δ and the whole grain is depleted, the ionosorbed oxygen may induce more conduction channel through the oxide. At last, the element refers to effective transduction of the target gas concentration information into a macroscopically accessible signal, that is, the change of the electrical resistance.

When the oxygen is ionosorbed on the metal oxide surface, which can act as electron acceptors due to their relative energetic position concerning the Fermi level (Fig. 1b). The required electrons for this process originating from donor sites, which are extracted from the conduction band and are trapped at the surface, leading to an electron depleted surface region, the so-called electron depletion layer. The negative surface charge leads to band bending and generates a potential barrier height. The height and depth of the band bending depend on the surface charge, which is determined by the amount and type of adsorbed oxygen. At the same time, the electron depletion layer depends on the Debye length, which is a characteristic of the semiconductor material and can fulfill the following equation (Yamazoe et al. 2003):

$$L = \sqrt{\frac{\varepsilon_0 \varepsilon k_B T}{e^2 N_d}}$$

where k_B is Boltzmann's constant, ε is the dielectric constant, ε_0 is the permittivity of free space, T is the operating temperature, e is the electron charge, and N_d is the carrier concentration, which corresponds to the donor concentration assuming full ionization.

For the polycrystalline sensing materials, electronic conductivity occurs along percolation paths via grain-to-grain contacts and therefore depends on the value of eV_s of the adjacent grains. In here, eV_s represent the Schottky barrier. The conductance G of the sensing material, in this case, can be written as (Madou and Morrison 2012):

$$G \approx \exp\left(\frac{-eV_s}{k_B T}\right)$$

In addition, atmospheric oxygen can dissociate into different forms at different temperatures, which can be classified as O_2^-, O^-, and O^{2-}, respectively. As shown in Fig. 1c, the more probable mechanisms of oxygen interaction with the sensing layer are the physisorption or chemisorption of O_2 in the molecular form in the lower operating temperature range ($T < 420$ K). In this temperature range (420 K $< T <$ 670 K), the atmospheric oxygen molecular and the adsorbed oxygen can dissociate into O^-, which can cause a fast change of resistance. In the higher temperature range ($T \geq 670$ K), the sensing mechanism becomes bulk controlled. When the adsorbed oxygen can dissociate into $O-$, the $O-$ also can dissociate into O^{2-}, and oxygen desorption becomes predominant (Neri et al. 2007).

2.1.3   Oxygen-Vacancy Model

This model concerns the oxygen vacancies at the surface, which are the critical factor for the chemiresistance gas sensor. For the oxygen-deficient sensing material, oxygen vacancies act as electron donors. Once the reducing target gas contacts with the sensing material, the partial reduction further induce more oxygen vacancies and thereby injection of free electrons in its conduction band, that is, increased conductivity. More specifically, the target gas removes oxygen from the surface of the lattice and thereby producing an oxygen vacancy. Meanwhile, the ionized vacancy introduces electrons into the conduction band and increasing the conductivity. When the reducing target gas is removed, the surface is reoxidized filling the

vacancy, one or more electrons are taken from the conduction band, which results in the decrease in conductivity (Tricoli et al. 2010). The role of oxygen-vacancy for oxidizing target gas is strongly material and temperature-dependent and requires further understanding. Furthermore, the slow kinetics of oxygen exchange at the surface and the immobilized ionization of oxygen vacancies at the operating temperatures are often not considered or avoided by the nonrealistic experimental conditions. To summarize, oxygen vacancies are elusive, almost invisible, but crucial entities on the surface of oxides. While the vast majority of the regular surface sites may not be involved in chemical reactions at room temperature, the defect sites, and the oxygen vacancies are among the most reactive sites of the whole surface (Pacchioni 2003). Thus, new perspectives should be proposed for the understanding of the oxygen-vacancy model, such as the characterization of spherical aberration coefficient-corrected transmission microscopy in the aspect of atoms.

2.1.4   Evaluation Criteria

The use of nanofiber-based gas sensor has potential advantages compared to conventional thin-film devices due to the large surface-area-to-volume ratio, tailored pore structures, large stacking density, and ease of surface modification of nanofibrous membranes, which can achieve high sensitivity, selectivity, fast response and recovery times, and good reversibility. The performance of gas sensors can be evaluated by different parameters like sensitivity, selectivity, stability, response and recovery time, reversibility. Here, we will introduce the definition and understanding of several most important evaluation criteria for nanofiber-based gas sensors.

2.1.5   Sensitivity

Sensitivity is one of the most essential indicators which describes the activity of a sensing material to target gas. Sensitivity can be defined as R_a/R_g for reducing gases and R_g/R_a for oxidizing gases, where R_a is the resistance of the gas sensor in the clean carrier (usually air) and R_g stands for the resistance of the sensor in the target gas (Wetchakun et al. 2011). Sometimes the concepts of sensitivity and response are ambiguous since both describe the variation degree of the resistance change. Therefore, other expression forms of sensitivity are also expressed by $[(R_a - R_g)/R_a]*100\%$ (Patil et al. 2015).

2.1.6   Selectivity

Selectivity is the ability of the gas sensors to detect a specific gas in a mixture of gases. It is well-known that gas sensors are normally sensitive to more than one target gas and usually show cross-sensitivities. Thus, selectivity is a measurand, which can be estimated by comparing the effects of different gases on a sensor. A major shortcoming of some semiconductor oxide as sensing material is its low selectivity, due to the presence of a wide range of adsorption sites on its surface that cannot distinguish the contribution of each type of gaseous molecules to the total electrical signal. The surface modification of a highly dispersed oxide matrix with clusters of transition metals or their oxides may affect the electrophysical and chemical properties of the surface, which can improve the selectivity. In addition, some materials can show different selectivity profiles at different operating temperatures due to the disproportionate sensitivity change toward various gases during the change in temperature. Take CO gas as an example, the typical behavior of the sensor response to a target gas as a function of the operating temperature as shown in Fig. 2. The unsatisfactory sensitivity at low temperature can be attributed to the insufficient dissociative adsorption of oxygen and less CO molecules adsorb in the surface and conversion to CO_2 molecules. When the sensor is working at high temperature, the poor sensitivity can be attributed to the insufficient desorption of CO and chemisorbed oxygen. This means that gases possessing a specific combination of adsorption/desorption parameters and reaction rates might have different temperature profiles of sensor response and therefore these gases could be selectively recognized (Morrison 1987).

2.1.7   Stability

The stability presents the performance of the gas sensor to maintain its output signal over a long period and to target systematically varying concentrations of a target gas. The baseline resistance shift as well as decreasing the sensitivity of a sensor. The growth of grain at the high operating temperature and the crystallization are the major reasons, which can cause the sensing performance gradually degenerate in the process of long-term measurement. Studies show that each size of crystallite has its

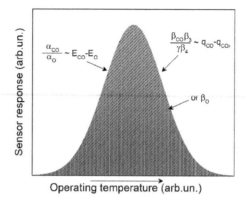

Fig. 2 Adsorption/desorption parameters controlling temperature dependence of oxides gas response to CO: α_{CO} and α_O are the adsorption coefficients of CO and O_2; β_{CO} and β_{CO_2} are the desorption coefficients of CO and CO_2; β_3 and β_4 are the coefficients of charging and neutralization of CO_2; E_O and E_{CO} are the adsorption activation energies of O_2 and CO, q_{CO} and q_{CO_2} are the desorption activation energies of CO and CO_2. Reproduced from ref. (Korotcenkov and Cho 2013). Copyright 2013, Elsevier

threshold temperature, above which a tendency of grain size growing occurs. The studies present the related rule of the size of grains and threshold temperature, which can be described by the expression (Korotcenkov et al. 2005):

$$T_{st} = 420(lg\,D)^{3/4}(°C)$$

where T_{st} is threshold temperature, D is grain size. Therefore, the stability is controlled by some of the energy of grains (energy of vacancy formation) characterizing the thermodynamic stability of crystallite. In addition, the factors which might be responsible for instability are structural transformation, phase transformation, poisoning, degradation of contacts and heaters, bulk diffusion, error in design, change in humidity in the surrounding atmosphere and interference effect (Korotcenkov and Cho 2017). But there is no uniform approach to increase the stability of the metal oxide sensors up to now.

2.1.8 Response and Recovery Time

Regarding the response and recovery times of the sensor, the response times is defined as the time at which the response value reaches 90% of its maximum one after the injection of target gas, while the recovery time is defined as the time at which the response value is decreased to its 10% of its maximum one after target gas was replaced by air (Liu et al. 2019a, b, c, d). The response kinetics is mostly determined by gas diffusion processes and chemical reactions between the solid surface and gaseous molecules. Generally, higher porosity sensing materials provide more channels for gas diffusion and ensure faster response and recovery than bulk materials. Increasing the operating temperature can also speed up the response and recovery behaviors, which can improve the reaction rate (Liu et al. 2018a, b, c, d; Zhou et al. 2019a, b).

2.1.9 Reversibility

Reversibility is described as whether a sensor signal returns to its original state when gas concentration returns to a normal state. More particularly, target gas and sensing materials interaction are considered reversible if removal of the target gas from the ambient air results in its complete desorption from the sensing materials, with no permanent physical or chemical change has occurred. On the other hand, the slow reversible process at room temperature should be distinguished with the process of truly irreversible (Wu et al. 2018). Note that the sensor signal is fundamentally different for the process of reversible and irreversible. For a reversible sensor, the frequency or amplitude change is proportional to target gas concentration. In this case, sensitivity is determined by the equilibrium value of the signal, relative to its baseline; upon removal of the target gas, the sensor signal returns to its original baseline value. For an irreversible sensor, however, exposure to the target gas produces a change in signal from which the sensor does not recover; repeated exposures produce a corresponding permanent shift in the baseline value. Thus, excellent reversibility is essential when considering scaling up to large-scale production of commercial devices in the actual application.

2.1.10 Application Fields

Nanofiber-based gas sensors have been applied in many fields including environmental monitoring, public safety, process control, and medical diagnostics. Nanofibers can be very useful components for the highly sensitive gas sensors due to their ultra-small size, high surface to volume ratio, surface reactivity, and low power consumption. Figure 3 shows a schematic summary of some rapidly emerging sensing concepts for public and workplace safety personalized medicine. Kim et al. designed bimetallic Pt-based nanoparticles supported mesoporous WO_3 nanofibers that demonstrate unprecedented sensing performance for detecting different target biomarkers in highly humid exhaled breath, which can clearly distinguish between the breath of simulated biomarker and healthy controls by assembling sensor arrays (Kim et al. 2017a, b). Lichtenstein et al. proposed chemically modified nanofibers-based devices arrays enable the supersensitive discriminative detection of explosive species, which allow for the rapid detection of explosives down to the parts-per-quadrillion concentration range from air collected samples, and these detection results represent the first generation of analytical platform for the ultra-trace detection and identification of a broad range of explosives species (Lichtenstein et al. 2014). Liu et al. reported the $Pt@In_2O_3$ core-shell nanofiber-based sensor, which can function as an

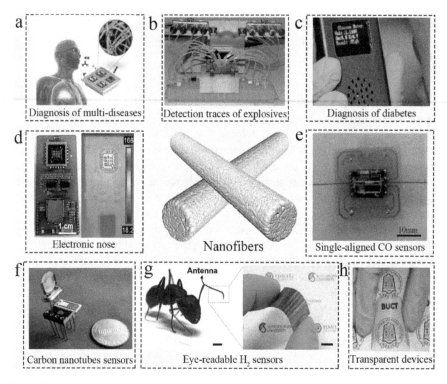

Fig. 3 Schematic summary of nanofiber gas sensors for practical application. For example, **a** Diagnosis of multi-diseases sensor array (Kim et al. 2017a, b), Copyright 2017, Wiley-VCH. **b** Detection traces of explosives (Lichtenstein et al. 2014), Copyright 2014, Springer Nature Publishing. **c** Diagnosis of diabetes (Liu et al. 2018a, b, c, d), Copyright 2018, Springer Nature Publishing. **d** Electronic nose (Moon et al. 2016), Copyright 2016, American Chemical Society. **e** Single-aligned CO sensors (Nikfarjam et al. 2017), Copyright 2017, American Chemical Society. **f** Carbon nanotubes sensors (Lu et al. 2015), Copyright 2015, Elsevier. **g** Eye-readable H₂ sensors (Han et al. 2017), Copyright 2017, Wiley-VCH. **h** Transparent devices (Wan et al. 2015). Copyright 2015, Wiley-VCH

ultrasensitive sensing platform for the real-time detection of diabetes biomarkers in exhaled breath by portable sensing device (Liu et al. Liu et al. 2018a, b, c, d). Moon et al. displayed different nanofibers based on three metal oxide materials to obtain electrical responses to targeted gases, the sensing properties were carefully measured in a chemiresistive electronic nose condition to exhaled breath, such as H_2S, NH_3, and NO, known as biomarkers for human diseases (Moon et al. 2016). Nikfarjam et al. fabricated a single-aligned TiO_2 nanofiber using a novel electrospinning procedure equipped with secondary electrostatic fields on highly sharp triangular and rectangular electrodes provided for gas sensing applications. The results exhibited that the rational design highly impact sensitivity improvement of the CO sensor in the range of ppt, and it can be employed for different related applications. Lu et al. investigated single-walled carbon nanotube-coated sensor arrays for the detection of volatile organic compounds on both chemiresistors and quartz crystal microbalance, which provided insight into the inter-tube interlocks and conductivity modulation of acidified single-walled carbon nanotube via a hydrogen bond (Lu et al. 2015). Han et al. presented a geometry-switchable and highly H_2-reactive Janus nanofiber array inspired by the structural features of the arthropod sensilla, which allowed simple structural change with the naked eye without requiring additional electrical apparatus. It can potentially be employed in wetting-controllable, H_2-reactive nanoactuators and in hazard warning devices that use an H_2-sensitive fastener (Han et al. 2017). Wan et al. showed carbon nanotube networks coated with hierarchically nanostructured polyaniline nanorods assembled a flexible, transparent, chemical gas sensor. The device demonstrated excellent sensitivity at room temperature towards ammonia gas without any obvious decrease in performance after 500 bending and extending cycles. It is anticipated that the research can be extended to other conducting polymer-containing nanocomposites, which can create opportunities for developing a low-cost, sensitive, reliable handheld, wearable, and transparent gas sensor electronic devices (Wan et al. 2015).

2.2 Nanofiber Gas Sensors: Composition and Morphology and Sensing Performance

2.2.1 Composition and Morphology

Nanofibers are featured with ultra-small size, high surface to volume ratio, large surface area, and small pore size. Various materials such as inorganic material, organic material, and carbon material can be used to fabricate uniform nanofibers with well-controlled sizes, compositions, and morphologies. Different nanofiber compositions and morphologies can be obtained via control of the processing conditions, which can improve the sensor performance with binary oxides heterojunctions, noble metal modification, and specific elements doping. To date, a variety of specific fiber structures originating from the complex self-assembly processes have been fabricated by carefully controlling the preparing conditions, such as ribbon nanofiber, helix nanofiber, porous nanofiber, necklace-like nanofiber, core-shell nanofiber, and hollow nanofiber. The following section will highlight the key microstructure categories that control gas sensing properties.

2.2.2 Inorganic Nanofibers

A great variety of inorganic nanofibers such as ZnO, SnO_2, TiO_2, In_2O_3, and WO_3 have proven to be excellent gas sensing materials for the detection of both reducing and oxidizing gases, including H_2, ethanol, NO_2, acetone, and toluene. Different compositions and morphologies states of inorganic nanofibers have been studied to be used in nanofiber gas sensors. It is reported that polycrystalline self-supporting ZnO nanofibers with an average diameter of 50 nm exhibit excellent sensing properties towards NO_2 at an operating temperature of 100 °C, including a lower detection limit (about 25 ppb), and high sensitivity (Aziz et al. 2018). Facile synthesis of bimodal pores-loaded tubular SnO_2 nanostructure functionalized by protein-templated Pt catalysts via advanced electrospinning. Plenty of mesopores and macropores are formed on the thin-wall of SnO_2 nanotubes as shown in Fig. 4a, which can facilitate gas diffusion into sensing layers. The combined synergistic effect of porous nanotubular morphology and uniform distribution of Pt nanoparticles on SnO_2 nanotubes enabled high sensitivity, superior selectivity against other interfering gases, and very low limit of detection (10 ppb) to simulated diabetic acetone molecules (Jang et al. 2016). By the coaxial electrospinning method, the TiO_2-SnO_2 core-shell nanofibers were successfully synthesized. Compared to pure TiO_2 and SnO_2, the TiO_2-SnO_2 core-shell nanofibers illustrated that the excellent response to acetone, selectivity to acetone and response/recovery processes, which can be attributed to hierarchical heterostructure provide more adsorbed oxygen species and electron depletion layer on the surface of the core-shell nanofibers (Li et al. 2017). Besides binary composites for TiO_2, ternary composites of Pt@ZnO-TiO_2 nanotubes can be designed as a high sensitivity toluene gas sensor. Figure 4b presents the small-sized and well-dispersed Pt NPs that by assembling into the ZnO nanoparticles fixed on the surface of TiO_2 nanotubes. The result exhibited that the ternary composites exhibit higher response, lower operating temperature and faster response/recover speed during the dynamic measurement towards toluene gas (Liu et al. Liu et al. 2018a, b, c, d). For In_2O_3 nanofibers, the studies show that the doping of noble metal and rare earth can change the morphology structure and effectively increase the specific surface area. As shown in Fig. 4c, the porous and ultra-length Pt-doped In_2O_3 nanofibers were obtained through the electrospinning and soaking method and the process of annealing, which can provide more gas diffusion paths and active sites for the gas sensor. It demonstrated superior performance in selectivity, reversibility, time stability with a period longer than 50 days and anti-humidity properties towards acetone gas (Liu et al. 2019a, b, c, d). Xu et al. prepared highly porous In_2O_3 nanotubes with doping rare earth, which is all with a controllable diameter of 80 nm and a wall thickness of 15 nm. The rare-earth doping has great influences on the structure and room-temperature gas-sensing properties of In_2O_3 nanotube sensors toward H_2S. It is important to observe that the lattice constants of the cubic In_2O_3 nanotube gradually decreased with the varied rare-earth dopant, which exhibited the excellent room-temperature response sensitivity to 20 ppm H_2S and response value reached as high as 1241 (Xu et al. 2010).

Despite many efforts for the improvement of the gas sensing property, the pure inorganic nanofibers still exist many limitations, such as the excellent selectivity. In order to overcome the drawbacks of inorganic nanofiber-based sensors, the functional group modified nanofiber-based sensing devices have been developed to an excellent candidate towards the selectivity sensing performance. Scattered researches have investigated the performances of sensors based on functional group modified nanofibers. As shown in Fig. 4d, the report presented a sub-ppm level NO_2 gas sensor

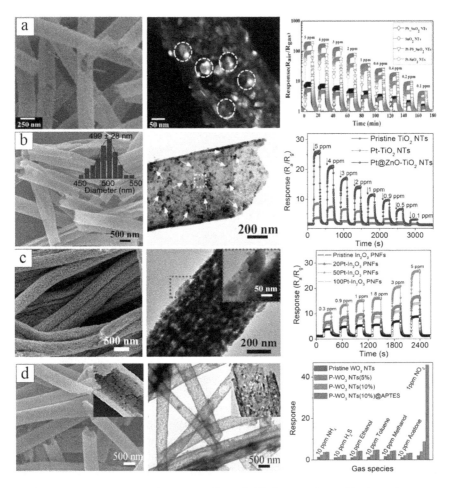

Fig. 4 **a** SEM and TEM images of the porous thin-wall of SnO₂ nanotubes and the dynamic response curve towards acetone gas. Reproduced from ref. (Jang et al. 2016). Copyright 2016, Wiley-VCH. **b** SEM and TEM images of Pt@ZnO-TiO₂ nanotubes and the dynamic response curve towards toluene gas. Reproduced from ref. (Liu et al. 2018a, b, c, d). Copyright 2018, Elsevier. **c** SEM and TEM images of Pt functionalized porous In₂O₃ nanofibers and the dynamic response curve towards acetone gas. Reproduced from ref. (Liu et al. 2019a, b, c, d). Copyright 2019, Elsevier. **d** SEM and TEM images of the APTES functionalized porous WO₃ nanotubes and the dynamic response curve towards NO₂ gas sensor Reproduced from ref. (Liu et al. 2018a, b, c, d). Copyright 2018, Royal Society of Chemistry

based on porous WO₃ nanotubes, which were chemically surface-modified with 3-aminopropyltriethoxysilane (APTES). The concept is to incorporate two substantial requirements of practical gas sensors, sensitivity, and selectivity, the rational design porous thin-walled WO₃ nanotubes as a scaffold providing larger surface area, more gas diffusion channel towards the improvement of sensitivity, and the surface-functionalized APTES with amino groups that selectively react with NO₂.

The amino groups modified WO_3 nanofibers sensor demonstrated a low detection limit of 10 ppb, quick response and recovery times of 11 s and 12 s with a response value of 184 (10 ppm NO_2), and long-term stability even at high humidity. This structure and method can be extended to design other types of highly selectivity gas sensors and will also provide an economical yet powerful tool for developing portable and flexible sensors, especially for environmental air quality monitoring (Liu et al. Liu et al. 2018a, b, c, d).

2.2.3 Organic Nanofibers

During the last decade, organic nanofibers have also been investigated for the detection of the gas sensor as alternatives to inorganic nanofibers, that is, conducting polymer nanofibers. The most commonly conducting polymers applied to the gas sensor has been reported for polyaniline (PANI), polypirrole (PPy), and polythiophene (PTh) and their derivatives. The advantages of using conductive polymers in gas sensors regard the high sensitivity and the possibility to perform detection at room temperature. The principle of the sensitivity of conducting polymer-based sensors is based on the chemical reactions occurring between the conducting polymer and the analyte, or by changes in the physical properties of the system. The conducting polymer sensing layer can interact chemically or physically with the analyte of interest and then generate a signal which may be qualitative or quantitative.

As shown in Fig. 5a, the study successfully fabricated PANI/polyacrylonitrile uniaxially aligned coaxial nanofiber yarns with a novel electrospinning method and followed by in situ solution polymerization. The core-shell nanofiber-constructed single yarn combined high mechanical performances of polyacrylonitrile with high conductivity of PANI, and it was long and robust enough to handle, which are capable of being manufactured into different textile architectures. The yarn sensor exhibited a high and fast sensing response towards NH_3 in the range of 10–2000 ppm at room temperature. These results demonstrate that the yarn sensor consisting of coaxial PANI/polyacrylonitrile nanofibers can be used as an alternative sensing element for wearable gas sensors (Wu et al. 2017). For PTh nanofibers, Fig. 5b showed the PTh sensing thin films with an interconnected nanofiber network and exhibited different morphology and by chemical bath deposition method on a glass substrate. This study highlighted the crucial role of monomer concentration to optimize highly porous interconnected PTh nanofibers network with precise control of pore size. Moreover, the present PTh sensing thin films successfully demonstrated for the gas sensing properties, which can display a high response and selectivity, towards NO_2 gas at room temperature (Kamble et al. 2017). Zhang et al. reported that a novel sensor based on the PPy nanofiber arrays by electropolymerization in the anodic aluminum oxide as shown in Fig. 5c, which presented a significantly high surface-to-volume ratio, high density, and small diameter (about 50 nm). The PPy nanofiber sensor was sensitive to ammonia at room temperature and showed a relatively high response in low concentration and comparatively short response and recovery time (Zhang et al. 2009).

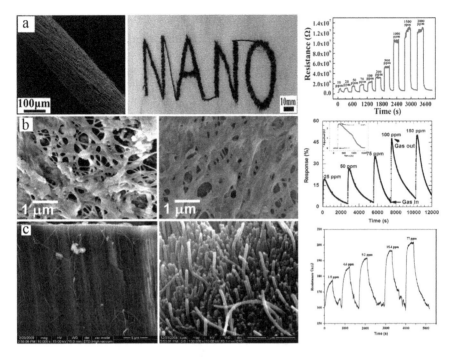

Fig. 5 **a** SEM images of PANI/polyacrylonitrile uniaxially aligned coaxial nanofiber yarns and the resistance transients towards ammonia at room temperature. Reproduced from ref. (Wu et al. 2017). Copyright 2017, Elsevier. **b** SEM images of PTh nanofiber and the dynamic response of NO_2 gas sensor at room temperature. Reproduced from ref. (Kamble et al. 2017). Copyright 2017, Elsevier. **c** SEM images of PPy nanowires and the time-dependent change of the resistance towards ammonia at room temperature. Reproduced from ref. (Zhang et al. 2009). Copyright 2009, Elsevier

2.2.4 Carbon Nanofibers

Carbon nanofibers are one of the advanced functional materials and possess many desirable mechanical and chemical properties towards gas detection, which super-sede many of the advanced materials of today. Generally, the sensing function of the carbon nanofiber composites is realized by testing the variation of electrical properties that have resulted from the change of the external conditions. The elec-trical conductivity of the carbon nanofiber composites can be reversibly changed by several orders of magnitude with the reversible change of the external conditions. However, commercialization of these potential applications remains elusive mainly due to the lack of control in the synthesis of specific morphology, diameter, and length of carbon nanofiber, which influences the device performance. The hybrid nanofiber-based sensing devices have been developed to combine the advantages of metal oxides nanoparticle and noble metal modified carbon nanofiber-based sensors.

For the metal oxides nanoparticle modified carbon nanofiber-based sensors, Seekaew et al. presents a highly sensitive room-temperature gas sensor based on titanium dioxide/graphene-carbon nanotube fabricated by chemical vapor deposition and sparking methods as shown in Fig. 6a. The results indicated that the optimal sparking time of 60 s led to an optimal sensor response of 42%–500 ppm at room temperature and exhibited substantially higher toluene response, sensitivity, and selectivity towards toluene gas (Seekaew et al. 2019). Lee et al. fabricated ultrafine ZnO and SnO_2-decorated hybrid carbon nanofibers by a single-nozzle co-electrospinning process using a phase-separated mixed polymer composite solution and heat treatment. The decoration morphology of the metal oxide nanoparticles could be well-controlled and these ultrafine hybrid carbon nanofibers were applied to a dimethyl methyl phosphonate gas sensor at room temperature with excellent sensitivity and minimum detectable level (0·1 ppb) (Lee et al. 2011). For the noble

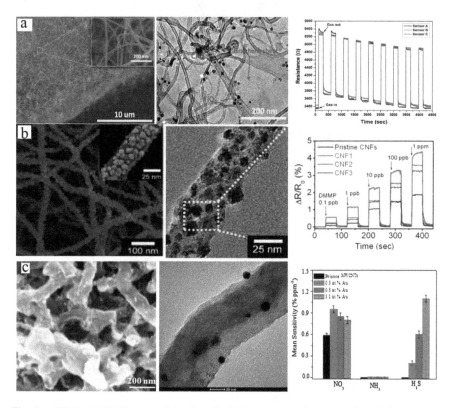

Fig. 6 **a** SEM and TEM images of titanium dioxide/graphene-carbon nanotube and the response curve towards toluene at room temperature. Reproduced from ref. (Seekaew et al. 2019). Copyright 2019, Elsevier. **b** SEM and TEM images of ultrafine ZnO and SnO_2-decorated hybrid carbon nanofibers and the resistance transients towards dimethyl methyl phosphonate gas at room temperature. Reproduced from ref. (Lee et al. 2011). Copyright 2011, American Chemical Society. **c** SEM and TEM images of multiwalled carbon nanofibers modified with Au nanoparticles and the mean sensitivity towards NO_2, NH_3, and H_2S at a sensor temperature of 150 °C. Reproduced from ref. (Dilonardo et al. 2016). Copyright 2016, Elsevier

metal modified carbon nanofiber-based sensors, Dilonardo et al. presented multi-walled carbon nanotube-based gas sensors that were decorated by electrophoretic deposition of electrochemically preformed Au nanoparticles with controlled size and loading. The measurement result revealed that the highest NO_2 response up to the sub-ppm level by using multiwalled carbon nanotubes functionalized by the lowest Au content instead, and the highest Au loading exhibited the worse NO_2 response. Thus, fine- tuning of the surface concentration of deposited Au nanoparticles permits to control the gas sensor properties.

2.2.5   Sensing Performance

Up to now, many efforts have been made to improve the sensor performance of nanofiber-based gas sensors including the enhancement of sensor sensitivity, low detection limit, sensor life, response–recovery time, selectivity as well as lowering the working temperature (Zhang et al. 2016). A typical method used for these purposes is the surface modification of the semiconductor metal oxide with various dopants, mostly noble metals such as Au, Ag, Pt, and Pd or different metal oxides. Despite the synthesis of noble metal sensitized nanofiber gas sensors as well as their advanced applications being widely reported, the sensing mechanism is always ambiguous. In order to understanding the effect of noble metals more systematically, Liu et al. reported that a sensing array composed of different noble metal nanoparticles (Au, Ag, and Pt) modified In_2O_3 nanofiber-based gas sensor sensitive layers, which can simultaneous detection of multiple exhaled breath biomarkers. It is exciting to observe that the introduction of noble metal nanoparticles can not only enhance the sensing performance, but also effectively adjust the selectivity. The Au−, Ag− and Pt-modified In_2O_3 nanofiber sensors exhibit excellent selectivity to hydrogen sulfide, formaldehyde, and acetone biomarkers, respectively. In addition, the density functional theory (DFT) results indicated that the enhancement of sensitivity and selectivity can be attributed to the strong adsorption energies between the noble metal nanoparticles and the stable molecular configuration towards certain target gases at the operating temperature of 300 °C. Furthermore, the sensor array was further used to the real-time detection of disease biomarkers in simulated exhaled breath. This work provides give a deeper understanding of the noble metal modifying effect and a promising technique for noninvasive diagnosis of different diseases (Liu et al. 2019a, b, c, d) (Fig. 7).

On the other hand, environmental humidity affects the sensing performance of nanofiber-based gas sensors. Physisorbed and chemisorbed water molecules product more OH^- ions at high temperature, which may react with surface oxygen vacancy thereby decreasing the baseline resistance and hence gas response. Adsorption of water molecules also prevents the chemisorption of oxygen on the surface of the sensor, which is the prime requirement for the sensor response. Thus, the maximized improvement of sensing performance need solves the effect of environmental humidity. Many attempts at eliminating the humidity effect have been made in some

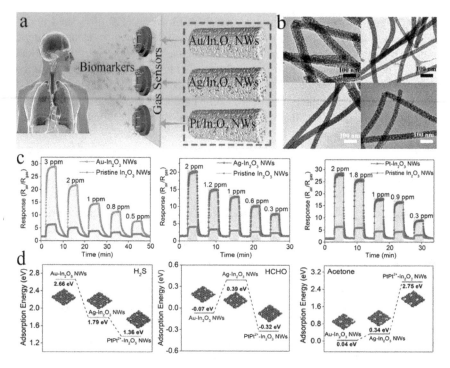

Fig. 7 **a** Schematic diagram of the sensor array. **b** TEM images of Au-, Ag- and Pt-modified In₂O₃ nanofiber. **c** Dynamic response curves, and **d** Adsorption energy towards hydrogen sulfide, formaldehyde, and acetone of Au−, Ag−and Pt-modified In₂O₃ nanofiber sensors. Reproduced from ref. (Liu et al. 2019a, b, c, d). Copyright 2019, Royal Society of Chemistry

works. For example, Konvalina et al. introduced humidity compensation and cross-reactive array methods, which can yield more accurate target gas values at various humidity levels (up to approximately 80% relative humidity) (Konvalina and Haick 2012). Both Nishibori and Mondal et al. added some extra dehydrating components to decrease the influence of the humidity. However, these methods also have many side-effects, resulting in sensing devices that are much larger and more complicated (Nishibori et al. 2009; Mondal et al. 2011). Liu et al. proposed a type of acetone sensor with ultra-high sensitivity with Pt@In₂O₃ core-shell nanofibers as a sensing layer and moisture resistance with SBA-15 molecular sieve as a dehumidifying layer. The In₂O₃ nanowires with a controllable Pt were designed and prepared by the electrospinning method. The sensor exhibits a high sensitivity, fast dynamic process, selectivity, and long-term stability towards acetone gas. The detection limit can be as low as 10 ppb, which is much lower than the concentration level of 1·8 ppm in the exhaled breath of diabetic patients. The influence of the humidity is greatly weakened by using the SBA-15 molecular sieve as a moisture filter layer, and improved response difference between breath samples obtained from healthy people and people with diabetes. The Pt@In₂O₃ core-shell nanofibers acetone sensor with the moisture filter

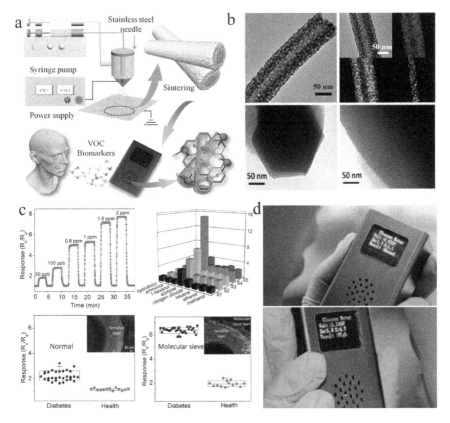

Fig. 8 **a** Schematic diagram of Pt@In$_2$O$_3$ core-shell nanofibers gas sensor device. **b** TEM images of Pt@In$_2$O$_3$ core-shell nanofibers and SBA-15 molecular sieve. **c** The acetone sensing performance of Pt@In$_2$O$_3$ core-shell nanofibers and the dehumidifying effect of SBA-15 molecular sieve. **d** The measurement of the handheld device towards the diabetic patient and healthy person. Reproduced from ref. (Liu et al. 2018a, b, c, d). Copyright 2018, Springer Nature Publishing

layer has been applied into a simple portable sensing device, which can distinguish healthy people and diabetic patient and provide a real-time measurement perform (Liu et al. 2018a, b, c, d) (Fig. 8).

3   Conclusion

In summary, nanofiber-based gas sensors are becoming more and more important in the future, which can apply in numerous modern industrial and medical applications, including early diagnosis and disease, public safety and environment monitoring. The continuous advances in nanoscience and nanotechnology, of course, also will boost the development of the nanofiber-based gas sensor. Nanofiber with well-defined material composition and morphology provides a better chance for us to construct stable and reproducible gas sensors. Various gas sensors comprising nanofibers of inorganic metal oxide, conducting polymer, and carbon composites were successfully fabricated with ultrahigh sensitivity, selectivity, very short response and recovery time, and good reversibility based on various sensing techniques and principles. In addition, gas sensing mechanisms have been reviewed for better understanding of their working principles. Then, the influence factors have been described in detail, such as composites, porous nanostructure, and doping on the nanoscale levels. By considering those influencing factors on nanoscale, novel nanofibers will be developed and then nanofiber-based gas sensing properties will be further improved. Thus, there are still many subjects of research that are worth challenging in order to progress the innovation of nanofiber-based gas sensors.

References

A. Aziz, N. Tiwale, S.A. Hodge, S.J. Attwood, G. Divitini et al., ACS Appl. Mater. Interfaces. **10**, 43817 (2018)

M.E. Franke, T.J. Koplin, U. Simon, Small **2**, 36 (2006)

H. Han, S. Baik, B. Xu, J. Seo, S. Lee et al., Adv. Funct. Mater. **27**, 1701618 (2017)

A. Gurlo, R. Riedel, Angew. Chem., Int. Ed. **46**, 3826 (2007)

A. Tricoli, M. Righettoni, A. Teleki, Angew. Chem., Int. Ed. **49**, 7632 (2010)

J.-S. Jang, S.-J. Choi, S.-J. Kim, M. Hakim, I.-D. Kim, Adv. Funct. Mater. **26**, 4740 (2016)

Y.H. Kim, S.J. Kim, Y.-J. Kim, Y.-S. Shim, S.Y. Kim et al., ACS Nano **9**, 10453 (2015)

S.-J. Kim, S.-J. Choi, J.-S. Jang, H.-J. Cho, I.-D. Kim, Acc. Chem. Res. **50**, 1587 (2017a)

S.-J. Kim, S.-J. Choi, J.-S. Jang, H.-J. Cho, W.-T. Koo et al., Adv. Mater. **29**, 1700737 (2017b)

G. Konvalina, H. Haick, A.C.S. Appl, Mater. Interfaces **4**, 317 (2012)

G. Korotcenkov, V. Brinzari, M. Ivanov, A. Cerneavschi, J. Rodriguez et al., Thin. Solid Films **479**, 38 (2005)

J.S. Lee, O.S. Kwon, S.J. Park, E.Y. Park, S.A. You et al., ACS Nano **5**, 7992 (2011)

A. Lichtenstein, E. Havivi, R. Shacham, E. Hahamy, R. Leibovich et al., Nat. Commun. **5**, 4195 (2014)

W. Liu, Y. Xie, T. Chen, Q. Lu, S. Ur Rehman et al., Sens. Actuators B **298**, 126871 (2019)

W. Liu, L. Xu, K. Sheng, X. Zhou, B. Dong et al., NPG Asia Mater. **10**, 293 (2018a)

W. Liu, L. Xu, K. Sheng, C. Chen, X. Zhou et al., J. Mater. Chem. A **6**, 10976 (2018b)

Y. Liu, R. Wang, T. Zhang, S. Liu, T. Fei, J. Colloid Interface Sci. **541**, 249 (2019a)

W. Liu, X. Zhou, L. Xu, S. Zhu, S. Yang et al., Nanoscale **11**, 11496 (2019b)

W. Liu, J. Sun, L. Xu, S. Zhu, X. Zhou et al., Nanoscale Horiz. **4**, 1361 (2019c)

H.-L. Lu, C.-J. Lu, W.-C. Tian, H.-J. Sheen, Talanta **131**, 467 (2015)

R. Lu, W.-W. Li, B. Mizaikoff, A. Katzir, Y. Raichlin et al., Nat. Protoc. **11**, 377 (2016)

M. J. Madou, S. R. Morrison (ed.), Chemical sensing with solid state devices (Elsevier, 2012)

H.G. Moon, Y. Jung, S.D. Han, Y.-S. Shim, B. Shin et al., ACS Appl. Mater. Interfaces. **8**, 20969 (2016)

S.R. Morrison, Sens. Actuators **12**, 425 (1987)

A. Nikfarjam, S. Hosseini, N. Salehifar, A.C.S. Appl, Mater. Interfaces **9**, 15662 (2017)

G. Pacchioni, Chem. Phys. Chem. **4**, 1041 (2003)

S.J. Patil, A.V. Patil, C.G. Dighavkar, K.S. Thakare, R.Y. Borase et al., Front. Mater. Sci. **9**, 14 (2015)

N.S. Ramgir, Y. Yang, M. Zacharias, Small **6**, 1705 (2010)

W. Liu, L. Xu, K. Sheng, X. Zhou, X. Zhang et al., Sens. Actuators, B **273**, 1676 (2018)

J. Wu, S. Feng, Z. Li, K. Tao, J. Chu et al., Sens. Actuators, B **255**, 1805 (2018)

G. Neri, A. Bonavita, G. Micali, G. Rizzo, N. Pinna et al., Sens. Actuators, B **127**, 455 (2007)

L. Zhang, F. Meng, Y. Chen, J. Liu, Y. Sun et al., Sens. Actuators, B **142**, 204 (2009)

M. Nishibori, W. Shin, N. Izu, T. Itoh, I. Matsubara, Sens. Actuators, B **137**, 524 (2009)

K. Wetchakun, T. Samerjai, N. Tamaekong, C. Liewhiran, C. Siriwong et al., Sens. Actuators, B **160**, 580 (2011)

S. P. Mondal, P. K. Dutta, G. W. Hunter, B. J. Ward, D. Laskowski et al., Sens. Actuators, B **158**, 292 (2011)

G. Korotcenkov, B. K. Cho, Sens. Actuators, B **188**, 709 (2013)

E. Dilonardo, M. Penza, M. Alvisi, C. Di Franco, R. Rossi et al., Sens. Actuators, B **223**, 417 (2016)

G. Korotcenkov, B. K. Cho, Sens. Actuators, B **244**, 182 (2017)

F. Li, X. Gao, R. Wang, T. Zhang, G. Lu, Sens. Actuators, B **248**, 812 (2017)

S. Wu, P. Liu, Y. Zhang, H. Zhang, X. Qin, Sens. Actuators, B **252**, 697 (2017)

D. B. Kamble, A. K. Sharma, J. B. Yadav, V. B. Patil, R. S. Devan et al., Sens. Actuators, B **244**, 522 (2017)

Y. Liu, X. Gao, F. Li, G. Lu, T. Zhang et al., Sens. Actuators, B **260**, 927 (2018)

T. Zhou, X. Liu, R. Zhang, Y. Wang, T. Zhang, Sens. Actuators, B **290**, 210 (2019)

Y. Seekaew, A. Wisitsoraat, D. Phokharatkul, C. Wongchoosuk, Sens. Actuators, B **279**, 69 (2019)

P. Wan, X. Wen, C. Sun, B.K. Chandran, H. Zhang et al., Small **11**, 5409 (2015)

C. Wang, L. Yin, L. Zhang, D. Xiang, R. Gao, Sensors **10**, 2088 (2010)

L. Xu, B. Dong, Y. Wang, X. Bai, J. Chen et al., J. Phys. Chem. C **114**, 9089 (2010)

N. Yamazoe, G. Sakai, K. Shimanoe, Catal. Surv. Asia **7**, 63 (2003)

J. Zhang, X. Liu, G. Neri, N. Pinna, Adv. Mater. **28**, 795 (2016)

T. Zhou, S. Cao, R. Zhang, J. Tu, T. Fei et al., ACS Appl. Mater. Interfaces. **11**, 28023 (2019)

Sapphire-Derived Fibers and Optical Fiber Sensing

Fufei Pang, Zhifeng Wang, Huanhuan Liu, Sujuan Huang, and Tingyun Wang

Abstract The optical fiber being optical sensors exhibit several advantages including the immunity to the electromagnetic interference, simple structure, easy to carry, and large dynamic range. Especially, the possibility of working under harsh environment marks the fiber sensors that are superior to the electrical sensor counterparts. To achieve new breakthroughs in high-temperature sensing technology, exploring high-temperature-resistant optical fibers with good mechanical properties is necessary. The sapphire-derived fiber (SDF), a high-concentration alumina-doped silica fiber based on a single-crystal sapphire rod, has good mechanical strength, high-temperature resistance, etc. Such SDF shows great potential in high-temperature sensing and distributed strain sensing. This chapter mainly introduces the SDF for high-temperature sensing application. Firstly, the fabrication and the characterization of SDF have been introduced. Secondly, the SDF-based optical devices have been discussed including Fabry–Perot interferometer, Bragg grating, Mach–Zehnder interferometer, long-period fiber grating. Then, the advantages of SDF fiber in the field of high temperature are proved by introducing the performance of various sensors based on SDF.

F. Pang (✉) · Z. Wang · H. Liu · S. Huang · T. Wang
Key Laboratory of Specialty Fiber Optics and Optical Access Networks, Joint International Research Laboratory of Specialty Fiber Optics and Advanced Communication, Shanghai Institute for Advanced Communication and Data Science, Shanghai University, Shanghai 200444, China
e-mail: ffpang@shu.edu.cn

Z. Wang
e-mail: Clairewangzf@shu.edu.cn

H. Liu
e-mail: hhliu@shu.edu.cn

S. Huang
e-mail: sjhuang@shu.edu.cn

T. Wang
e-mail: tywang@shu.edu.cn

© Springer Nature Singapore Pte Ltd. 2020
L. Wei (ed.), *Advanced Fiber Sensing Technologies*,
Progress in Optical Science and Photonics 9,
https://doi.org/10.1007/978-981-15-5507-7_9

155

Keywords Sapphire-derived fiber · Refractive index modulation ·
High-temperature sensor · Bragg grating · Long-period grating · Fabry–perot
cavity · Mach–zehnder interferometer

1   Introduction

Optical fiber sensors have superiorities of simple structure, easy to carry, large
dynamic range, stable operation in harsh environment, etc. (McCague et al. 2014;
Leung et al. 2015; Lee 2003). It has broad application prospects in aerospace,
biochemistry, etc., and becomes the current research hotspots in the sensor field
(Li et al. 2014; Zheng et al. 2013; Liao et al. 2013). Traditional fiber-optic sensors
for monitoring high-temperature and high-pressure environments are made of standard silica fibers (Tu et al. 2017; Zhang et al. 2018; Huang et al. 2010). Nevertheless,
limited by the low-concentration doped core, the stability of the sensor device and
the mechanical strength of the sensing fiber at high temperatures will deteriorate. As
a result, it is difficult to apply the silica fiber sensor at higher temperatures. Therefore, exploring a high-temperature-resistant fiber with good mechanical properties is
of great significance to break through the traditional high-temperature fiber sensing
technology.

The sapphire-derived fiber (SDF) is a high-concentration alumina-doped specialty
fiber, which can survive at temperatures above 1000°C (Liu et al. 2019; Wang et al.
2019; Zhang et al. 2018, 2019). The SDF was drawn successfully for the first time by
the rod-in-tube method in 2012 (Dragic et al. 2012). It is a special optical fiber made
of single-crystal sapphire (SCS) as core rod and a pure silica tube as sleeve under high
temperature, and the melting point is about 2045°C (Merberg and Harrington 1993).
Once the drawing temperature is above the melting point of the SCS, the silica tube
is in a molten state and SCS melts. Therefore, two elements of alumina and silica
undergo thermal diffusion (Ballato et al. 2009), thereby forming a high-concentration
alumina–silica co-doped glass core. SDF has a very low Brillouin gain coefficient
and becomes smaller as the alumina content in the fiber increases (Dragic et al.
2012). This characteristic provides a strong support for the realization of a single
parameter Brillouin strain distribution measurement sensor system (Ballato et al.
2009; Dragic et al. 2013). In 2014, a Bragg grating capable of withstanding 950°C
was written on the SDF using femtosecond (fs) laser technology, which proved the
high-temperature resistance of SDFs (Elsmann et al. 2014). In addition, due to high-
concentration alumina-doped core, mullite crystals with high refractive index (RI)
will be precipitated in the core of SDFs during re-heating and cooling treatment.
Based on crystallization effect of SDF, a Fabry–Perot interferometer (FPI), long-
period grating (LPG), and Mach–Zehnder interferometer (MZI) can be fabricated for
high-temperature sensing (Liu et al. 2019; Hong et al. 2017; Xu et al. 2017; Zhang
et al. 2018). Previous studies have shown that SDF optical devices can withstand up
to 1600°C (Liu et al. 2019).

This chapter mainly introduces the basic characteristics of the SDF, and various optical devices based on the SDF, such as Fabry–Perot interferometer (FPI), fiber Bragg grating (FBG), Mach–Zehnder interferometer (MZI), and long-period grating (LPG). Moreover, the SDF-based optical devices for temperature sensing have been demonstrated.

2 Fabrication of SDF

The high-concentration alumina-doped SDF is drawn by using the rod-in-tube technique (Dragic et al. 2012). A single-crystal sapphire (SCS) rod is sheathed into a pure silica tube and sent together into a drawing tower as shown in Fig. 1. In view of the melting point of SCS (2045°C) that is higher than the melting state of silica tube (1750°C), the temperature of drawing tower should be higher than 2050°C to make them fully heated. The drawing of alumina-doped fiber is an unsteady-state diffusion process, and the concentration of alumina in the preform will change with time at a certain temperature (Kobelke et al. 2017). Due to the thermal diffusion effect, alumina in the optical fiber will change the core composition through diffusion, thereby affecting the refractive index (RI) distribution of the fiber. The RI distribution has a direct impact on the transmission performance of the fiber the transmission performance of optical fiber. Therefore, it is necessary to control the diffusion of alumina in the fiber. During the drawing process of SDF, the diffusion of alumina is closely related to concentration and viscosity of alumina, temperature of drawing tower, drawing speed, and feed rate (Kobelke et al. 2017; Ma et al. 2019). Table 1 shows the drawing parameters of the SDF with a cladding diameter of 125–170 μm. Taking into account the drawing effect of the cladding and the core, the set drawing temperature is much higher than the melting temperature of the silica tube, resulting in the decrease of drawing tension of cladding.

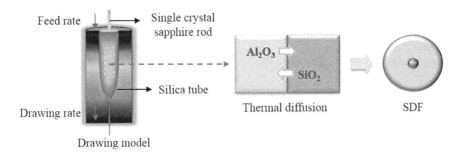

Fig. 1 Schematic drawing of a SDF by rod-in-tube method

Table 1 Drawing parameters of a SDF (Dragic et al. 2012)

Parameters	Values
Diameter of single-crystal sapphire rod	2.8 mm
Dimensions of silica tube (inner/outer)	3 mm/30 mm
Drawing temperature	2100°C
Drawing rate	20 m/min

3 Performance Characterization of SDF

3.1 Optical Performance

The RI profile, geometrical dimensions, and numerical aperture of optical fibers are the main specifications that affect the sensing characteristics of optical fibers. Therefore, the accurate measurements of RI and geometrical dimensions are of great significance for optical fiber manufacturing and application. The geometrical dimensions and core composition of SDF can be measured by an OXFORD scanning electron microscope (SEM). As an example, Fig. 2a shows the SEM results of SDF with core and cladding diameters of 16 μm and 125 μm, respectively. The aluminum element is only concentrated in the core, while the silicon element is mainly distributed in the cladding and a small amount is distributed in the core.

The RI profile of an alumina-doped 35 mol% SDF tested with a commercial SHR-1602 RI profiler is shown in Fig. 2b. The test result proves that the high-concentration alumina-doped core can introduce a high refractive index difference (RID) between the core and the cladding. Figure 2c is a simulation result of the effective RI obtained by using the finite element method analysis software COMSOL based on the parameters in Table 2. Due to the high RID and large core diameter, SDF supports many modes.

The optical fiber transmission loss is also an important parameter reflecting the performance of optical fiber, which is generally obtained by cut-back method. Figure 2d shows the transmission loss of the SDF with the length of 1.4 m at room temperature, which is about 2.06 dB/m. At present, the lowest loss of SDF at 1550 nm is about 0.2 dB/m (Dragic et al. 2012). The transmission loss of SDF could be further improved by the drawing process, the purity of the sapphire rod and preform, etc.

3.2 Crystallization Performance

The core of a SDF is a highly doped alumina glass material. According to the liquid-phase immiscibility in the binary system, SDFs undergo phase separation during re-heating and cooling treatment, thereby precipitating crystals in the core of SDFs. When a SDF is spliced with single-mode fiber (SMF) by arc discharge as shown in Fig. 3a, the current density distribution of the arc discharge is Gaussian along the

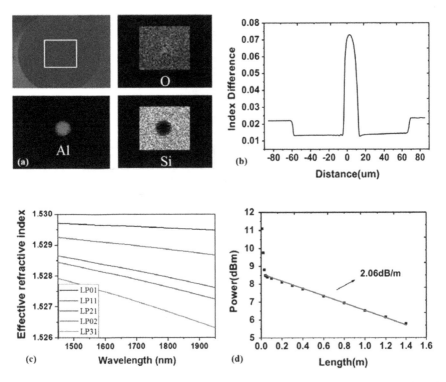

Fig. 2 Composition and optical properties of SDF. **a** Element test results and **b** the RI profile of SDF. **c** The simulation of effective RI of various modes in SDF. **d** Transmission power variation of SDF with the different length at room temperature

Table 2 Simulation parameters of COMSOL

Parameters	Values
Diameter of SDFs core	16 μm
Diameters of SDFs cladding	125 μm
RI of SDFs core	1.53
RI of SDFs cladding	1.46

fiber axis. Since the current density distribution is proportional to the temperature distribution in the arc area, the temperature of the heat treatment can be controlled by adjusting the discharge parameters (Llewellyn-Jones and Hinrichs 1967). As shown in Fig. 3b, because the temperature in the central area of the arc is higher than the upper temperature limit of the metastable phase, no crystallization will occur in the core area near the fusion point. Meanwhile, due to the high temperature, the aluminum element will diffuse into the cladding, which will expand the core diameter. As the distance from the fusion point increases, the temperature on the optical fiber gradually decreases until it enters the metastable phase separation, and the crystallization effect

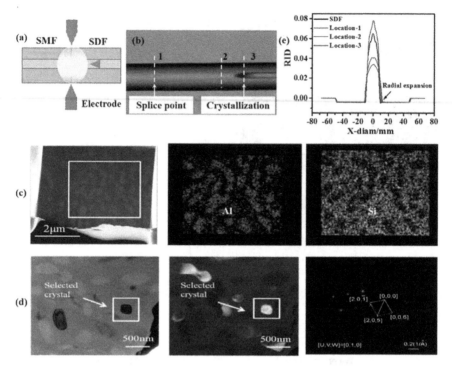

Fig. 3 Crystallization effect introduced by arc discharge on SDF. **a** Schematic diagram and **b** microscopy image of the crystallized SDF by using arc discharge. **c** Composition analysis results and **d** electron diffraction analysis results of crystallized SDF at Location 3. **e** RI profile of the crystallized SDF at typical positions with Location 1 (purple), 2 (blue), and 3 (red)

occurs in the core area. As the temperature further decreases, the temperature of the metastable phase separation is not reached. Since the SiO_2 and Al_2O_3 in the core have relatively high viscosities, phase separation cannot occur, that is, no crystallization effect is generated.

Figure 3c shows the composition analysis results of the crystallization area (Location 3) of the SDF by a transmission electron microscope (TEM). In the test area, the pixels represent the content of the element. The more pixels distributed, the higher the element content. Al element is mainly concentrated in the microcrystalline particle position, which is obviously higher than that in the substrate. The content of Si element in the substrate is higher than that in the microcrystalline particle. According to the diffraction pattern of the microcrystalline particles in Fig. 3d, it can be determined that the microcrystalline particles are mullite crystal. Since the mullite crystal has high-temperature resistance, high-temperature fiber sensors can be developed by using the crystallization effect of SDFs.

The core of the SDF is in a glassy state, and its molecular arrangement is disordered. During the crystallization process, the molecules are rearranged and transformed into an ordered structure. Therefore, the RI of the crystallization area of the

SDF also changes. Figure 3e shows the RI profile of the crystalline area of the SDF. The crystallization area can be formed by a FSM-80S Fujikura fusion splicer, and the "MM-MM" mode is selected as the discharge parameter. Three typical positions in Fig. 3b are selected to manifest the effect of crystallization on the RI, marked as Location 1, Location 2, and Location 3. Attributed to the diffusion effect, the higher the temperature near the fusion point, the faster the diffusion rate (Unger et al. 2011). Since the aluminum element near the fusion point (Location 1) diffuses to the cladding, the diameter of core expands. Therefore, the RID between expanded core and the cladding is lower than that of SDF without crystallization. As the distance from the fusion point deviates significantly, the effect of the thermal effect becomes weak, and the diffusion rate of alumina is reduced. Thus, the RI at Location 2 is higher than that at Location 1. Location 3 is the crystallized area, which has a higher RID between crystallized core and the cladding than that of SDF without crystallization.

At present, the reported the RID between the crystallized core and core without crystallization is about 0.03. Thus, crystallization-induced RI modulation can reach about 0.015. The crystallization of a SDF by arc discharge is a new method of RI modulation. The use of crystallization-induced optical fiber RI modulation not only has a simple processing method, but also has high RI modulation. High RI modulation can be used to make all-fiber devices more efficiently. Meanwhile, such a device has more stable performance under high temperature and high RI environments. Therefore, this technology will provide important support for the development of high-temperature optic devices.

4 High-Temperature Transmission Performance

The SDF has good light transmission from 600 to 2400 nm as shown in Fig. 4a. Because SDF supports many modes, the transmission spectrum of SDF has inter-mode interference. High temperature will not only change the viscosity and RI of fiber materials, but also release the fiber stress. It can be seen from Fig. 4b that SDF' transmission loss increases with the increase of temperature. At higher temperature, the light transmission performance of the fiber is affected by Rayleigh scattering. When the temperature reaches 1000 °C, the loss increases up to 10 dB. In the cooling process, the viscosity and RI of SDF will return to the original value with the decrease of temperature. Therefore, the light loss will be reduced as shown in Fig. 4c. Compared with the spectra of heating to 400 °C and cooling to 400 °C, as shown in Fig. 4d, the two spectra are only slightly different, which may be caused by thermal stress release.

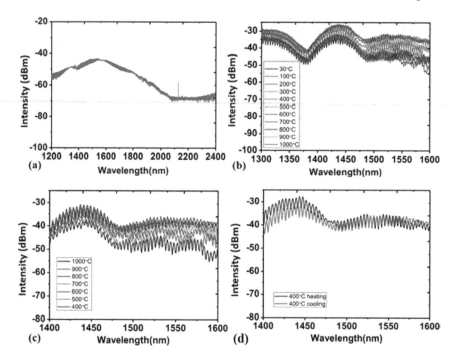

Fig. 4 Transmission spectra of SDF at **a** room temperature, **b** the heating, and **c** cooling processes. **d** Comparison of transmission spectra between heating and cooling processes

5 SDF-Based Optical Devices

In order to achieve SDF-based optical devices, various structures have been written in the SDF, such as FPI, FBG, MZI, and LPG. Figure 5 shows the schematic diagrams of the SDF-based devices. The preparation of these devices will be discussed following.

6 SDF-FPI

The formation of FPI needs two highly reflective surfaces. Figure 5a is the structure of SDF-FPI with an air cavity (Chen et al. 2015). Since the core material of the high-concentration alumina-doped SDF is not compatible with that of the low-concentration germanium-doped SMF, when these two types of optical fibers are fused by a commercial fiber fusion splicer, there may be fusion defects. The viscosity of the core of SDF is greater than that of the cladding, so the degree of deformation of the core is less than that of the cladding. Therefore, an air cavity will be generated between SDF and SMF.

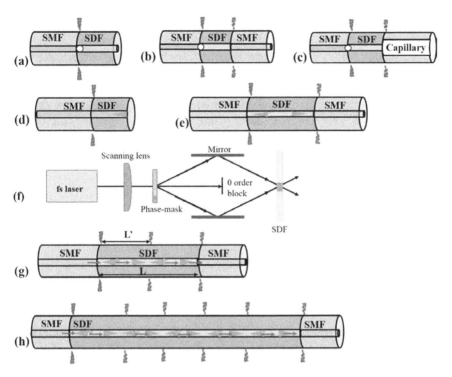

Fig. 5 Various interference structures prepared based on SDF. Schematic diagram of **a** and **b** and **c** Air cavity sensor based on SDF. **d** FPI with a single crystallized reflection surface; **e** FPI with two crystallized reflection surfaces; **f** FBG fabricated by a two-beam phase mask interference system (Elsmann et al. 2013); **g** cascaded SDF-MZI with four crystallized areas; **h** LPG based on crystallized SDF

Based on this principle, a FPI with an air cavity can be developed. Firstly, a SMF and SDF can be spliced by arc discharge. Then, the SDF is cut to a micron length. Because the RI of SMF and SDF are both higher than that of air, the two ends of the air cavity and the end surface of the SDF are three reflective surfaces, thus forming a composite FP cavity. When the end surface of the SDF directly contacts the external environment, it is susceptible to infection, so a structure shown in Fig. 5b can be designed to solve it (Ma et al. 2018). In addition, the air cavity can also be obtained by chemically etching (Zhang et al. 2019). Because the fiber' core is doped with germanium, the chemical corrosion rate of the core is faster (Klini et al.1998). Therefore, a concave cavity can be formed at the end of the SMF. Then, a SMF with a concave cavity is spliced with SDF by arc discharge as shown in Fig. 5c.

Due to the high RI of the crystallization area in SDF, when light in a SMF is coupled into the crystallization area, a part of the light will be reflected, and a part of the light is coupled into the higher-order mode at the crystallization area to continue transmitting forward. Therefore, the crystallization area of the SDF can be used as both a reflective surface and a mode converter.

Figure 5d is a FPI with single crystallized reflection surface. When a SDF is spliced with the standard SMF by the fiber fusion splicer, the arc discharge method produces the crystallization effect at a distance of several hundred microns from the fusion point. The interface between SDF and SMF, and the two interfaces of the crystallization area constitute three reflective surfaces, thereby forming a FPI. The crystallization area contains many crystal particles. If the interface of the crystallization area directly contacts the external environment, the structure of the crystallized area may be easily damaged. Therefore, a double-crystallized structure shown in Fig. 5e can be designed to form the FP cavity.

7   SDF-FBG

A FBG is a periodic or aperiodic reflective structure (Mihailov et al. 2010). Through the grating region, the forward-transmitting core mode and the backward-transmitting core mode will be coupled, so that the energy of the forward-transmitting core mode is transmitted to the backward-transmitting core mode, forming reflection of the incident wave.

FBG is usually produced by exposing the fiber to a strong optical interference pattern, as shown in Fig. 5f. The RI of SDF is related to the alumina concentration. The higher the alumina concentration, the higher the RI. Laser or mechanical gratings are used to change the morphology of the fiber core and affect the distribution of alumina, thereby inducing the RI distribution. In addition, due to the good light transmission of SDF at high temperature, the ordinary grating inscribed on SDF will have high-temperature resistance.

8   SDF-MZI

Because the crystallization region can be used as a mode converter and mode coupler, an MZI can be prepared. The two crystalline areas in Fig. 5e can also function as mode splitting and combined, respectively. At the first crystalline area, part of the guided mode in the core continues to propagate forward, and part of it stimulates higher-order modes. When passing through the second crystallization area, the guided mode interferes with the higher order. Therefore, a MZI can be fabricated based on the crystallization characteristics of SDF. However, the extinction ratio (ER) of the interference spectrum formed by the two crystallization areas is relatively small. In order to improve the ER of the MZI spectrum, a cascaded MZI can be fabricated as shown in Fig. 5g. With the new crystallization area, the higher-order mode can be excited to transmit in the fiber again, and the high-order mode and the guided mode of the core are coupled at the last crystallization. Thus, a cascaded MZI with L', L-L' and L cavity length is formed.

9 SDF-LPG

LPG is a periodic grating commonly used in communications, fiber filters, and other applications. It has the characteristics of low back reflection, high sensitivity, wide bandwidth, and polarization independence. Fiber grating is a kind of fiber device which is made of periodic index modulation (Davis et al. 1999; Albert et al. 2013). Based on crystallization-induced RI modulation, it can be used to fabricate a LPG. Because the crystallization of SDF is related to the temperature and duration of heat treatment, it is necessary to adjust the discharge current and time by arc discharge. Figure 5h is a schematic diagram of a LPG fabricated by arc discharge. A short segment of SDF is spliced between two standard SMFs by a commercial fiber fusion splicer. When the discharge current is large, there will be a pair of separate crystallizing cones in SDF, and the core between them is still transparent. According to the Al_2O_3-SiO_2 binary system, crystallization occurs when the temperature is lower than the liquidus temperature. When the temperature is higher than that of the metastable phase, and the temperature drops rapidly, there will be no phase separation or crystals will not precipitate.

Figure 6a is a micrograph of a LPG obtained through a fiber fusion splicer with a discharge current of 13.5 mA and a discharge time of 400 ms (Hong et al. 2017). A short segment of SDF is spliced with two standard SMFs by a fiber fusion splicer. There will be a pair of separate crystallizing cones in SDF when the setting discharge current is large, and the core between them is still transparent. According to the Al_2O_3-SiO_2 binary system, crystallization occurs when the temperature is lower than the liquidus temperature. When the temperature is higher than that of the metastable phase, and the temperature drops rapidly, there will be no phase separation or crystals will not precipitate.

Figure 6c shows the spectrum response of a SDF-LPG as the number of discharge points increases. When the number of discharges gradually increases, the spectrum shows a red shift. After four times of discharge, the spectrum achieves strong coupling

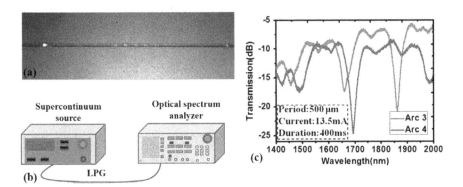

Fig. 6 LPG with a period of 500 μm based on SDF. **a** Microscopic images of SDF-LPG, **b** the experimental setup, and **c** spectrum response of LPG

with the ER of about 16 dB. When the SDF-LPG sensor is placed in different RI liquid, the transmission spectrum has no shift. It indicates that such a SDF-LPG has very good stability in different RI liquid.

Traditional LPG can excite cladding mode, which is sensitive to external environment. However, the coupling mode of the SDF-LPG is the coupling between the core modes. Since these modes are transmitted in the core, the SDF-LPG is not sensitive to externally changing refractive indices. In addition, the working principle of LPG is similar to that of MZI, which is based on inter-mode interference for sensing. Therefore, such a SDF-LPG has potential for high temperature and high-temperature strain applications.

10   SDF-Based Optical Devices for Sensing Applications

10.1   SDF-FPI for Temperature Sensing

FPI is a commonly used interference sensor with the advantages of small cavity length, high sensitivity, good stability, diverse preparation methods, etc. When a FPI experiences temperature variation, due to thermal expansion and thermo-optic effects, its cavity length and RI will change, which will cause the operating wavelength of the sensor to shift.

The FPI experiment system in Fig. 7 is used to heat up the FPI sensor and record the reflection spectrum at different temperatures. The FPI structure in Fig. 5a can be obtained through a fiber fusion splicer with a discharge current of 100 bits and a discharge time of 500 ms (Chen et al. 2015). According to the temperature measurement results of such a FPI sensor, it can be found that the spectrum shifts to longer wavelengths with the increase of temperature. Meanwhile, the reflected spectrum still maintains a good spectrum after experiencing a high temperature of 1000 °C. In order to avoid the direct contact between the fiber end face and the external environment, an interference structure in Fig. 8a can be designed. Such a new FPI with an air cavity can have the ER of 18 dB and keep good optical signal at 1100 °C as shown in Fig. 8b, c (Ma et al. 2018). Due to the different composition of the media at

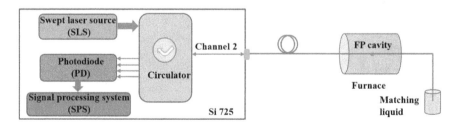

Fig. 7   High-temperature experimental setup of a SDF-FPI

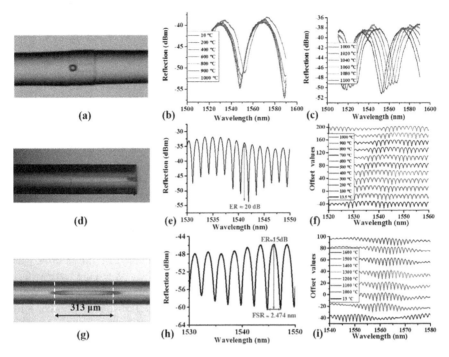

Fig. 8 Microscopic images of SDF-FPIs and the corresponding temperature response. **a** FPI with an air cavity; **b** FPI with a single crystallized reflection surface; and **c** two crystallized reflection surfaces. The corresponding interference spectra and temperature response are shown in the right figures

both ends of the air cavity and the cavity, the deformation and RI changes caused by heating are different. Thus, when the temperature is below 500 °C, the spectrum has almost no shift. While the temperature is higher than 500 °C, the spectrum begins to shift directionally. When the temperature reaches 1000 °C, the spectrum shifts significantly toward longer wavelength due to thermal expansion effect of optical fiber.

Figure 8d shows a SDF-FPI with a single crystallized reflection surface, which has a high ER shown in Fig. 8e, about 20 dB. Benefiting from the high-temperature stability of mullite crystals, such a crystallized SDF-FPI sensor can work stably at 1000 °C as shown in Fig. 8f. When the end face of the fiber is directly exposed to the air, the end face will inevitably be damaged or polluted, which will affect the transmission of optical signals. Figure 8f is the optimized structure of the crystallized FPI (Liu et al. 2019). The reflection spectrum of the crystallized SDF-FPI with two crystallized reflection surfaces in Fig. 8h shows the free spectral range (FSR) of 2.474 nm and the ER of about 15 dB. The FSR of the FPI can be expressed as:

$$\Delta\lambda_{\text{FPI}} = \frac{\lambda_F^2}{2 \cdot n_F \cdot L_F},\qquad(1)$$

where λ_F is the resonant wavelength, $\Delta\lambda_{FPI}$ is the FSR, referring to the wavelength difference between adjacent interference dips. n_F is the effective RI of the FP cavity, about 1.53. Thus, the cavity length, L_F, is calculated to be about 317 μm by Eq. (1), which is approximately equal to the measured length by the microscope.

The reflection spectrum of the crystallized SDF-FPI with two crystallized reflection surfaces is measured from 13 to 1600 °C with a step of 100 °C, as shown in Fig. 8i. The optical signal of such a FPI at 1000 °C shows almost no degradation. Meanwhile, the reflected spectrum is red-shifted as the temperature increases. At present, 1000 °C is the limit temperature of most fiber sensors (Hua et al. 2015; Liu et al. 2018). The traditional temperature optical fiber sensor is made of a commercial SMF. In order to study the high-temperature properties of the FPI based on a SDF and SMF at temperatures higher than 1000 °C, a SMF sensor with high-temperature resistant needs to be prepared. It has been reported that a SMF-FPI sensor fabricated by the fs laser can work up to 1100 °C (Wang et al. 2010; Wei et al. 2008). The high-energy irradiation of the fs pulse can induce the RI modulation of the fiber, thereby creating a reflective surface in the fiber. When the SMF-FPI fabricated by the fs laser was heated to 1000 °C, only a slight decrease in the ER. However, when the SMF-FPI is heated to 1200 °C, the spectrum of the SMF sensor is completely distorted. Different from SMF-FPI sensor, the SDF-FPI with two crystallized reflection surfaces can still maintain a good spectral when heated to 1600 °C.

In addition, the stability of the sensor at high temperature can also be reflected by the change of the ER of the spectrum. The ER of FPI based on SMF decreases sharply when the temperature is higher than 800 °C, while the ER of crystallized SDF-FPI with two crystallized reflection surfaces only started to decrease at 1600 °C. The SDF-FPI sensor has unmatched optical performance of SMF at high temperature, which is attributed to the stable material properties of the reflected mirror made of the mullite crystal. According to the Al_2O_3–SiO_2 binary system, the silica–mullite eutectic temperature is about 1600 °C (Aramaki and Roy 1959). Mullite crystals will dissolve at a temperature of 1600 °C or higher, thereby reducing the reflectivity of the crystallization area and reducing the ER of the reflected spectrum. Thus, such a SDF-FPI with two crystallized reflection surfaces provides more options for some harsh environments.

10.2   SDF-FBG for Temperature Sensing

With the development of laser devices and laser technology, FBG sensors have become more and more widely used in the field of high-temperature sensing (Canning 2008; Mihailov et al. 2006; Bennion et al. 1996). Fs lasers are widely used in the preparation of FBGs due to short pulse width and high peak power, which can induce larger RI modulation s in optical fibers. Fiber material and the RI modulation induced by the writing method both affect the high-temperature stability of fiber gratings. According to the high-temperature stability performance, there are two types of fiber gratings written by fs laser, one is Type I and the other is Type II (Liao et al. 2010).

Table 3 Performance comparison of FBG-SDF

Alumina concentration in SDF(mol%)	Diameter of SDF (core/cladding) (μm)	Working temperature (°C)	Temperature sensitivity	Reference
49.4	21/125	950	12.4 pm/K	Elsmann et al. (2014)
4	8/125	900	13.0 pm/°C	Grobnic et al. (2015)
30	30/150	700	14.7 pm/°C	

The RI modulation mechanism of Type I FBG is similar to that of a SMF induced by UV laser, so it is easy to be erased at a lower temperature. The RI modulation of Type II FBG is based on the permanent damage of fiber, which has high RI modulation and high-temperature stability.

Type II Bragg sensors can be fabricated by a two-beam phase mask interference system, as shown in Fig. 5f (Elsmann et al. 2013). Firstly, an ultrafast laser is divided into two coherent beams. Then, the two coherent beams are combined into a holographic interference pattern by using two rotating mirrors. When a SDF is placed in the interference area, the holographic interference light will introduce periodic RI distribution into a SDF. Due to the difference of RI between core and cladding and the larger diameter of core, the reflection spectrum of a Type II FBGs written on a SDF will exhibit significant multi-mode characteristics. Thus, Type II FBGs can achieve a highly multi-mode grating response similar to that in pure sapphire rods.

Table 3 lists the temperature resistance of FBGs that have been prepared based on different SDFs. The FBG based on SDF has a linear relationship between operating wavelength and temperature. When the same type FBG is written on SDFs doped with different concentrations of alumina, the temperature sensitivities of the FBG sensors are slightly different. It indicates that the temperature resistance of the FBG temperature sensor is related to the doping concentration of alumina in a SDF. Since the RI profile of SDF is related to its core diameter and the alumina concentration of the doped core, the performance of FBG may be related to the core diameter of SDF.

The Bragg grating can also be written by point-by-point (Yang et al. 2017) or line-by-line scanning (Huang and Shu 2016). For different grating writing methods, the RI modulation induced on the fiber is different. Therefore, the performance of FBGs based on SDF may be further improved based on other grating writing methods.

10.3 SDF-MZI for Temperature Sensing

MZI is a structure of dual-beam interferometer. Due to simple preparation, small size, high sensitivity, large dynamic range, etc., it attracts wide attention in many industrial fields (Xu et al. 2012; Wen et al. 2014). A MZI with two crystallization areas can be obtained by splicing a centimeter-long SDF between two SMFs. When the connected SDF is discharged again at a distance L' from one of the fusion points, a cascaded SDF-MZI with four crystallization areas can be obtained. Figure 9a shows the spectra of a SDF-MZI with two crystallization areas and a cascaded SDF-MZI with four crystallization areas. It can be seen that the cascaded interference structure can not only increase the ER of interference fringes, but also increase the fringe pitch.

The fast Fourier transform of the interference spectra of the two types of SDF-MZI is used to achieve the frequency domain spectrum as shown in Fig. 9b. There is only one main peak (f_1) in the frequency domain of a SDF-MZI with two crystallization areas. It shows that the power is mainly concentrated in a dominant mode, and the interference spectrum is mainly obtained by coupling the fundamental mode and the specific mode. The cascaded SDF-MZI with four crystallization areas has three main peaks, f_1, f_2 and f_3. f_2 and f_3 are corresponding interference cavity with cavity length of $L', L - L'$, respectively. When $L' = L/2, f_2 = f_3$. The spatial frequency of the MZI, f_{MZI}, can be expressed as (Choi et al. 2007; Chen et al. 2017),

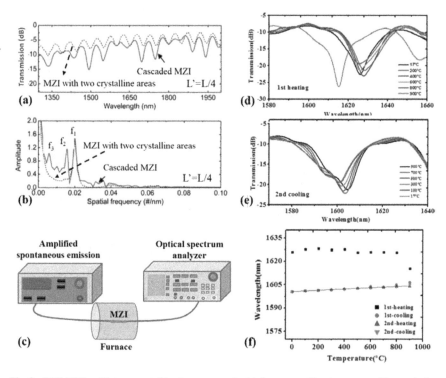

Fig. 9 SDF-MZIs with two crystallization areas and with four crystallization areas. **a** Transmission spectra; **b** Fast Fourier transform of the interference spectra. Temperature test system of **c** first heating process and **d** second cooling process; **f** Temperature response of a cascaded SDF-MZI with four crystallization areas

$$f_{MZI} = \frac{\Delta n_{eff} \cdot L_M}{\lambda_M^2},   \qquad (2)$$

where L_M is the interference cavity length of SDF-MZI. λ_M is the resonant wavelength, Δn_{eff} is the effective RID between core guide mode and excited higher-order mode. The effective RID is calculated to be about 2.02×10^{-3} by Eq. (2), which corresponds to the RID between LP_{01} (guided mode) and LP_{11} (excited higher-order mode) in the SDF.

The experimental setup for observing and recording the spectrum of the MZI is shown in Fig. 9c. During the first heating process shown in Fig. 9d, when the temperature is raised to 800 °C, the monitored resonance peak fluctuates only slightly, until the temperature is higher than 800 °C, the resonance peak shifts toward the shorter wavelength. This can be caused by the introduction of residual stress during the drawing (Mohanna et al. 1990). Therefore, before calibration and use of the sensor, an annealing process should be performed to release the residual stress of the optical fiber. It can be found from Fig. 9f that the temperature sensing performance of the cascaded SDF-MZI with four crystallization areas is basically stable after the first 900 °C heat treatment, and the transmission spectrum has good repeatability during the second heating and cooling process. The experiment results demonstrate that such a cascaded SDF-MZI with four crystallization areas has the temperature sensitivity of about 3.7 pm/°C, and it can perform high-temperature sensing at 900 °C. The cascaded SDF-MZI with four crystallization areas shows good temperature insensitivity in high-temperature environment. In addition, since the interference modes LP_{01} and LP_{11} have similar thermal expansion coefficients (Li et al. 2005), temperature variation has little effect on the interference modes' effective RID. Therefore, the temperature sensitivity of cascaded SDF-MZI with four crystallization areas is small.

Based on these characteristics, SDF-MZI can solve the problem of temperature cross-sensitivity in optical fiber sensor and reduce the cross talk between various parameters. It shows advantages in fields requiring high-temperature resistance and low-temperature cross-sensitivity.

Table 4 lists the types and temperature resistance of interference sensors that have been prepared based on SDF. Both SDF-FPI with air cavity and with crystallization area can work at 1000 °C and above. When the end face of the sensing fiber is directly exposed to the air, its end face may be damaged or polluted. When both ends of the sensing fiber are spliced with a SMF by a fiber fusion splicer, the thermal stability performance of SDF-FPI will be improved. Especially for the SDF-FPI with crystallization area, the maximum working temperature can reach 1600 °C. At present, the high-temperature performance of SDF-FBGs written on SDFs by a two-beam phase mask interference system is weaker than that of SDF-FPI. In order to further improve the high-temperature performance of FBG, there are two schemes. One is to use other femtosecond laser writing techniques to prepare better FBGs in SDF, such as the point-by-point method or line-by-line scanning method, and the

Table 4 Performance comparison of various sensors prepared based on SDF

Structure	Working temperature (°C)	Sensing type	Reference
FPI with air cavity (Fig. 5a)	Up to 1000	Temperature	Chen et al. (2015)
FPI with air cavity (Fig. 5b)	500–1100	Temperature	Ma et al. (2018)
FPI with air cavity (Fig. 5c)	Up to 1000	Temperature and strain	Zhang et al. (2019)
FPI with a single crystallized reflection surface	Up to 1000	Temperature and RI	Zhang et al. (2018)
FPI with two crystallized reflection surfaces	Up to 1600 Up to 1000 (stain)	Temperature and strain	Liu et al. (2019) Wang et al. (2019) (strain)
FBG	Up to 950	Temperature	Elsmann et al. (2014)
MZI	Up to 900	Temperature	Xu et al. (Xu et al. 2017)

other is to prepare a crystallized FBG. The principle of SDF-MZI is based on inter-mode interference. Although SDF can support multi-mode transmission, the high-order mode loss at high temperature affects the interference spectrum. Therefore, high-temperature stability of a SDF-MZI is weaker than the former two sensors.

11   Conclusion

Sapphire-derived fiber (SDF) is a glass fiber with an alumina-doped core drawn by a rod-in-tube technology, which has a high refractive index core. Due to high-concentration alumina-doped core, when a SDF is treated by arc discharge, the temperature variation introduced by arc discharge will cause devitrification of Al_2O_3-SiO_2 co-doped cores. As a result, mullite crystals will be precipitated in the core of SDF. The high refractive index of mullite crystals can be used to prepare highly reflective surfaces, periodic gratings, and mode beam splitters and mode couplers. Meanwhile, the high-temperature resistance of mullite crystals can make crystallized devices based on SDFs to have good stability at high temperature. However, aeronautical engines, blades, and other equipment will introduce large vibrations during work. This requires that the optical fiber sensors embedded in high-temperature equipment have high mechanical properties at high temperature, which will be a development direction of SDF in the future. At present, optical devices such as FBG, FPI, LPG, and MZI can be fabricated based on SDF. Due to good optical properties and good thermal properties, it has broad application prospects in aerospace, oil extraction, metallurgy, and other fields.

References

J. Albert, L.Y. Shao, C. Caucheteur, Laser Photonics Rev. **7**, 1 (2013)

S. Aramaki, R. Roy, Nature **184**, 4686 (1959)

J. Ballato, T. Hawkins, P. Foy, B. Kokuoz, R. Stolen, C. McMillen, M. Daw, Z. Su, T.M. Tritt, M. Dubinskii, J. Zhang, T. Sanamyan, M.J. Matthewson, J. Appl. Phys. **105**, 5 (2009)

I. Bennion, J.A.R. Williams, L. Zhang, K. Sugden, N.J. Doran, Opt. Quant. Electron. **28**, 2 (1996)

J. Canning, Laser Photonics Rev. **2**, 4 (2008)

C.H. Chen, W.T. Wu, J.N. Wang, Technology **23**, 2 (2017)

H.Y. Choi, M.J. Kim, B.H. Lee, Opt. Express **15**, 9 (2007)

D.D. Davis, T.K. Gaylord, E.N. Glytsis, S.C. Mettler, Electron. Lett. **35**, 9 (1999)

P. Dragic, T. Hawkins, P. Foy, S. Morris, J. Ballato, Nat. Photonics **6**, 9 (2012)

P.D. Dragic, J. Ballato, S. Morris, A. Evert, R.R. Rice, T. Hawkins, Fiber Optic Sens. Appl. X **8722**, 8 (2013)

T. Elsmann, T. Habisreuther, A. Graf, M. Rothhardt, H. Bartelt, Opt. Express **21**, 4 (2013)

T. Elsmann, A. Lorenz, N.S. Yazd, T. Habisreuther, J. Dellith, A. Schwuchow, J. Bierlich, K. Schuster, M. Rothhardt, L. Kido, H. Bartelt, Opt. Express **22**, 22 (2014)

D. Grobnic, S.J. Mihailov, J. Ballato, P.D. Dragic, Optica **2**, 4 (2015)

L. Hong, F.F. Pang, H.H. Liu, J. Xu, Z.Y. Chen, Z.W. Zhao, T.Y. Wang, IEEE Photonics Technol. Lett. **29**, 9 (2017)

L.W. Hua, Y. Song, J. Huang, X.W. Lan, Y.J. Li, H. Xiao, Appl. Opt. **54**, 24 (2015)

B. Huang, X.W. Shu, Opt. Express **24**, 16 (2016)

Y. Huang, Z. Zhou, Y.N. Zhang, G.D. Chen, H. Xiao, I.E.E.E. Trans, Instrum. Meas. **59**, 11 (2010)

A. Klini, T. David, E. Bourillot, J. Lightwave Technol. **16**, 7 (1998)

J. Kobelke, K. Schuster, J. Bierlich, S. Unger, A. Schwuchow, T. Elsman, J. Dellith, C. Aichele, A. R. Fatobene, H. Bartelt, Adv. Electr. Electronic En. **15**, 1 (2017)

B. Lee, Opt. Fiber Technol. **9**, 2 (2003)

C.K.Y. Leung, K.T. Wan, D. Inaudi, X.Y. Bao, W. Habel, Z. Zhou, J.P. Ou, M. Ghandehari, H.C. Wu, M. Imai, Mater. Struct. **48**, 4 (2015)

Q. Li, C.H. Lin, P.Y. Tseng, H.P. Lee, Opt. Commun. **250**, 4–6 (2005)

W. Li, Z.F. Hu, X.Y. Li, W. Fang, X. Guo, L.M. Tong, J.Y. Lou, Opt. Commun. **32**, 5 (2014)

C.R. Liao, Y.H. Li, D.N. Wang, T. Sun, K.T.V. Grattan, IEEE Sens. J. **10**, 11 (2010)

C.R. Liao, D.N. Wang, Y. Wang, Opt. Lett. **38**, 5 (2013)

D.J. Liu, Q. Wu, C. Mei, J.H. Yuan, X.J. Xin, A.K. Mallik, F.F. Wei, W. Han, R. Kumar, C.X. Yu, S.P. Wan, X.D. He, B. Liu, G.D. Peng, Y. Semenova, G. Farrell, J. Lightwave Technol. **36**, 9 (2018)

H. Liu, F. Pang, L. Hong, Z. Ma, L. Huang, Z. Wang, J. Wen, Z. Chen, T. Wang, Opt. Express **27**, 5 (2019)

F. Llewellyn-Jones, C.H. Hinrichs, Am. J. Phys. **35**, 5 (1967)

C. McCague, M. Fabian, M. Karimi, M. Bravo, L.R. Jaroszewicz, P. Mergo, T. Sun, K.T.V. Grattan, J. Lightwave Technol. **32**, 5 (2014)

G.N. Merberg, J.A. Harrington, Appl. Opt. **32**, 18 (1993)

S.J. Mihailov, D. Grobnic, H.M. Ding, C.W. Smelser, J. Broeng, IEEE Photonics Technol. Lett. **18**, 17 (2006)

S.J. Mihailov, D. Grobnic, C.W. Smelser, Opt. Lett. **35**, 16 (2010)

Y. Mohanna, J.M. Saugrain, J.C. Rousseau, P. Ledoux, J. Lightwave Technol. **8**, 12 (1990)

P.F. Chen, F.F. Pang, Z.W. Zhao, L. Hong, N. Chen, Z.Y. Chen, T.Y. Wang, in *Proc. SPIE* **9634** (2015)

T.Y. Tu, F.F. Pang, S. Zhu, J.J. Cheng, H.H. Liu, J.X. Wen, T.Y. Wang, Opt. Express **25**, 8 (2017)

Z.W. Ma, F.F. Pang, H.H. Liu, Z.Y. Chen, T.Y. Wang, OFS **WF48** (2018)

S. Unger, J. Dellith, A. Scheffel, J. Kirchhof, Phys. Chem. Glasses-B **52**, 2 (2011)

Y. Wang, Y. Li, C. Liao, D.N. Wang, M. Yang, P. Lu, IEEE Photonics Technol. Lett. **22**, 1 (2010)

Z. Wang, H. Liu, Z. Ma, Z. Chen, T. Wang, F. Pang, Opt. Express **27**, 20 (2019)

T. Wei, Y. Han, H.L. Tsai, H. Xiao, Opt. Lett. **33**, 6 (2008)

X. Wen, T. Ning, C. Li, Z. Kang, J. Li, H. You, T. Feng, J. Zheng, W. Jian, Appl. Opt. **53**, 1 (2014)

F. Xu, C. Li, D. Ren, L. Lu, W. Lv, F. Feng, B. Yu, Chin. Opt. Lett. **10**, 7 (2012)

J. Xu, H.H. Liu, F.F. Pang, L. Hong, Z.W. Ma, Z.W. Zhao, N. Chen, Z.Y. Chen, T.Y. Wang, Opt. Mater. Express **7**, 4 (2017)

S. Yang, D. Hu, A.S. Wang, Opt. Lett. **42**, 20 (2017)

Y.M. Zhang, X.D. Ding, Y.M. Song, M.L. Dong, L.Q. Zhu, Meas. Sci. Technol. **29**, 3 (2018)

P.H. Zhang, L. Zhang, Z.P. Mourelatos, Z.Y. Wang, Appl. Opt. **57**, 30 (2018)

P.H. Zhang, L. Zhang, Z.Y. Wang, X.Y. Zhang, Z.D. Shang, Opt. Express **27**, 19 (2019)

J. Zheng, X.Y. Dong, P. Zu, J.H. Ji, H.B. Su, P.P. Shum, Appl. Phys. Lett. **103**, 18 (2013)

Z.W. Ma, H.H. Liu, Y.N. Shang, F.F. Pang, Z.F. Wang, Z.Y. Chen, X. Gong, T.Y. Wang, in *Proc. SPIE*, **11048** (2019)

Thermoelectric Fibers

Ting Zhang, Haisheng Chen, and Xinghua Zheng

Abstract Flexible thermoelectrics enables a direct and green conversion between heat and electricity to power or refrigerate flexible and wearable electronics. Organic polymer-based flexible thermoelectric materials are particularly fascinating because of their intrinsic flexibility, affordability, and low toxicity, but low thermoelectric performance limits their development. The other promising alternatives of inorganic-based flexible thermoelectric materials that have high energy-conversion efficiency, large power output, and stability at relatively high temperature, yet it is impeded by inferior flexibility. Hence, researchers propose a paradigm shift in material research as flexible thermoelectrics requires the material of which the device is made to simultaneously have inorganic semiconductor-like high thermoelectric performance and organic material-like mechanical flexibility. In this chapter, this dilemma is tackled, on both material level and device level, by introducing a new kind of flexible thermoelectric fibers, which overcomes the problems that thin film-based thermoelectrics can only be bent in one direction and lack essential wearable properties such as air permeability. Herein, the state-of-the-art in the development of flexible thermoelectric fibers and devices is summarized, including organic conducting polymer thermoelectric fibers, fully inorganic flexible thermoelectric fibers, and inorganic TE materials hybridized with organic polymer fibers. Finally, the remaining challenges in flexible thermoelectric fibers are discussed in conclusion, and suggestions and a framework to guide future development are provided, which may pave the way for a bright future of fiber-based flexible thermoelectric devices in the energy market.

Keywords Thermoelectric (TE) fibers · Wearable electronics · Seebeck coefficient · Thermal conductivity · Electrical conductivity · Power factor · ZT

T. Zhang (✉) · H. Chen · X. Zheng
Institute of Engineering Thermophysics, Chinese Academy of Sciences,
Beijing 100190, China
e-mail: zhangting@iet.cn

H. Chen
e-mail: chen_hs@iet.cn

X. Zheng
e-mail: zhengxh@iet.cn

© Springer Nature Singapore Pte Ltd. 2020
L. Wei (ed.), *Advanced Fiber Sensing Technologies*,
Progress in Optical Science and Photonics 9,
https://doi.org/10.1007/978-981-15-5507-7_10

175

value · PEDOT: PSS fiber · Thermal drawing · Wet-spinning · Electrospun · Mechanical flexibility · Conductive polymers · Power output

1 Introduction

In fossil fuel combustion, typically only $\sim 34\%$ of the resulting energy is used efficiently, while the remainder is lost to the environment as waste heat (Fitriani et al. 2016). Broad societal needs have focused attention on technologies that can reduce ozone depletion, greenhouse gas emissions, and fossil fuel usage. Thermoelectric (TE) devices, which are semiconductor systems that can directly convert between thermal and electrical energy, are increasingly being seen as having the potential to make important contributions to reducing CO_2 and greenhouse gas emissions and providing cleaner forms of energy. TE devices exhibit many advantages, such as having no moving parts, no moving fluids, no noise, easy (or no) maintenance, and high reliability, which has been commercially used for a variety of applications, including thermal cycles for DNA synthesizers, car seat cooler/heaters, laser diode coolers, certain low-wattage power generators, and radioisotope thermoelectric generators, etc. (Du et al. 2018).

TE materials offer a way to convert low-grade waste heat energy into electrical power, based on the Seebeck effect (Fig. 1a) (Du et al. 2018). This effect was discovered in 1821 by German scientist Thomas Johann Seebeck and can be used in a wide range of energy-conversion applications. The TE energy-harvesting mechanism of a material is that when a temperature gradient (ΔT) is applied, the charge carriers (electrons for n-type materials or holes for p-type materials) from the hot side diffuse to the cold side. As a result, an electrostatic potential (ΔV) is induced (Li et al. 2010). The electrostatic potential generated by a single n- or p-type TE legs is very low (from several μV to mV depending on context). Therefore, to achieve

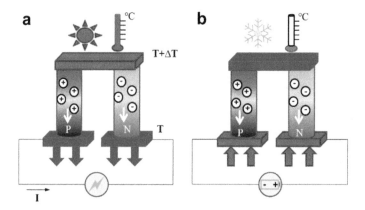

Fig. 1 Schematic illustrations of a TE module for **a** power generation (Seebeck effect) and **b** active refrigeration (Peltier effect). Reproduced with permission. (Du et al. 2018) Copyright 2018, Elsevier

high output voltage and power, TE generators are typically made of dozens, or even hundreds, of TE couples. TE materials can also convert electrical power into thermal energy (i.e., cooling or heating) based on the Peltier effect (Fig. 1b) (Du et al. 2018), discovered in 1834 by French scientist Jean Charles Athanase Peltier. The Peltier effect is essentially the inverse of the Seebeck effect. The efficiency of TE devices is strongly associated with the dimensionless figure of merit (ZT) of TE materials, defined as $ZT = (S^2\sigma/\kappa)\,T$, where S, σ, κ and T are the Seebeck coefficient, electrical conductivity, thermal conductivity, and absolute temperature (Du et al. 2018). High electrical conductivity (corresponding to low Joule heating), a large Seebeck coefficient (corresponding to large potential difference) and low thermal conductivity (corresponding to a large temperature difference) are therefore necessary in order to realize high-performance TE materials. The ZT value is also a very convenient indictor for evaluating the potential efficiency of TE devices. In general, good TE materials have a ZT value of close to unity. However, ZT values of up to three are considered to be essential for TE energy converters that can compete on efficiency with mechanical power generation and active refrigeration (Li et al. 2010).

High-performance TE materials have been pursued since Bi_2Te_3-based alloys were discovered in the 1960s. Until the end of last century, moderate progress had been made in the development of TE materials. The benchmark of $ZT \approx 1$ was broken in the mid-1990s by two different research approaches: one by exploring new materials with complex crystalline structures, and the other by reducing the dimensions of the materials (Li et al. 2010). In the recently discovered TE compounds, such as skutterudites, clathrates, and Zintl compounds, the thermal conductivity can be reduced greatly while maintaining the electrical conductivity at a high level (Lan et al. 2010). In research on low-dimensional material systems, Dresselhaus et al. suggested that the power factor ($P = S^2\sigma$) can be enhanced through the use of quantum confinement effects. Dresselhaus's pioneering work has shed light on various low-dimensional systems, including superlattices, nanowires, and quantum dots (Dresselhaus et al. 2007). Venkatasubramanian et al. reported $Bi_2Te_3/Sb_2\,Te_3$ superlattices with a high-ZT value of up to 2.4 (Venkatasubramanian et al. 2001). Subsequently, Harman et al. reported PbTe/PbTeSe quantum dot superlattices with a ZT value of greater than 3.0 at 600 K (Harman et al. 2002). However, these high-ZT low-dimensional materials are difficult to be applied on general applications, and it is still desired to develop potential routes for improvement in the major TE bulk material systems, including Bi–Te alloys, skutterudites, Ag–Pb–Sb–Te or 'LAST', half-Heusler alloys and some high-ZT oxides (Snyder and Toberer 2008). In 2014, an unprecedented ZT value of 2.6 for bulk materials is realized in SnSe single crystals at 923 K (Zhao et al. 2014), and the hole doped SnSe single crystals exhibit ZT > 1.5 at 600–773 K while the highest value of 2.8 at 773 K for electron-doped SnSe single crystals (Chang et al. 2018). These attributes are generally associated with low thermal conductivity. Compared to other high-performance thermoelectrics, SnSe single crystals demonstrate that a high ZT also can be realized in simple layered, anisotropic and anharmonic systems, without nanostructuring.

Nowadays, TE materials chipped into devices are mostly prepared in the form of cube or cuboid blocks from a TE ingot by means of a top-down dicing process. A

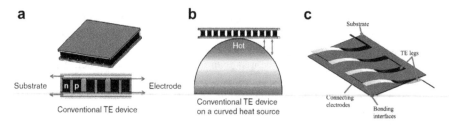

Fig. 2 **a** A conventional planar-structured TE device. **b** Scheme of power generation of the conventional TE generator on a curved heat source. **c** Schematic diagram of a typical module for flexible TE devices. **a, b** Reproduced with permission. (Park et al. 2016) Copyright 2016, Springer Nature. **c** Reproduced with permission. (Wang et al. 2019) Copyright 2019, Wiley-VCH

potential problem of this conventional procedure is a relatively high production cost due to the energy intensive processing for ingots such as zone-melting or hot-pressing as well as the post-processing due to shape control. The latter, in fact, suffers from another problem that attempts to realize any complicated shape other than a cube are technically impossible within the context of mass production. On the other hand, in the real-world applications, the minimization of heat loss due to incomplete contact between the surface of the heat source and the TE module is no less important than the figure-of-merit of materials. It is noted that the majority of heat sources for TE generators has irregular shapes, where the conventional planar-structured TE devices composed of cubic blocks should fail in achieving a desirable contact (Fig. 2a, b) (Park et al. 2016). One readily available solution to settling down the aforementioned issues is the development of flexible TE materials and devices.

Flexible TE materials, in this situation, are promising because their conformability enables effective contact with curved heat sources in order to maximize heat harvesting. These flexible TE materials can also be fabricated at much lower temperatures. Compared with other rigid devices, flexible TE devices are lightweight and can be conformably attached onto human skin, enabling the direct harvesting of electricity from body heat without recharging, but also preventing or minimizing heat losses during energy transfer. Such advantages make flexible TE devices promising power sources for wearable electronics. Figure 1c illustrates a typical flexible TE device, composed of a substrate, TE legs, connecting electrodes, and bonding interfaces (Wang et al. 2019). In such an flexible TE device, the substrates are flexible and insulating in order to provide the devices with flexibility, which guarantees interference-free carrier transport within the TE legs; the TE legs are in the form of thin films, which can be adapted for device flexibility; the electrodes are applied to connect n- and p-type TE legs in series; and the bonding interfaces stabilize the TE legs and electrodes (Chen et al. 2015). Moreover, the Seebeck effect is the primary enabling physical phenomenon that facilities the function of flexible TE devices as potential body temperature monitors by transferring body temperature changes into an output voltage signal (Yang et al. 2018). By the Peltier effect, flexible TE devices can also be applied in flexible temperature-control systems, which can be used in microclimate systems that maintain the body temperature in extreme conditions

(Zhang et al. 2018), medical devices such as cooling blankets (Delkumburewatte and Dias 2012), and emerging portable electronic devices that require cooling (He et al. 2015). Although their applications as power implantable devices are still controversial, with concerns about their toxicity, there is clearly a bright future of flexible TEs.

For flexible TE materials, the most common approaches are to use either fully organic TEs or inorganic/organic hybrids. Some conducting polymers exhibit relatively good TE properties, such as poly(3,4-ethylenedioxythiophene):poly(styrenesulfonate) (PEDOT:PSS), polyaniline, polypyrrole, and their derivatives, due to their intrinsic flexibility and conductivity (Wang and Zhang 2019). Additionally, conductive polymers are abundant, nontoxic, or low toxicity, and generally easy to shape and process for industrial applications. The major drawback of conductive polymers is their poor electrical transport properties, which can be improved by doping and secondary doping engineering (Chen et al. 2015). Nevertheless, higher TE performance (in terms of output power and/or efficiency) may be achieved in inorganic/polymer hybrid materials. For example, in inorganic/organic composite materials, the high electrical conductivity and Seebeck coefficient of the inorganic constituent can be integrated with the low thermal conductivity of the polymers, and thus achieve high TE efficiency. More complex approaches to inorganic/organic hybrid materials include incorporating TE fillers into conductive polymers via ex situ (Du et al. 2014) and in situ synthesis (Gao and Chen 2018), or by intercalating organic molecules into inorganic layered structures (Wan et al. 2015). In such organic/inorganic hybrids, inorganic fillers such as inorganic TE particles and carbon-based nanomaterials can provide extra current pathways and in turn induce energy-filtering effects in the flexible polymer matrix, while organic molecules can provide the desired flexibility for the inorganic host (Gao et al. 2017). For inorganic flexible TE thin films, continuous flexibility can be realized through either depositing inorganic TE thin films on flexible substrates using atomic deposition techniques (Yang et al. 2017), or applying carbon nanotube (CNT) scaffolds (Jin et al. 2019) or nanostructure tailoring (Paul et al. 2017) to develop freestanding inorganic flexible TE films. However, limited by the device size, small temperature gradient, and low energy-conversion efficiency of flexible TE materials, the output power of flexible TE film devices is only a few μW to several mW (Kim et al. 2016). Optimization of the design of devices is another key factor to achieving their high performance. For example, devices have been achieved in the form of fiber-based textile structures, which significantly enhances device performance and essentially meets the wearable requirement of air permeability, showing great promise for the development of TE clothes (Du et al. 2017).

Fiber-shaped electronic devices, adopting one-dimensional (1D) structure with small diameters from tens to hundreds of micrometers, have attracted broad interests for wearable electronic fields (Lan et al. 2010). Typically, fiber-shaped electronic devices can be prepared on the basis of coaxial, twisting, and interlacing architectures. These fiber-shaped electronic devices are lightweight and flexible, and they can adapt to various deformations like bending, distortion, and stretching. More importantly, it is feasible to weave the fiber-shaped electronic devices into flexible, deformable, and breathable textiles that can facilitate practical applications. Hence, quite lots of researchers have devoted themselves to enhancing device performances (Wang et al. 2020). To make full use of the advantages of fiber and textile electronics, stretchable, self-healing, or shape-memory, fiber-shaped electronic devices have been further developed (Lee et al. 1902). More importantly, people have to realize self-powering devices by integrating energy harvesting with energy storage (Zhu et al. 2020). Fiber-based flexible TE energy generators are 3D deformable, lightweight, and desirable for applications in large-area waste heat recovery, and as energy suppliers for wearable or mobile electronic systems in which large mechanical deformations, high energy-conversion efficiency, and electrical stability are greatly demanded. TE fiber devices can be manufactured at low or room temperature under ambient conditions by established industrial processes, offering cost-effective and reliable products in mass quantity, which has inspired the industry like electronic and clothing industry because fiber-shaped TE devices can be woven through the well-developed commercial textile technology.

Here, we provide a critical overview of the progress of fiber-shaped TE materials and devices. We aim to identify the existing benchmarks, point out current problems, and predict future directions for further improvement. This chapter covers the materials fabrication methods, device structures and potential applications of TE fibers. Discussions are presented on the opportunities and challenges of TE fibers in terms of their device structure and applications. Last, we provide our perspectives on future research directions of fiber-shaped TE devices. We hope that this report will encourage researchers to putting more effort to translate the progress made in research laboratories for developing TE fibers into practical applications.

2 Fiber-Shaped TE Materials and Devices

The selection of appropriate materials is extremely important for the design and fabrication of fiber-shaped TE materials and devices. These materials should possess high energy-conversion efficiency with high carrier mobility and appropriate carrier concentration. Meanwhile, these materials should have the expected thermal and mechanical properties, be flexible and eco-friendly, show stable performance, and be easy to fabricate at low or room temperature using low-cost materials. According to constituent materials, TE fibers can be summarized as organic TE fibers, inorganic TE fibers, and organic/inorganic composite TE fibers.

- **Organic TE fibers**

Conductive polymers are organic materials with intrinsically electrical conduction. Alan Heeger, Alan MacDiarmid, and Hideki Shirakawa first discovered conducting polyacetylene in 1970s. They were jointly awarded with the Nobel Prize in 2000 for this discovery. Like traditional polymers, conducing polymers have a low thermal conductivity when compared to inorganic TE materials, which is beneficial for high

ZT. Furthermore, conducting polymers also have good flexibility, low density, low cost and easy synthesis and processing into versatile forms; therefore, much attention has recently been paid to conducting polymers for TE applications. Aqueous PEDOT:PSS dispersions are commercially available. The properties of PEDOT:PSS vary with the product types and its electrical conductivity, as well as TE properties, can be significantly enhanced by adding secondary dopants or compositing with special semiconductors. Besides, PEDOT:PSS possesses excellent stability, flexible mechanical properties, and high transparency, making it a promising candidate for organic TE materials (Yue and Xu 2012). The maximum ZT at room temperature was 0.42, which was measured from the DMSO-doped PEDOT:PSS (Kim et al. 2013). In addition to the doping method, post-treatments, such as adding solvents or dedoping, have been explored. After treatment, although the ZT was only ≈ 0.1, the power factor was as high as $112 \ \mu Wm^{-1}K^{-2}$ (Park et al. 2014). Moreover, H_2SO_4 has proven to be the most effective solvent for conductivity enhancement. After treatment with H_2SO_4, the conductivity of PEDOT:PSS increased and reached ~4380 S cm^{-1} (Kim et al. 2014). These studies suggest that H_2SO_4 is a promising candidate to replace the organic solvents commonly used in the coagulation bath of PEDOT:PSS fiber spinning.

Razal et al. (2019) reported a facile one-step method to produce highly conducting PEDOT:PSS fibers that effectively removes the insulating PSS component within seconds, thereby enabling their fabrication in a fast one-step process. The highest electrical conductivity is 3828 S cm^{-1} for a ~15-micron fiber. PEDOT:PSS fibers were fabricated at room temperature using a custom-made bottom-up wet-spinning apparatus (Fig. 3). Typically, the spinning formulation was loaded into a syringe

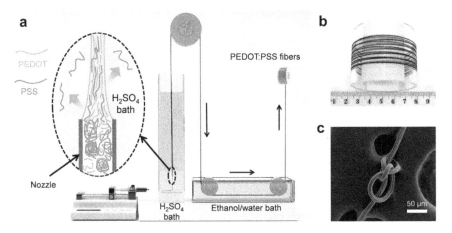

Fig. 3 **a** Schematic illustration of the modified setup used in wet-spinning PEDOT:PSS fibers. Inset shows the schematic illustration of the alignment of PEDOT:PSS domains during fiber formation and the outward diffusion of excess PSS to H_2SO_4 coagulation bath. **b** Sample spool containing over 100 meters of continuous PEDOT:PSS fiber. **c** Representative scanning electron microscopy (SEM) image of a PEDOT:PSS showing great flexibility through tight knots. Reproduced from Ref. (Zhang et al. 2019) with permission from The Royal Society of Chemistry

and then mounted on a syringe pump. The formulation was extruded into H_2SO_4 coagulation bath through a blunt needle from the bottom of the coagulation bath. The fiber was passed through a washing bath of ethanol/water mixture to remove H_2SO_4 from fiber. PEDOT:PSS fibers were dried under tension using a halogen light as heating source and collected by winding onto a spool. Multifilament PEDOT:PSS fibers were produced using the same method by replacing the single needle with a 7-hole spinneret. These fibers can withstand mechanical mistreatment. Testing has been conducted in various environments including sonication and exposure to boiling water for extended periods. The study on the mechanism of conductivity enhancement shows that the spinning method efficiently removes the PSS component during fiber formation and improves orientation of the PEDOT chains, which facilitates efficient intramolecular and intermolecular charge transport, leading to the enhancement in electrical properties.

Xu and his co-workers (Liu et al. 2018) reported that a highly conductive p-type PEDOT:PSS fiber was produced by gelation process, which was three orders of magnitude higher than that of previous hydrogel fibers. In brief, a certain volume of PEDOT:PSS suspension (Clevios, PH1000) containing 0.05 M H_2SO_4 was sealed in a poly(tetrafluoroethylene) capillary (inner diameter = 1.0 mm), which was kept at 90 °C for 3.0 h. To optimize the synthesis procedure, vacuum evaporation was employed to condense the solid content of PEDOT:PSS in solution. After concentration, 5.0 vol.% dimethyl sulfoxide (DMSO) was added into the as-condensed solution following ultrasonic treatment for 10 min. After the formation of hydrogel, the as-obtained fibers were released into isopropyl alcohol (IPA), ethyl alcohol, or acetone bath from capillary by pumping air. They were cleaned by deionized water repeatedly and dried under 60 °C in vacuum for 1.0 h. Then, the resultant fibers were immersed in EG and DMSO (Fig. 4a–c). Finally, the as-prepared fibers were washed with deionized water again for several times and dried at 80 °C for 30 min. Surprisingly, a posttreatment with organic solvents such as EG and DMSO tripled their electrical conductivity with an only 5% decreased Seebeck coefficient, consequently

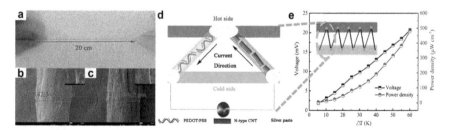

Fig. 4 **a** Digital photographs of as-fabricated PEDOT:PSS fiber. SEM images of PEDOT:PSS fibers **b** before and **c** after ethylene glycol (EG) posttreatment. **d** Schematic diagram for TE device and **e** the output voltage and power density of five couples of legs at various temperature differences ranging from 5 to 60 K. The inset is a scheme for the fiber device consisting of five pairs of p-n legs consisting of EG-treated p-type PEDOT: PSS fibers and n-type CNT fibers. Reproduced with permission. (Liu et al. 2018) Copyright 2017, American Chemical Society

leading to an optimized thermoelectric power factor. Furthermore, they assembled a p-n-type thermoelectric device connecting five pairs of p-type PEDOT:PSS fibers and n-type carbon nanotube fibers (Fig. 4d). This fiber-based device displayed an acceptable output voltage of 20.7 mV and a power density of 481.2 $\mu W \cdot cm^{-2}$ with a temperature difference of ~60 K (Fig. 4e), which might pave the way for the development of organic TE fibers for wearable energy harvesting.

A prerequisite for TE fibers is wash and wear resistance, which is necessary to withstand both textile manufacture and daily use. Integration of the conducting organic material into the fiber, which can be achieved through either incorporation during fiber spinning or impregnation of an existing fiber, is likely to result in a more durable functionalization because a surface coating is prone to delamination during bending and abrasion during wear. The use of natural materials such as polysaccharide-based cotton and protein-based silk offer a number of advantages because they can be derived from renewable sources, are biocompatible, and, being traditional textile materials, are readily compatible with existing manufacturing routines. A lasting wear and wash resistant finish can be realized through functionalization with materials that carry ionizable groups, which is a common feature of acid dyes that are used by the textile industry to color natural fibers such as wool, mohair, and silk. Similarly, water-soluble conjugated polyelectrolytes that carry negatively charged acetate or sulfonate counterions can tightly bind to fiber through electrostatic interactions with cationic sites, provided that the pH is adjusted during the dyeing process to ensure polypeptides with a net positive charge (Müller et al. 2011). Ryan et al. demonstrated that PEDOT:PSS can be used akin to an acid dye. Impregnation of silk from the silkworm Bombyx mori allows realizing durable high-modulus fibers with a bulk electrical conductivity of 14 S cm^{-1} (Fig. 5). A high Young's modulus of approximately 2 GPa combined with a robust and scalable dyeing process results in up to 40 m long fibers that maintain their electrical properties when experiencing repeated bending stress as well as mechanical wear during sewing. Moreover, a high degree of ambient stability is paired with the ability to withstand both machine washing and dry cleaning. To illustrate the potential use for e-textile applications, they realized an in-plane thermoelectric module that comprises 26 p-type legs by embroidery of dyed silk fibers onto a piece of felted wool fabric (Ryan et al. 2017).

- **Inorganic TE fibers**

Conventional inorganic TE materials are rigid and brittle so that they are hard to be directly manufactured into flexible and freestanding fibers. Together with the insoluble feature, physical depositing methods such as thermal evaporation and magnetron sputtering were used for coating inorganic TE materials onto flexible fibers at first. In 2008, Yadav et al. prepared TE fiber by coating Ni and Ag layers onto flexible silica fibers via thermal evaporation and demonstrated power generation performance of the fiber-based TE generator (Yadav et al. 2008). Adjacent stripes of nickel and silver are deposited on one side of a silica fiber of diameter 710 μm by thermal evaporation under a vacuum of 5×10^{-7} Torr, as shown in Fig. 6a, b. First, a silver layer of thickness 120 nm is deposited through a mask; the mask is then moved

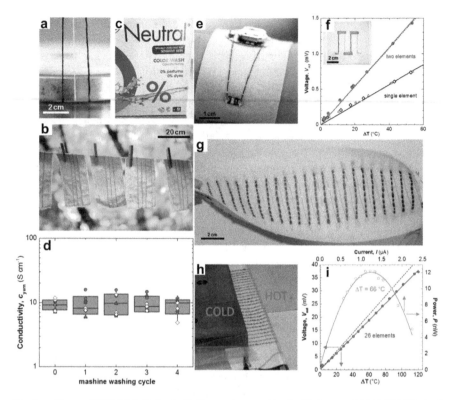

Fig. 5 **a** Neat and PEDOT:PSS dyed silk fibers submerged in distilled water; **b** PEDOT:PSS dyed silk fibers were sewn onto cotton fabrics and washed in a household Electrolux washing machine; **c** using Neutral COLOR WASH detergent; **d** electrical conductivity of PEDOT:PSS dyed silk fibers as a function of washing cycle (MeOH($\Diamond$), EG (blue triangle), DMSO (red circle)); **e** Image of an LED connected with PEDOT:PSS dyed silk fibers to a battery (embroidered on felted wool fabric); **f** in-plane textile TE device with two ~5 cm long p-type legs of PEDOT:PSS dyed silk fibers (10 fibers per leg) connected with silver wire (inset) and output voltage as a function of applied temperature gradient (single element (black line), two elements (red line), calculated output (dotted line)); **g** in-plane textile TE device with 26 p-type legs prepared in a similar way as the two leg module, and **h** the device in use, suspended between a hot and cold temperature reservoir; **i** electrical measurements of the 26 leg module, showing output voltage a function of temperature gradient (all elements (red line), calculated (dotted line) $V_{out}/\Delta T$; see (**f**)) and power output as a function of measured current. Reproduced with permission. (Ryan et al. 2017) Copyright 2017, American Chemical Society

to cover the deposited silver, and a nickel layer of thickness 120 nm is deposited. The junctions are formed by a 0.5 mm overlap between the two metals. To test the thermoelectric properties of the fiber, they heated alternate junctions, resulting in a temperature gradient that mimics the heating profile of the fiber inside a woven pattern which has a temperature gradient normal to the surface. As shown in Fig. 6c, they used electrically insulated posts with heating wire wrapped around them as heat

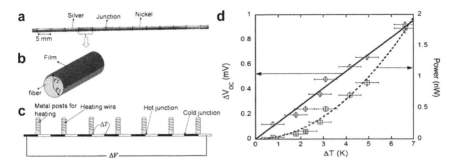

Fig. 6 **a** Picture of striped thin film TE fiber made with thermal evaporation of nickel and silver. **b** Illustration of fiber with thin film deposited on one side. **c** Schematic of experimental setup for applying a temperature gradient and measuring the induced open circuit voltage. **d** Net thermal voltage and maximum power output as a function of temperature applied for 7 couples. Reproduced with permission. (Yadav et al. 2008) Copyright 2007, Elsevier

sources, bringing alternate fiber junctions in contact with them. The unheated junctions were exposed to ambient (22 °C) and lose heat due to convection. Temperature differences between neighboring junctions are measured using micro thermocouples. Open circuit voltage versus temperature was measured, as well as the induced voltage across a matched load resistor. An open circuit voltage of 19.6 μV/K per thermocouple junction was measured for Ni-Ag fiber, and a maximum power of 2 nW for 7 couples at $\Delta T = 6.6$ K was measured (Fig. 6d).

Wang and his co-workers (Ren et al. 2016) explored the feasibility of a powder-processing-based method in the fabrication of inorganic Bi_2Te_3 TE fibers with large length-to-area ratio as shown in Fig. 7a, b. Powders were milled from Bi_2Te_3-based bulk materials and then mixed with a thermoplastic resin dissolved in an organic solvent. Through an extrusion process, flexible, continuous fibers with sub-millimeter diameters were formed. The polymer phase was then removed by sintering. Electrical resistivity of the sintered TE fibers was strongly affected by the sintering temperature such that increasing the sintering temperature led to decreased resistivity, which might be related to the residual porosity and grain boundary contamination. Electrical resistivity of the sintered fibers was as low as 14.1 mΩ·cm for p-type fibers sintered at 590 °C and 153 mΩ·cm for n-type fibers sintered at 580 °C, however, which were much higher than the corresponding bulk materials. The Seebeck coefficient of the fibers, on the other hand, exhibited comparable values to the bulk materials, implying that the Seebeck coefficient was not significantly affected by the sintering. The Seebeck coefficient is 179 μV/K and -207 μV/K for p-type $Bi_{0.5}Sb_{1.5}Te_3$ and n-type $Bi_2Te_{2.7}Se_{0.3}$ fiber, resulting the power factor of 1.7 μW/cmK^2 and 0.3 μW/cmK^2 for p-type fiber and for n-type fiber, respectively. Figure 7c is a picture of a prototype miniature uni-couple fabricated from the sintered fibers. High output power could be achieved by connecting multiple uni-couples either in series (with higher output voltage) or in parallel (with higher output current). For example, Fig. 7d shows the performance of two uni-couples connected in series, where the maximum output power was increased to 18 nW and the corresponding output voltage was

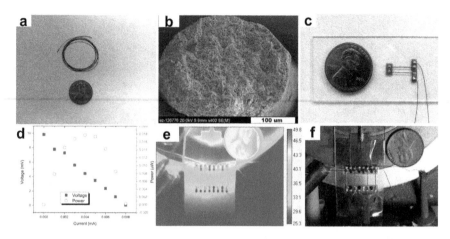

Fig. 7 **a** An as-extruded fiber, **b** the cross section of a p-type fiber sintered at 590 °C in vacuum for 4 h. **c** Prototype miniature uni-couple fabricated from sintered fibers and **d** I–V curve and power output of a module consisting of two uni-couples connected in series. **e** Infrared image of a prototype module consisting of two uni-couples showing the temperature distribution in the module at equilibrium, and **f** its corresponding digital picture. Reproduced with permission. (Ren et al. 2016) Copyright 2015, Springer Nature

increased to 4.8 mV. Figure 7e is the infrared image of this two uni-couple module showing the temperature distribution at thermal equilibrium; the hot size and cold side temperatures were approximately 42 and 30 °C, respectively. The corresponding picture of this module is shown in Fig. 7f. Prototype modules consisting of up to 4 uni-couples (8 fiber elements) were constructed and the power output and the voltage (or the current) were quadrupled compared to single uni-couples. Due to the large electrical resistivity, these prototype uni-couples and modules fabricated from such fibers showed poor performance in terms of power output. Therefore, further studies are highly desirable to seek improved sintering conditions to optimize the fiber compositions and to reduce their porosity.

Graphene fibers have been extensive researched in sensing devices, energy harvesters and supercapacitor on account of its superior mechanical properties and high electrical conductivity. In contrast with traditional wet- or dry-spinning techniques, self-assembly in tube module is a more accessible method to prepare graphene fibers (Xu et al. 2020). Furthermore, the semiconductor behavior of graphene is easily to be converted from p-type to n-type via doping of electron donating molecules, providing us an inspiration to prepare an integral p-n connected graphene TE fiber. Lin et al. (2019) obtained an integral p-n connected all-graphene fiber without any additional conductive adhesive and assembled a wearable fiber-based TE energy-harvesting device. As shown in Fig. 8, graphene oxide (GO) dispersion was prepared by oxidation of natural graphite powder via modified Hummers' method. Homogeneously mixed GO and ascorbic acid (AA) hybrid solution was sealed in polytetrafluoroethylene (PTFE) tube. The GO fiber was kept at 80 °C for 2 h. The GO fiber was further reduced with $N_2H_4 \cdot H_2O$ to achieve a higher TE performance, which

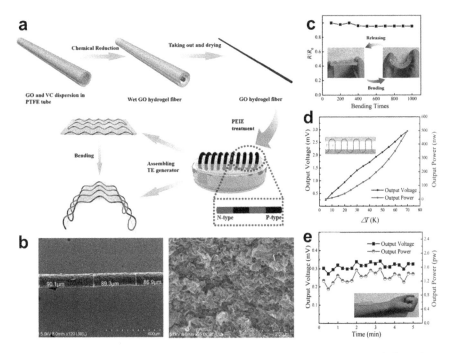

Fig. 8 **a** Schematic illustration of the fabrication process of an integral p-n connected all-graphene fiber. **b** SEM images of GO fibers and its surface morphology. **c** Digital photographs of TE generators. **d** The output voltage and output power of p-n connected TE fibers at various temperature differences ranging from 5 to 70 K. **e** The stability with time for output performance of TE generators. Reproduced with permission. (Lin et al. 2019) Copyright 2019, Elsevier

was noted as p-GO fiber. Then, to obtain an integral p-n connected all-graphene fiber, as-prepared GO fiber was rolled onto a plastic rod and partially immersed into polyethyleneimine ethoxylated (PEIE) solution, which was noted as p-n-GO fiber. Finally, the integral p-n connected all-graphene fiber was integrated into polydimethylsiloxane (PDMS) to achieve a flexible fiber-based TE device. Young's modulus of GO fiber was calculated as 18.4 GPa. After treated with $N_2H_4 \cdot H_2O$, Young's modulus increased about 40% to 25.8 GPa for p-GO fiber, which might be ascribed to the reductant-induced enhancement of π-π stacking interactions between graphene sheets. For n-GO fiber, Young's modulus decreased to 12.8 GPa compared to that of p-GO fiber. They assembled a wearable TE energy-harvesting device via integrating a p-n-GO fiber consisting of 20 pairs p-n legs into flexible PDMS substrate (Fig. 8c). Mechanical reliability is an important indicator for wearable device. They tested the electrical resistance variation of this fiber-based device for about 1000 times of bending-releasing circle. The initial electrical resistance was measured as about 75 kΩ and the electrical resistance maintained almost the same with about less than 5% variation, indicating a good mechanical reliability of this device (Fig. 8c). The output voltage and power of this fiber-based device consisting of 4 pairs of p-n legs were calculated under the temperature gradient ranging from 0 to 70 K. As shown in Fig. 8d, one can see that the maximal output voltage and output power were calculated 0.75 mV and 124.1 nW per p-n leg under 70 K temperature difference. Furthermore, Fig. 8e recorded a practical measurement results of the output voltage and power for several minutes at temperature difference between human body and air temperature ($\Delta T = 10 \pm 0.5$ K), indicating a stable TE energy-harvesting performance.

Most CNT-based TE generators have been fabricated via alternative stacking/or printing of p- and n-type CNT films/or inks with metal deposition as an electrode between p- and n-type units. However, in the case of highly electrically conductive CNT, the contact resistance between metal electrode and CNT is higher than the resistance of metal or CNT itself, thus decreasing the output power density of flexible TE generators. Choi et al. (2017) reported a robust CNT fiber with excellent electrical conductivity of 3147 S cm^{-1} which is due to increased longitudinal carrier mobility derived from highly aligned structure. On the basis of highly conductive CNT fiber, they fabricated an all-carbon TE generator with superior TE performance. Figure 9a shows a schematic of flexible TE generator based on CNT fiber without metal electrodes. The synthesized CNT fiber was wound onto a flexible supporting unit, and alternatively doped into n- and p-types using polyethylenimine (PEI) and FeCl$_3$ solutions, respectively, with undoped material between the two doped regions. The CNT fiber between the doped regions was used as electrodes to minimize the circuit resistance because it has excellent electrical conductivity. The CNT fiber used in TEG fabrication was continuously produced by direct spinning after synthesis. Thousands of individual double-walled CNTs compose a CNT fiber of 30 μm in diameter as shown in the scanning electron microscopy (SEM) images in Fig. 9b. Highly integrated CNT fibers have a density of ∼1.0 g/cm3 and high specific strength of ∼1 GPa/(g·cm^{-3}). Furthermore, the CNT fiber has excellent specific electrical conductivity on the order of 103 S·cm^2/g, which is comparable to some metals (Fig. 1c). On the basis of mechanical strength and shape advantage of CNT fiber for modular fabrication, various types of flexible TE generators were prepared. A prototype flexible TE generator with 60 pairs of n- and p-doped CNT fiber shows a maximum power density of 10.85 and 697 μW/g at temperature differences (ΔT) of 5 and 40 K, respectively, which are the highest values among reported TE generators based on flexible materials (Fig. 9d–f). Furthermore, the flexible TE generator based on 240 pairs of n- and p-doped CNT fiber can successfully power a red LED using an energy-harvesting circuit at $\Delta T = 50$ K (Fig. 9g–i).

- **Organic/inorganic composite TE fibers**

As discussed in the previous sections, the TE properties of conducting polymers have been significantly improved; however, they are still much lower than those of inorganic thermoelectric materials. However, for flexible TE devices, the use of inorganic materials is in general hampered by their inherent rigidity. Overcoming

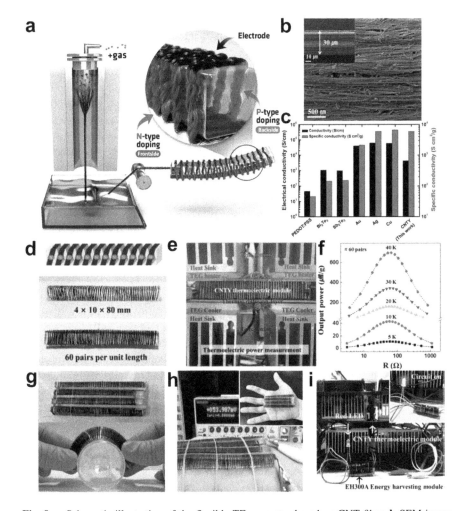

Fig. 9 a Schematic illustration of the flexible TE generator based on CNT fiber. **b** SEM image of the CNT fiber. **c** Electrical conductivity and specific conductivity of the CNT fiber and typical conducting materials. **d** Photograph of fiber-based TE generator with 60 PN pairs. **e** Photograph of power measurement system. **f** Output power density of fiber-based TE generator with 60 PN pairs as a function of load resistance. **g** Photographs showing the practical application of flexible fiber-based TE generator. **h** Output voltage of fiber-based TE generator obtained from the temperature difference between body heat and atmosphere. **i** Photograph showing a red LED powered using fiber-based TE generator at $\Delta T = 50$ K. Reproduced with permission. (Choi et al. 2017) Copyright 2017, American Chemical Society

this issue is to prepare organic/inorganic composite TE fibers. The most obvious manners are to deposit inorganic materials onto a flexible organic fiber substrate, or embed inorganic materials into a flexible organic fiber substrate.

Lee and his co-workers proposed the fabrication of flexible, woven, and knitted textiles based on electrospun organic polymer nanofiber cores that are coated with inorganic n- and p-type semiconductor sheaths (Bi_2Te_3 and Sb_2Te_3, respectively) and twisted into flexible TE fibers (Lee et al. 2016). Separate n-type and p-type fibers were prepared by depositing Bi_2Te_3 and Sb_2Te_3, respectively. Highly aligned sheets of polyacrylonitrile (PAN) nanofibers, with average nanofiber diameter of $\approx$ 600 nm, were electrospun on two parallel wire collectors (Fig. 10a). A PAN solution in the syringe is extruded through the metal needle. A high voltage supply is connected to the spinning tip and to two counter electrodes, so that highly aligned PAN nanofiber sheets are collected. Tiger-type fibers were fabricated by first sputtering alternating strips of Sb_2Te_3 and Bi_2Te_3 (Fig. 10b) on both sides of the electrospun PAN nanofiber sheet using a stencil mask which produces sheath-core nanofiber structures containing $\approx$ 50% by volume of active material and then interconnecting these TE strips using narrow sputtered gold strips. The thicknesses of these coatings were 110 nm for the TE materials and 75 nm for the gold. The thereby coated nanofiber sheets were then twisted into yarns using an electric motor, typically using a twist density of $\approx$ 630 turns per meter, relative to the initial sheet length, followed by a thermal annealing at 200 °C to crystallize the semiconductors, resulting in a strong, flexible sheath-core TE fiber. Figure 10c demonstrates that the initial tiger structure is retained during twist insertion and subsequent complete fiber untwist. The power output of 3D fiber-based TE generators assembled in different fabric structures was also investigated. Figure 10d–i displays the schematic illustrations and photographs of zigzag-stitch, garter-stitch, and plain-weave TE generators using p-type Sb_2Te_3 fibers and n-type Bi_2Te_3 fibers. At temperature difference of 55 K, the power output per area and per couple were 0.11 W m^{-2} and 0.24 μW for the zigzag-stitch TE generator (Fig. 10j), and 0.09 W m^{-2} and 0.21 μW for the garter-stitch TE generator (Fig. 10k), respectively, while the plain-weave TE generator (Fig. 10l) presented the highest power output per area of 0.62 W m^{-2} and 1.01 μW per couple, respectively. Besides, the power output was proportional to ΔT^2. When the temperature difference increased to 200 K, the generated power greatly increased to 8.56 W m^{-2} and 14.1 μW per couple, respectively.

Inversely, Sun et al. (2020) presented and developed a π-type inorganic CNT fiber-based TE module using organic PEDOT:PSS and oleamine doping combined with electrospray technology. Figure 11a schematically exhibited the fabrication process of TE modules. First, CNT fiber was p-hybridized by dipping into a commercial PEDOT:PSS solution. Subsequently, n-type CNT fiber was achieved in an equal interval using polypropylene mask by oleamine doping combined with electrospray technology. Therefore, a TE fiber with alternatively doped n- or p- segment at the distance of (L-4 mm)/2 was formed, where L (mm) is the length of p-n repeat unit of the fiber. The 2 mm long undoped section is electrically conducting and treated as

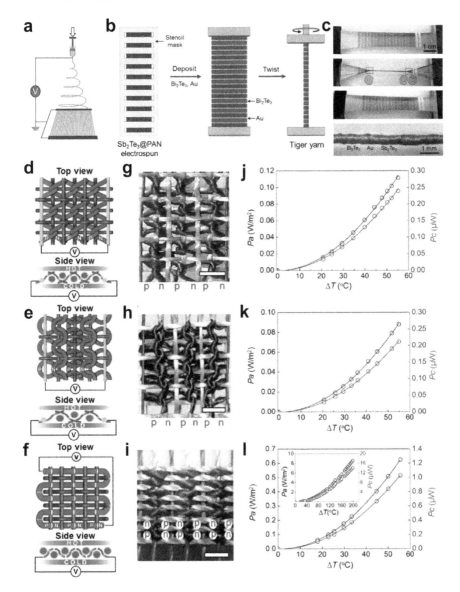

Fig. 10 a Illustration of the electrospinning process used to make PAN nanofiber sheets for conversion into a yarn. **b** Illustration of the stencil mask-based method used to provide alternating p-Au-n segments for tiger fibers by sequentially sputtering Sb_2Te_3-gold-Bi_2Te_3 on both sides of a PAN nanofiber sheet. **c** Illustration of the twist spinning process used to convert the nanofiber sheet into a tiger fiber. **d–f** Schematic illustrations and **g–i** corresponding photographs of the fabricated zigzag-stitch, garter-stitch, and plain-weave fiber-based TE generators. The scale bars are 2 mm. **j–l** The power output per textile area (Pa) and per TE couple (Pc) of the fabricated zigzag-stitch, garter-stitch, and plain-weave fiber-based TE generators as a function of temperature difference. Reproduced with permission. (Lee et al. 2016) Copyright 2016, Wiley-VCH

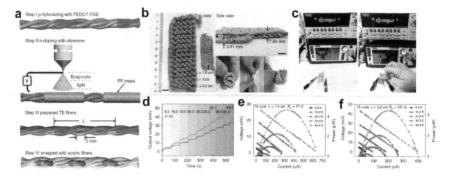

Fig. 11 **a** p-Hybridized CNT and n-Doped CNT TE fibers fabricated by an electrospray technique wrapped with acrylic fibers. **b** Photograph of TE generator composed of 15 units (3 × 5) with different TE repeat length L. Scale bars, 1 cm. **c** The output voltage display of TE generator (15 units) with repeat length of 16 mm before and after fingertip touching. **d** The output voltage contrast of the two TE generators at various steady temperature differences given by Peltier elements. **e, f** Power output of the TE generators with L of 16 mm/32 mm at various steady temperature differences. R_0 is the internal resistance of TE generators. Reproduced with permission. (Sun et al. 2020) Copyright 2020, Springer Nature

an electrical interconnection between the two doped sections. Furthermore, to avoid a short circuit, the doped CNT fiber was wrapped with acrylic fibers using a cover spun technique, except the electrode segment, which was exposed to maximize the thermal contact between the TE generator and the human body.

A high tightness δ of 10.6 is given and the TE repeat length L of TE generator with the same amount of units was adjusted from 32 to 16 mm to optimize power density. A dramatically decreased occupied area can be seen in Fig. 11b. Besides, the thickness of the TE generator reduced from 7.8 to 3.8 mm, but the ~40° standing angles of TE legs were maintained. Figure 11b also shows the ability of the prepared TEG to adapt to various loading conditions, for example, bending, twisting, and folding. Furthermore, a 3.5 mV voltage can be immediately detected when human fingertip touches the small TE generator as shown in Fig. 11c. The output voltage at various steady temperature differences are displayed in Fig. 11d. When a through thickness temperature difference is applied, an obvious Seebeck voltage is immediately created. After reducing the repeat length to 16 mm, the voltage decreases by ~26%, in accordance with the positive dependence of temperature difference ΔT across TE legs on repeat length L. Conversely, the power increases after reducing the repeat length to 16 mm as shown in Fig. 11e, f. Using 44.4 K temperature difference, the power enhances from 4.40 to 4.64 µW, accompanying an short circuit current increasing from ~450 to ~700 µA. A 4-fold rise in power density is obtained after reducing the repeat length from 32 to 16 mm. Giving 44.4 K temperature difference, the power density increases from 14 µWm^{-2} with $L = 32$ mm to 69 µWm^{-2} with $L = 16$ mm, which is the highest output reported for a flexible organic TE generator.

Recently, thermal drawing technique was developed to produce inorganic TE fibers such as $Bi_{0.5}Sb_{1.5}Te_3$, Bi_2Se_3, $(Te_{85}Se_{15})_{45}As_{30}Cu_{25}$, In_4Se_3, and Bi_2Te_3 fibers

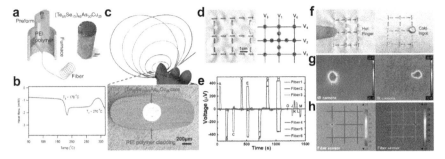

Fig. 12 **a** Schematic diagram of the drawing process of a TE fiber from the semiconducting glass rod and PEI polymer. **b** Glass transition temperature (T_g) and crystallization temperature (T_x) of the semiconducting glass measured by differential scanning calorimeter (DSC). **c** Single TE fiber with good flexibility and the optical microscope image of fiber cross section. **d** Photograph of the fabricated 3 × 3 TE fiber array embedded into a flexible fabric and the marked points for finger touching. **e** Corresponding output voltages of the TE fiber array when touching the marked points. **f** Finger heating (33 °C) and ingot cooling (22 °C) on the TE fiber array. Corresponding **g** thermal images recorded by IR camera and **h** reconstructed thermal maps using the data of TE fiber. Reproduced with permission. (Zhang et al. 2019) Copyright 2019, American Chemical Society

(Loke et al. 2020; Zhang et al. 2017, 2019). The process required a cladding material with glass transition temperature slightly higher than the melting point of TE materials. The TE fibers fabricated by this method usually had high TE properties but general mechanical flexibility. For example, Zhang et al. (2019) prepared flexible $(Te_{85}Se_{15})_{45}As_{30}Cu_{25}$ fiber was prepared by thermal drawing. Figure 1a schematically shows the preform-based fabrication of TE fibers. A semiconducting glass rod of $(Te_{85}Se_{15})_{45}As_{30}Cu_{25}$ shown in the inset of Fig. 12a is first synthesized by a standard sealed-ampoule technique (Lucas et al. 2013), and this semiconducting glass offers two superior properties in constructing flexible thermal sensors. One is that its Seebeck coefficient is large enough to ensure the sensitivity of thermal sensing. The other is that its glass transition temperature (T_g) is 170 °C and the crystallization temperature (T_c) is 270 °C, as measured in Fig. 12b, offering a wide glass stability range of 100 °C, which makes it particularly suitable for the thermal fiber drawing process with the polymer matrix. Then, a hollow rectangle-shaped PEI polymer tube with a softening temperature of 217 °C is used to support and confine $(Te_{85}Se_{15})_{45}As_{30}Cu_{25}$, while providing excellent mechanical properties. The synthesized semiconducting glass is reshaped and filled in the hollow PEI block, and then the entire assembly is thermally sealed in a vacuum oven to form an all-solid macroscopic preform. The preform is further loaded in a fiber draw tower with a narrow heating zone and thermally drawn at the temperature of 250 °C, where the viscosity of the PEI cladding is much larger than that of the TE glass core, thus supporting the whole structure and yielding hundreds of meters of TE fibers. Figure 12c shows the photograph of the resulting TE fiber and its optical microscope image of the cross section. Owing to the presence of polymer cladding, the resulting fiber exhibits high mechanical flexibility to be easily bent, coiled, and woven. The continuous semiconducting glass core and polymer cladding are well maintained in a round and rectangular shape, respectively, and the interfaces between the core and cladding are clearly defined. The required TE fiber dimension can be precisely controlled via adjusting the feeding speed of the preform and the drawing speed of the fiber.

The fiber displayed not only high TE power factor but also excellent thermal sensing performance, as illustrated in Fig. 12d–h. When a temperature difference was applied between the two fiber ends, thermoelectric voltage was generated because of the whole diffusion. The single TE fiber-based thermal sensor showed temperature resolution higher than 0.05 °C. Additionally, it was capable of localizing the position of thermal source, because the output voltages were different when touching different positions along the fiber. More importantly, the TE fiber showed good mechanical flexibility. The bending radius was only ~2.5 mm even for the fiber with thickness of 0.6 mm. Nonetheless, it was hard to simultaneously detecting the temperature and position of thermal source in the single TE fiber-based thermal sensor. The authors further assembled a 3×3 TE fiber array and embedded it into a flexible fabric (Fig. 12d). Figure 12e recorded the output voltages of the TE fiber array when finger touching the marked points. The distinguishable voltage signals enabled the simultaneous detection of the temperature and position of thermal source. Figure 12f gave two examples for temperature and position identification, one is finger heating (33 °C) and another is ingot cooling (22 °C). The corresponding thermal images recorded by IR camera and reconstructed thermal maps using the data of TE fiber array were shown in Fig. 12g, h, respectively.

Similarly, Zhang et al. (2017) prepared flexible p-type $Bi_{0.5}Sb_{1.5}Te_3$ and n-type Bi_2Se_3 fibers by thermal drawing and assembled them into textile to construct TE generator (Fig. 13a-c). The prepared p-type $Bi_{0.5}Sb_{1.5}Te_3$ fiber exhibits power factor of 3520 $\mu Wm^{-1}K^{-2}$ and n-type Bi_2Se_3 fiber with power factor of 650 $\mu Wm^{-1}K^{-2}$, respectively. Due to the p-type $Bi_{0.5}Sb_{1.5}Te_3$ with the melting point of 873 K and n-type Bi_2Se_3 with the melting point of 983 K as the TE fiber cores, so the borosilicate glass tube with the glass transition temperature of 1100 K is the best choice for the cladding material, as schematically shown in Fig. 13a. To prevent oxidation of the inserted TE rod during the drawing process, the empty space inside the glass tube is filled up with borosilicate glass rods, and the entire assembly is sealed from both ends under vacuum. The preform is then loaded in a fiber draw tower, fed into the furnace, and thermally drawn at the temperature of 1323 K, where the viscosity of the glass cladding is large enough to offer a structural support to the liquid TE cores, thus yielding hundreds of meters of continuous TE fibers. A polymer coating layer with the thickness of 50–150 μm depending on the resulting fiber diameter is then applied to the fiber outer surface for further mechanical protection. Figure 13b shows the photograph of a single bent TE fiber with the length of one meter and its SEM image of cross section. The fiber possesses good circularity and a clean interface between the concentric TE core and the glass cladding. The diameters of TE core and glass cladding are 33 μm and 400 μm, respectively, which has nearly the same diameter ratio of the TE rod (0.8 mm) and glass tube (10 mm) in the original preform, proving that such a thermal drawing is a proportional size-reduction process.

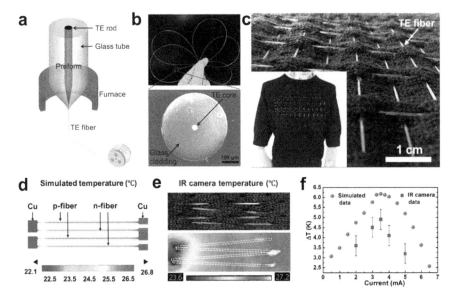

Fig. 13 a Schematic diagram of the thermal drawing process. **b** Single TE fiber with a length of one meter showing a good flexibility and its cross-sectional SEM image. **c** TE fibers are woven into a large-area fabric to construct a wearable TE device. **d** Simulated temperature profile of TE cooling with two pairs of p-n TE fibers at an input current of 2 mA by finite element modeling. **e** Corresponding experimental temperature profile recorded by IR camera. **f** Simulated and experimental temperature differences as a function of input current. Reproduced with permission. (Zhang et al. 2017) Copyright 2017, Elsevier

To engage the TE fibers for flexible and wearable electronics on large areas, they constructed the TE fibers into a two-dimensional fabric (Fig. 13c). These fibers possess many extraordinary characteristics such as thin, lightweight, and flexible to deform under small external force, so the TE fiber woven fabrics can be sufficiently sparse to maintain breathability of the covered areas, which is a critical requirement of long-term comfort around curvilinear human skins, and can be used as a TE cooler to cool the human body via a low impressed current. The cooling performance of TEG with two pairs of TE fibers was evaluated by finite element modeling and experimental characterizations. In Fig. 13d, the simulated temperature reduction was 22.1 °C with the temperature difference of 4.7 °C when the hot side was fixed at 26.8 °C under an input current of 2 mA. The measured temperature reduction was 23.6 °C with the temperature difference of 3.6 °C recorded by IR camera (Fig. 13e). The difference between simulation and test results might be induced by the measurement accuracy of IR camera as well as the unoptimized condition of TE generator. Figure 13f displays both the simulated and experimental temperature differences as a function of input current. Maximum temperature reduction of ~5 °C was obtained when the input current was 3.5 mA.

3   Conclusion

In this chapter, an overview and review of state-of-the-art flexible TE fiber materials and devices were presented with respect to the materials mechanism, fabrication methods, device structures, and application performance. Organic TE fibers such as PEDOT;PSS are flexible with high electrical conductivity and processed at low or room temperature via solution routes (wet-spinning, electrospinning, and dyeing process). Inorganic TE fibers fabricated by coating, extruding, and printing approaches have relatively high performance, but the high cost, limited flexibility, and complicated fabrication limit their development. However, organic/inorganic composite TE fibers fabricated by surface modification (physical/chemical deposition) or embedding approaches (thermal drawing) simultaneously have inorganic semiconductor-like high TE performance and organic material-like mechanical flexibility, which shows great potentials to power and refrigerate wearable electronic devices.

Although flexible TE fibers exhibit enormous application potential, TE fiber materials and devices in the early stage of development. The studies on TE generators made up of TE fibers are very few and the device performance is much lower than that of bulk TE materials based TE generators. To achieve more advancements of TE fiber, the performance and stability of the fiber should be firstly improved significantly. Then, the effects such as the device structure resulted from different textile techniques and the size of TE legs on the device performance of fiber-based TE generators should be further explored. Moreover, instruments to measure the thermal conductivity, Seebeck coefficient, and electrical conductivity of the TE fiber are homemade in most cases and lack of standardized measurement at present. Finally, current fabrication methods are mostly limited to the laboratory and are difficult to realize in mass production, which should be further explored. In a word, TE fibers have presented promising applications in smart wearable electronics, and there is huge room for the development and practical applications of fiber-based TE generators.

References

C. Chang et al., Science **360**, 778 (2018)

Y. Chen, Y. Zhao, Z. Liang, Energy Environ. Sci. **8**, 401 (2015)

J. Choi et al., ACS Nano **11**, 7608 (2017)

G.B. Delkumburewatte, T. Dias, J. Textile Inst. **103**, 483 (2012)

M.S. Dresselhaus, G. Chen, M.Y. Tang, R.G. Yang, H. Lee, D.Z. Wang, Z.F. Ren, J.-P. Fleurial, P. Gogna, Adv. Mater. **19**, 1043 (2007)

Y. Du, K.F. Cai, S. Chen, P. Cizek, T. Lin, ACS Appl. Mater. Interfaces. **6**, 5735 (2014)

Y. Du, K.F. Cai, S.Z. Shen, R. Donelsonand, J.Y. Xu, H.X. Wang, T. Lin, RSC Adv. **7**, 43737 (2017)

Y. Du, J. Xu, B. Paul, P. Eklund, Appl. Mater. Today **12**, 366 (2018)

L. Wang, K. Zhang, Energy Environ. Mater. 0, 1 (2019)

R. Fitriani, B.D. Ovik, M.C. Long, M. Barma, M.F.M. Riaz, S.M. Sabri, Said, R. Saidur, Renew. Sustain. Energy Rev. **64**, 635 (2016)

C. Gao, G. Chen, Small **14**, 1703453 (2018)

J. Gao, C. Liu, L. Miao, X. Wang, Y. Peng, Y. Chen, J. Electronic Mater. **46**, 3049 (2017)

T.C. Harman, P.J. Taylor, M.P. Walsh, B.E. LaForge, Science **297**, 2229 (2002)

W. He, G. Zhang, X. Zhang, J. Ji, G. Li, X. Zhao, Appl. Energy **143**, 1 (2015)

Q. Jin et al., Nature Mater. **18**, 62 (2019)

G.H. Kim, L. Shao, K. Zhang, K.P. Pipe, Nature Mater. **12**, 719 (2013)

N. Kim, S. Kee, S.H. Lee, B.H. Lee, Y.H. Kahng, Y.-R. Jo, B.-J. Kim, K. Lee, Adv. Mater. **26**, 2268 (2014)

S.J. Kim, H.E. Lee, H. Choi, Y. Kim, J.H. We, J.S. Shin, K.J. Lee, B.J. Cho, ACS Nano **10**, 10851 (2016)

Y. Lan, A.J. Minnich, G. Chen, Z. Ren, Adv. Functional Mater. **20**, 357 (2010)

J. Lee, B. Llerena Zambrano, J. Woo, K. Yoon, T. Lee, Adv. Mater. 32, 1902532 (2020)

J.A. Lee et al., Adv. Mater. **28**, 5038 (2016)

J.-F. Li, W.-S. Liu, L.-D. Zhao, M. Zhou, NPG Asia Mater. **2**, 152 (2010)

Y. Lin, J. Liu, X. Wang, J. Xu, P. Liu, G. Nie, C. Liu, F. Jiang, Compos. Commun. **16**, 79 (2019)

J. Liu et al., ACS Appl. Mater. Interfaces. **10**, 44033 (2018)

G. Loke, W. Yan, T. Khudiyev, G. Noel, Y. Fink, Adv. Mater. **32**, 1904911 (2020)

P. Lucas et al., J. Mater. Chem. A **1**, 8917 (2013)

C. Müller, M. Hamedi, R. Karlsson, R. Jansson, R. Marcilla, M. Hedhammar, O. Inganäs, Adv. Mater. **23**, 898 (2011)

H. Park, S.H. Lee, F.S. Kim, H.H. Choi, I.W. Cheong, J.H. Kim, J. Mater. Chem. A **2**, 6532 (2014)

S.H. Park et al., Nature Commun. **7**, 13403 (2016)

B. Paul, J. Lu, P. Eklund, ACS Appl. Mater. Interfaces. **9**, 25308 (2017)

F. Ren, P. Menchhofer, J. Kiggans, H. Wang, J. Electronic Mater. **45**, 1412 (2016)

J.D. Ryan, D.A. Mengistie, R. Gabrielsson, A. Lund, C. Müller, ACS Appl. Mater. Interfaces. **9**, 9045 (2017)

G.J. Snyder, E.S. Toberer, Nature Mater. **7**, 105 (2008)

T. Sun, B. Zhou, Q. Zheng, L. Wang, W. Jiang, G.J. Snyder, Nature Commun. **11**, 572 (2020)

R. Venkatasubramanian, E. Siivola, T. Colpitts, B. O'Quinn, Nature **413**, 597 (2001)

C. Wan et al., Nature Mater. **14**, 622 (2015)

Y. Wang, L. Yang, X.-L. Shi, X. Shi, L. Chen, M.S. Dargusch, J. Zou, Z.-G. Chen, Adv. Mater. **31**, 1807916 (2019)

L. Wang, X. Fu, J. He, X. Shi, T. Chen, P. Chen, B. Wang, H. Peng, Adv. Mater. **32**, 1901971 (2020)

T. Xu, Z. Zhang, L. Qu, Adv. Mater. **32**, 1901979 (2020)

A. Yadav, K.P. Pipe, M. Shtein, J. Power Sources **175**, 909 (2008)

C. Yang et al., Nature Commun. **8**, 16076 (2017)

L. Yang, Z.-G. Chen, M.S. Dargusch, J. Zou, Adv. Energy Mater. **8**, 1701797 (2018)

R. Yue, J. Xu, Synth. Met. **162**, 912 (2012)

T. Zhang, K. Li, J. Zhang, M. Chen, Z. Wang, S. Ma, N. Zhang, L. Wei, Nano Energy **41**, 35 (2017)

L. Zhang, S. Lin, T. Hua, B. Huang, S. Liu, X. Tao, Adv. Energy Mater. **8**, 1700524 (2018)

J. Zhang, S. Seyedin, S. Qin, P.A. Lynch, Z. Wang, W. Yang, X. Wang, J.M. Razal, J. Mater. Chem. A **7**, 6401 (2019a)

T. Zhang et al., ACS Appl. Mater. Interfaces. **11**, 2441 (2019b)

L.-D. Zhao, S.-H. Lo, Y. Zhang, H. Sun, G. Tan, C. Uher, C. Wolverton, V.P. Dravid, M.G. Kanatzidis, Nature **508**, 373 (2014)

Y.-H. Zhu, X.-Y. Yang, T. Liu, X.-B. Zhang, Adv. Mater. **32**, 1901961 (2020)

In-Fiber Breakup

Jing Zhang, Zhe Wang, and Zhixun Wang

Abstract In recent years, multifunctional multimaterial fibers based on the thermal drawing process have made considerable development, which enables various practical fiber devices with optoelectronics, photonics, acoustics, biomedicine, and energy harvesting functionalities. The future development of multifunctional fibers requests highly integrated ingenious in-fiber structures and excellent material properties. To meet these challenges, the technologies of using fluidic instabilities induced in-fiber breakup phenomena are presented, allowing us a way to modify the traditional axially invariant in-fiber structure and to achieve in-fiber material engineering. The post-drawing thermal treatment can soften the selective part of the functional fibers, induce perturbations on the interface between materials, and eventually break the continuous fiber inner structures to fabricate in-fiber functional structures. The in-fiber breakup process enables the fabrication of in-fiber structured particles by a variety of materials with a wide range of processing temperatures from 2400 to 400 K and material viscosity ratio of 10 orders. Moreover, the in-fiber breakup process provides a useful tool to form fiber-based functional devices. On the other hand, the fundamental understanding of the in-fiber breakup phenomena shall be contributing to optimizing the fiber thermal drawing process. By selecting suitable materials and suppressing the in-fiber breakup phenomena, the designed structure of the fiber preforms will be preserved in maximum. This chapter covers aspects of (1) the introduction and theory of thermal treatment induced in-fiber fluidic instabilities, (2) the in-fiber fabrications based on in-fiber breakup process, (3) potential applications of in-fiber breakup process, and (4) the future research directions.

J. Zhang (✉) · Z. Wang · Z. Wang
School of Electrical and Electronic Engineering, Nanyang Technological University,
50 Nanyang Avenue, Singapore 639798, Singapore
e-mail: jzhang048@e.ntu.edu.sg

Z. Wang
e-mail: zwang030@e.ntu.edu.sg

Z. Wang
e-mail: zhixun001@e.ntu.edu.sg

© Springer Nature Singapore Pte Ltd. 2020   199
L. Wei (ed.), *Advanced Fiber Sensing Technologies*,
Progress in Optical Science and Photonics 9,
https://doi.org/10.1007/978-981-15-5507-7_11

Keywords Thermal treatment · In-fiber breakup · Capillary instability · Structured particles · Material engineering · Optoelectronic fiber device

1 Introduction

Multifunctional multimaterial fibers have been well-developed recently based on the thermal drawing process. The preform with designed structures and selected materials can be heated and drawn into kilometers functional fiber with uniform dimensions and superior material qualities. In the past decades, this technology has successfully achieved distinctive classes of multifunctional fibers and integrated devices with different components (metals, insulators, semiconductors, glass, and polymers) and exquisite structures (single-core, multi-core, cylindrical shells, layer structures, and wires). Furthermore, plenty of smart wearable fabrics and textiles have been constructed with the electronic, optoelectronic, acoustic, and energy harvesting functional fibers that are developed from this technology.

Nevertheless, most of typical multifunctional fibers are held in an axially invariant structure without sophisticated ways to re-modify the in-fiber structures and change material properties, which greatly limit the multifunctional fibers to be further applied in wider areas. It is generally believed that the continuous development of multifunctional fibers requests more types of highly integrated ingenious structures in both of the fiber cross section axial and the length axial to achieve composite functions. Consequently, innovative approaches to precisely pattern the internal micro/nanostructure of fiber and to direct conduct in-fiber materials engineering are of the greatest importance.

In recent years, the approach of using thermal treatment induced in-fiber fluidic instabilities has been proven to be a feasible way for developing viable techniques to modify the in-fiber structure as well as to engineer the fiber materials properties. The thermal treatment process transfers multimaterial fibers from the solid state to the viscous state by heating, then induces a perturbation wave on the interfaces between materials. Interacted by the surface tension as well as the viscoelastic force between different materials, the perturbation of the fluidic instabilities further develops and leads the continuous fiber structure to breakup, which is called in-fiber breakup process. The in-fiber breakup process has been gaining increasing attention and showing plenty of promising applications, such as the generation of in-fiber spherical particles and fiber-shaped device fabrication with crystalline functional materials.

In this chapter, we will discuss the in-fiber breakup process in detail. First, the fundamental mechanism of the in-fiber fluidic instabilities is briefly reviewed. Second, we will introduce the utilizing of the in-fiber breakup phenomena as a post-drawing process to fabricate in-fiber micro/nano-particles, wires, and layered structures. In addition, we present some applications based on the in-fiber breakup process. Lastly, the current barriers and future opportunities of this technology are discussed at the end of this chapter.

2   Main Body Text

- In-fiber capillary instability
    - Introduction to capillary instability
    The classical capillary instability describes a cylindrical liquid to breakup into a train of spheres under the surface tension at the interface. This phenomenon is perhaps one of the most ubiquitous fluid instabilities. As shown in Fig. 1, when a fluid cylinder is surrounded by another kind of liquid or air, the surface tension will induce a sinusoidal perturbation at the interface of two materials and lead the cylinder to breakup into droplets in a spherical shape with the development of the perturbation wave amplitudes. This phenomenon frequently appears in daily life (Gennes et al. 2004). For example, the faucet dripping, the ink-jet printer system, and the wine tears on the glass are all typical capillary instability phenomena.
    As a common physical phenomenon, the study of capillary instability has a very long history. In 1849, Joseph Plateau proved experimentally that a vertically falling water column would breakup into some drops which have the same volume. Then, J. Plateau explained this phenomenon with an analysis of surface tension in balance. The reason for the breakup process to occur is also due to its result that can reduce the surface energy until the establishment of an equilibrium situation.
    Then, Lord Rayleigh studied the phenomenon theoretically by applying a linear stability analysis. His study showed that the breakup of a vertically falling water column needed to satisfy two conditions. Firstly, the falling water column will breakup into some drops if the length of the column exceeds its circumference.

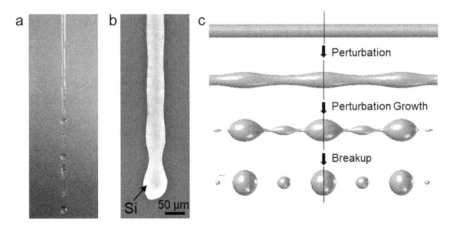

Fig. 1 In-fiber Plateau-Rayleigh capillary instability. **a** The capillary instability phenomenon in the faucet dripping (adapted from MIT lecture notes on Fluid Jets). **b** SEM image of the capillary instability perturbation in the in-fiber silicon cylinder core. **c** Simulation of the in-fiber capillary instability development based on COMSOL

Secondly, the drops result from some small disturbances that develop with time (Rayleigh 1878, 1879, 1882).

Based on these previous researches, Plateau-Rayleigh capillary instability theory, which is also called Rayleigh instability linear theory or PR linear instability theory, theoretically describes the physical phenomenon that a liquid column presents to breakup into small spheres which have the same volume. The conditions for the breakups to occur are (a) liquid column is surrounded by another kind of material, and (b) the liquid column's length is long enough (Rayleigh 1878).

Subsequently, S. Tomotika found that the viscosities' ratio between two kinds of materials, such as liquid-in-liquid or liquid-in-gas, is the critical factor of the capillary instability. At the same time, he furthermore improved Lord Rayleigh's linear instability theory. By the separation of time and space, he described that the development of small disturbance is affected by a factor named instability growth rate. This model also offers possibilities for conducting a numerical simulation to analyze the capillary instability (Tomotika 1935). Based on Tomotika's theory and related mathematic models, many additional phenomena have been inquired deeply (Kao and Hockham 1966; Gennes et al. 2004; Liang et al. 2011). It is worth to mentioning that the theory is also proved to be solid in sub-micron scale (Frolov et al. 2014).

- Thermal treatment induced in-fiber capillary instability

  Although not being found on solids, the capillary instability does take place in molten multimaterial fibers. Due to the surface tension and viscous dissipation force between the molten fiber's cladding and core (Chandrasekhar 1961), the core cylinder breaks up and forms a string of spheres that are frozen in situ upon cooling. Recent studies on in-fiber capillary instability phenomenon have promoted the development of fabrication and material engineering inside a fiber, especially for the fabrication of microspheres (Shabahang et al. 2011; Gumennik et al. 2013; Zhang et al. 2017), which is strongly related to the Plateau-Rayleigh (PR) capillary instability in a fiber. Fabrication of micro-spheres has been achieved based on fluidizing the pre-fabricated fibers either by the iso-thermal heating via furnace (Aktas et al. 2014; Khudiyev et al. 2014) or the axial thermal gradient heating via oxyhydrogen flame and laser (Gumennik et al. 2013; Zhang et al. 2017). The thermal treatment induced in-fiber breakup process has been applied in a variety of materials with a wide range of processing temperatures from 2400 to 400 K and material viscosity ratio of 10 orders.

- Instability wavelength and growth rate

  According to the PR capillary instability, for a perfect cylindrical liquid thread which is covered by another liquid, the perturbation wave occurs on the inter-face due to the competition of surface tension and viscoelastic force between two materials. The perturbation wave can be decomposed into a Fourier series with an amplitude ε and wavelength λ,

$$E^{\mathrm{surf}}(\lambda) = E_0^{\mathrm{surf}} + \frac{\pi\gamma}{2\lambda R_0}\left[(2\pi R_0)^2 - \lambda^2\right]\varepsilon^2, \tag{1}$$

where E_0^{surf} is the initial surface energy of the inner liquid cylinder, γ is the interfacial tension, and R_0 is the radius of the inner liquid cylinder. When the wavelength is smaller than $2\pi R_0$, the surface energy E^{surf} will increase which conflicts the actual situation and this perturbation are unsuitable. Instead, when the wavelength satisfies the criterion

$$\lambda > \lambda_{\mathrm{crit}} = 2\pi R_0, \tag{2}$$

the surface energy E^{surf} will decrease to minimize with the development of the perturbation wave and subjects to the capillary instability. Therefore, the perturbation with a wavelength above the critical wavelength λ_{crit} could be described as the amplitude function $\varepsilon(t, \lambda)$ showing the exponentially with time and growth rate q.

$$\varepsilon(t, \lambda) = \varepsilon(0)\exp(iqt), \tag{3}$$

The solution for the value of the perturbation growth rate q is a function of core viscosity η_{inner}, cladding viscosity η_{outer}, and wavenumber $x = ka$ (Tomotika 1935).

$$q = \frac{\gamma}{2a\eta_{\mathrm{outer}}}\Phi\left(x, \frac{\eta_{\mathrm{inner}}}{\eta_{\mathrm{outer}}}\right), \tag{4}$$

In Eq. 4, the growth factor function Φ is complicated Bessel functions depending on the wavelength x and viscosities of liquids.

- In-fiber filamentation instability

  - Introduction to filamentation instability
    The filamentation instability indicates that a viscous sheet transfer into an array of filaments (Fig. 2a), instead of particles. Differing from the PR capillary instability or the spinodal instability, the in-fiber breakups occur over some thin films and transfer them into an array of filaments instead of the particles. This fluidic instability phenomenon is identified as the filamentation instability. In this phenomenon, the fluidic instability develops along the transversal direction and instabilities along the longitudinal direction are suppressed. During the fiber thermal drawing process, the viscous sheet of the longitudinal direction is dragged by the gravity force as well as the fiber drawing force, which suppresses the development of the instability along the longitudinal direction.

  - Linear theory
    For the in-fiber filamentation instability, the curvature of the viscous sheet is negligible since the thickness of the viscous sheet is much smaller than the

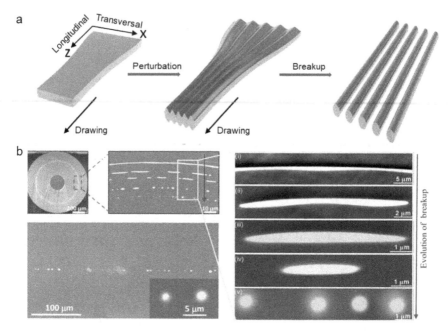

Fig. 2 Evolution and experimental results of in-fiber filamentation instability breakup. **a** Sketch of the in-fiber filamentation instability on a thermal drawing viscous sheet. **b** SEM images of the in-fiber filaments generation based on the filamentation instability breakup (Xu et al. 2019). Copyright 2019, American Physical Society. Reproduced with permission

radius. According to the linear theory of the filamentation instability during the fiber thermal drawing process (Deng et al. 2008, 2010, 2011; Xu et al. 2019), the growth rates along the longitudinal direction ($q_{\parallel}$) and the transverse direction ($q_{\perp}$) of the fiber cross section for the instability perturbation wave can be calculated according to the sheet thickness and thermal drawing speed. Here, the instability of the viscous sheet should be decided by the largest growth rates along both directions,

$$q_{\parallel}^{m} = \max\left(q_{\parallel}\left(z_{p}, k_{\parallel}\right)\right) \tag{5}$$

$$q_{\perp}^{m} = \max\left(q_{\perp}\left(z_{p}, k_{\perp}\right)\right) \tag{6}$$

where the z_{p} is the perturbation position and k is the perturbation wavenumber. Therefore, a filamentation instability phenomenon can appear when $q_{\perp}^{M} < q_{\parallel}^{M}$. Furthermore, the growth rate $q_{\perp}^{m}$ and the wavenumber $k_{\perp}^{m}$ of the perturbation can be calculated with the initial thickness H_{0} of the viscous sheet and the perturbation induced location z_{p}. The theoretical filamentation instability breakup wavelength can be described as $\lambda_{m} = 2\pi S(z_{m})/k_{\perp}^{m} S_{f}$, where the S_{f} is the attenuation rate at $z = L$.

- Utilizing in-fiber breakup for fabricating mico/nano-particles, wires, and layered structures

  - Fabrication of micro/nano-particles based on the capillary instability
    The need for scalable fabrication of spherical particles at both micro- and nanoscales is well-documented, and achieving this would enable a variety of applications ranging from microelectronics (Van Buuren et al. 1998), photonics (Wang et al. 2012), health care (Park et al. 2009), biology (Watts et al. 2013), chemical sensing (Ward and Benson 2011), to cosmetics. Indeed, the thermal treatment induced in-fiber capillary instability is excellent for fabricating micro- and nanoscale structured particles. This approach seamlessly links the full spectrum of particle radii and introduces various material systems into the production of size-homogeneous structured particles.

Iso-thermal and gradient-thermal heating method
Fabrication of micro- and nanoscale spherical particles has been achieved by fluidizing the pre-fabricated fibers either by the iso-thermal heating via furnace (Aktas et al. 2014; Khudiyev et al. 2014) or the axial thermal gradient-thermal heating via oxyhydrogen flame or laser (Gumennik et al. 2013; Zhang et al. 2017). Iso-thermal heating method induces the in-fiber capillary instability by providing a wide homogeneous temperature distribution heating source (furnace or oven). The key characteristics of this approach are (1) wide and homogeneous heating profiles; (2) satellite spheres generation; and (3) a high core-to-sphere amplification ratio. As the heating area is wide enough, a long section of fiber core is heated and breakup into spheres simultaneously. The transformation from a continuous cylinder to a chain of spheres starts from the formation of the necks, then the necks undergo pinch-off (Mansour and Lundgren 1990). Satellite and subsatellite drops are generated owing to self-repeating of the breakup mechanism in every neck and pinch-off section. Meanwhile, as the heating area becomes wider, the core-to-sphere amplification ratio $r_{am} = D_{sphere}/D_{core}$ is relatively large. Therefore, the iso-thermal approach is suitable to generate meso/micro-scale particles in large amounts (Fig. 3).
Gradient-thermal heating method induces the in-fiber capillary instability by providing a smaller and gradient temperature distribution heating source such as laser. The characteristics of this approach are (1) a small-scale and precisely controlled heating profile; (2) a gradient heating profile; and (3) a small core-to-sphere amplification ratio. As the heating area is significantly small and even comparable with the breakup wavelength, the short length of fiber is heated to form the spheres. The limited-size heating zone avoids the self-repeating of the breakup mechanism in the neck and pinch-off section and eliminates the formation of satellite spheres. Meanwhile, as the limited heating area is very small, the amplification ratio r_{am} can be relatively small. Indeed, by feeding the fiber through the heat zone continuously, the gradient-thermal heating method achieves mass production of in-fiber particles. This approach is suitable for the fabrications of micro- to nanoscale particles with a highly uniform size (Fig. 4).

Fig. 3 In-fiber capillary instability breakup based on iso-thermal heating method. **a** Development of the in-fiber capillary instability breakup based on the iso-thermal heating method. **b** The magnified image of the satellites and the subsatellites generated from the iso-thermal heating method (Shabahang et al. 2011). Copyright 2011, AIP Publishing. Reproduced with permission

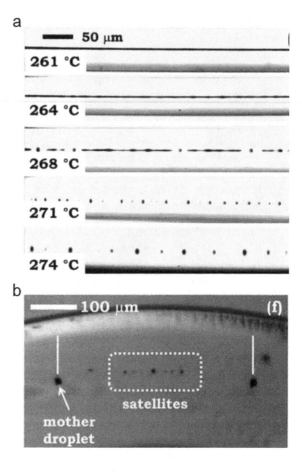

Structured particle generation

With the thermal treatment induced in-fiber breakup process, solid functional cylindrical cores embedded in cladding materials can be transferred to uniformly sized spherical particles. More significantly, the particles with multiple materials and complex structures (hollow-sphere, core-shell, multi-shell, Janus, beach ball, etc.) have been achieved (Fig. 5), enabling more products and functionalities.

First, by structuring the fiber cores during the preform preparations, the in-fiber particles with complex structures could be achieved from the core. As the fiber preforms are assembled in macroscale, complex geometries can be easily constructed into the fiber core by judiciously structuring the morphologies in the radial and azimuthal directions (Kaufman et al. 2012; Tao et al. 2016). As shown in Fig. 5a, the modification in the radial directions provides a way to generate Janus and beach ball particles and the modification in the azimuthal directions provides a way to generate multilayer-structured particles. On this basis, the in-fiber breakup fabrication approach has the advantages of exquisitely designable structures.

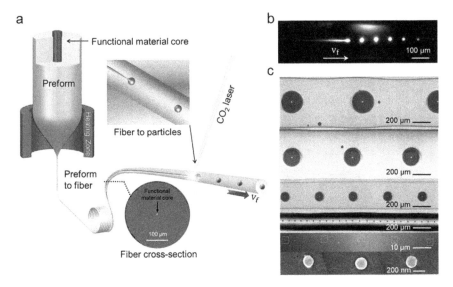

Fig. 4 In-fiber capillary instability breakup based on gradient-thermal heating method. **a** Schematic of the in-fiber breakup based on the laser-induced in-fiber capillary instability. **b** Optical micrograph of the laser-induced in-fiber capillary instability process. **c** Spheres generated from laser-induced in-fiber capillary instability method with sizes from 250 μm to 200 nm (Zhang et al. 2017). Copyright 2017, WileyVCH. Reproduced with permission

Second, benefiting from the diverse thermal modulation, the applicable material library for the in-fiber fabricated particles has been expanded to metal (Au, Ag, Cu, and Pt), semiconductor (Si and Ge), chalcogenide glass, polymers, and functional materials (biocompatible, porous, piezoelectric, triboelectric, and thermoelectric materials). The availability of applicable materials guarantees the powerful functions and broad applications of the in-fiber breakup generated particles.

Furthermore, through the design of multi-core structures and materials, the in-fiber breakup process creates a new way to fabricate in-fiber structures. As shown in Fig. 5b, the in-fiber breakup process can help to generate multiple-sphere particles with different materials. Moreover, by using liquid material to fill the hollow core, the biocompatible capsule particles can also be achieved with in-fiber breakup process. These capsule particles could be surface-functional particles by the surface modification technology with optical fluorescence functions as well as biomedicine functions (Kaufman et al. 2013, 2016).

In-fiber device assembly

By precisely tuning the surface tension and viscosity of different materials at time and spatial domains, the in-fiber breakup method proved a new method for in-fiber device fabrication using multi-core fiber and particle assemblage.

Selective-breakup: Firstly, differing from breaking all continuous fiber cores into spheres, only selected fiber cores are broken into spheres, and the other fiber cores of the same fiber are remained in the continuous cylinder shape by utilizing the

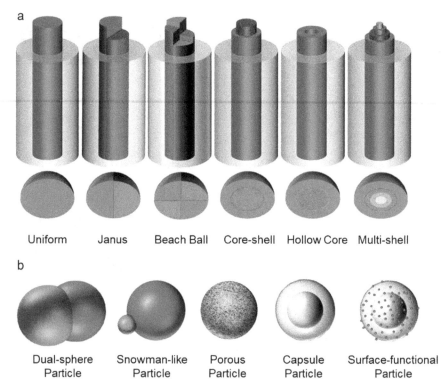

Fig. 5 Structured particles fabricated from the in-fiber capillary instability breakup approach. **a** Schematic diagram of the fiber preforms and particle cross sections for the uniform, Janus, Beach ball, core-shell, hollow core, and multi-shell particles. **b** Schematic diagram of dual-sphere, snowman-like, porous, capsule, and surface-functional particles. Different colors represent different materials

distinct melting points and viscosity features of different materials (Fig. 6a). For example, when inducing the breakup process into a metal-semiconductor (e.g., Platinum–Germanium) cores fiber, the thermal heating process will transfer the functional material into a viscous liquid statue. Because of the distinct melting temperatures between the Ge and Pt, only semiconductor core (e.g., Ge) is treated to break and form a chain of spheres, and the metal core (e.g., Pt) will hold the continuous cylinder shape. Secondly, the spatial selective-breakup can also be achieved by the laser heating method. For a much smaller heating spot size, the laser beam can heat a very short length of the fiber and transfer the selected part of the fiber core into micro/nanoscale spherical particles (Fokine et al. 2017). The spatial selective-breakup of the multimaterial fiber has many applications, such as the Bragg grating.

Breakup-to-contact: As the diameter of generated particles is usually larger than the original diameter of fiber core, the breakup would create the breakup-to-contact assemble process. A series of silicon-in-silica PN molecules is achieved

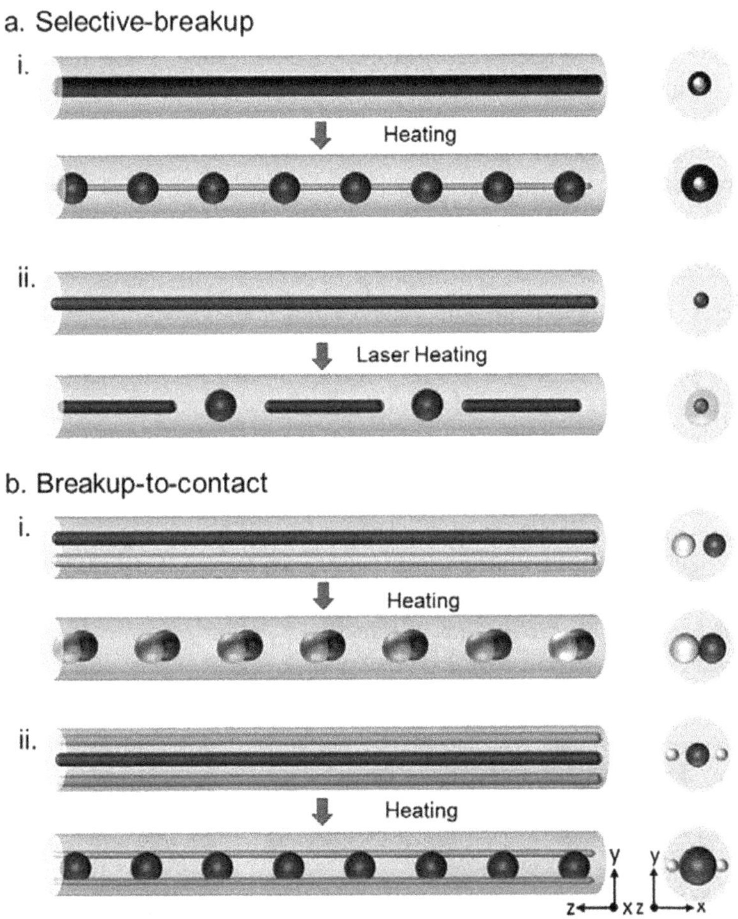

a. Selective-breakup

b. Breakup-to-contact

Fig. 6 Schematic diagram of the in-fiber device assembly. **a** Selective-breakup process. (i) In-fiber selective-breakup shape because of the distinct melting points and viscosity of materials. (ii) In-fiber selective-breakup shape based on the laser limited heating zone. **b** Breakup-to-contact process. (i) Breakup-to-contact process for dual-sphere particle generation. (ii) In-fiber functional device fabrication based on selective-breakup and break-to-contact process

by this technology (Gumennik et al. 2013). The direct contact of p-type and n-type semiconductor cores is not possible in one treatment of the completely melted thermal drawing process. Therefore, separated dual-core semiconductor fibers with a p-type core and an n-type core are fabricated as instead. The micro/nano-spherical particles in-fiber generation based on in-fiber breakup process (Fig. 6b). As the diameter of generated spheres is larger than the diameters of the original core, the two chains of spheres contact with each other, forming in-fiber PN junction molecules.

In-fiber material engineering

As mentioned, the in-fiber breakup process is based on the in-fiber thermal treatment, which can also be developed for in-fiber material engineering. By precisely controlling the thermal treatment parameters such as the heating temperature, the heating profile, and the cooling rate, a precisely controlled resolidification process that induces recrystallization on the particle material structure can be defined, eventually distribution of built-in stress and material structures can be modified (Gumennik et al. 2017; Zhang et al. 2019). Suitable heating parameters are beneficial to control the growth of the crystal grains, the microstructure of the particles, and the possibility of the supercooling in the cladding materials.

As shown in Fig. 7, the laser-induced in-fiber breakup process can be used to induce gigapascal level built-in stress onto the in-fiber particles during the resolidification, which is sufficiently strong that capably modify the bandgap of the semiconductor and to enable inelastic deformations on these semiconductor particles (Healy et al. 2014; Zhang et al. 2019). Therefore, the in-fiber breakup process can generate encapsulated and free-standing stressed semiconductor particles. Furthermore, by precisely control the laser heating parameters, such as the laser power, the laser spot size, and the laser beam's moving speed, this in-fiber built-in stress can be accurately controlled (as shown in Fig. 7b), which opens a new route for the further in-fiber fabrication and device assembly.

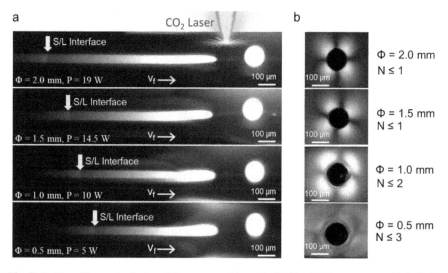

Fig. 7 In-fiber silicon spherical particles generated from the in-fiber breakup process with built-in stress. **a** In-fiber silicon spheres generation based on the laser-induced in-fiber capillary instability. **b** The birefringence intensity distribution around the silicon spheres through a transmission optical microscope in crossed polarizer configuration (Zhang et al. 2019) Copyright 2019, American Chemical Society. Reproduced with permission

– Fabrication of micro/nanowires based on the filamentation instability
  Basically, there are two methods to fabricate micro/nanowires in a fiber. One is based on the in-fiber filamentation instability (Deng et al. 2011), and another is based on a multiple-times thermal drawing process (Khudiyev et al. 2014). Here, we mainly introduce the approach based on the in-fiber filamentation instability. During the thermal drawing process of multilayer fibers, researchers have observed that breakups occur over some films with an extremely small thickness. This phenomenon was triggered by the filamentation instability, which will impact the resulting lay geometry while the preform is heated to fluid state as shown in Fig. 2. The viscous sheet embedded in the fiber cladding material preserved stability along the longitudinal direction and breakup due to the instability along the transverse direction, resulting in the production of micro/nanowires. According to the linear theory of filamentation instability, the final diameter of the micro/nanowires is determined by the sheet thickness (H) and the viscosity ratio between the viscous thin sheet and the cladding materials. Therefore, the duplication of multilayer has been studied over and been finely controlled to fabricate micro/nanowires during the fiber thermal drawing process.

– Fabrication of in-fiber layered structure by suppressing in-fiber fluid instability
  For the in-fiber micro/nano-particles and wires fabrication, the in-fiber instability is a marvelous tool. However, for the uniform layered structure fiber, the in-fiber instability will lead the layers to breakup and destroy the designed structure, which is counterproductive and should be suppressed. Fortunately, the stability criteria of the fluidic instability theories suggest some conditions that help to maintain the layered structure during the fiber thermal drawing process (Deng et al. 2011). Considering the complexity of the thermal drawing process, these conditions are necessary although not sufficient.
  As mentioned in the linear theory, the instability time scale (τ_{inst}) is decided by the structure geometry (radius and thickness of the shell) and the material properties (viscosity and surface tension). And the dwelling time ($\tau_{dwelling}$) is the time of materials in the viscous state before being frozen in-fiber during the thermal drawing process. The stability criterion of the layered structure is defined as

$$\tau_{inst} > \tau_{dwelling} \qquad (7)$$

which implies that the in-fiber instability is suppressed, and the layered structure can be retained. Instead, when the instability time scale τ_{inst} is short than the dwelling time $\tau_{dwelling}$, the instability in the shell structure will have enough time to develop and eventually break the in-fiber layered structure.

Therefore, by giving the viscosity and surface tension of cladding/sheet materials, the stability criteria can suggest the reliability of the designed layered in-fiber structure. For example, in the case of $\tau_{dwelling} \approx 100\,s$, the As_2Se_3-PES,

PVDF-PC, and Se-PSU are suitable material pairs and Se-PE is unsuitable pair (Deng et al. 2011; Liang et al. 2011).

- Structured multifunctional fiber devices and their applications

  - Surface-modified and fluorescence biomedicine particles
    With a doping material and digitally design structures, the in-fiber generated spheres gain more functionalities such as emitting fluorescence and biomedicine applications. In addition, coating the spheres with antibodies and antigens would make these spheres useful in an immunoassay (Kaufman 2014). Fluorescence structured particles and particles with surface-functionalities were achieved by structured core multifunctional fibers. Notably, particle structures with higher complexities may be generated by ingeniously structuring the multimaterial fiber core. These functional particles have demonstrated the potentials to be used in bio-imaging, targeted drug delivery, and gene therapies. Conventional ways to synthesize these functional micro/nano-particles majorly depending on chemical reactions in liquid organic solvent, making the cost high and the throughput relatively low. Therefore, directly fabricating biomedicine particles based on the in-fiber engineering approaches shows outstanding features such as mass production compatibility, high consistency of particles, and non-toxic, which is much more suitable for practical applications.
  - Optical resonator arrays
    Spheres fabricated from the breakup method features ultra-smooth surface roughness, crystal structure owing to surface tension and large built-in stress, which is an extraordinary platform for developing optical resonators (Vukovic et al. 2013; Aktas et al. 2014; Suhailin et al. 2015). Therefore, the generation of micron and nano-spherical particles based on in-fiber capillary instability has opened a new route for fabricating microcavity resonator arrays (Aktas et al. 2014). In general, optical materials, such as As_2Se_3 and Si, are made into the rod and are drawn into the fibers with suitable cladding materials by a thermal drawing process. Following the in-fiber breakup process, the continuous functional core breaks into chains of spheres due to the in-fiber Plateau-Rayleigh instability (Fig. 8). These spheres have an ultra-smooth surface, perfectly spherical shape, crystal structure, and uniform diameters, which consisted high-Q factor whispering gallery mode (WGM) microresonator array. For example, by using this technology, the Q factor of the As_2Se_3 WGM resonator can reach 3.1×10^5 and the Q factor of the Silicon WGM resonator can reach 7.1×10^5.
  - Optoelectronic fiber devices
    Optoelectronic and photodetecting fiber devices can be fabricated based on the selective amplification of in-fiber capillary instability (Wei et al. 2017). Based on a precisely defined metal-semiconductor-metal trial cores fiber, a fully functional fiber-shaped optoelectronic device is generated by employing in-fiber breakup process. As mentioned above, the thermal heating process helps us to selectively break functional material core (e.g., Germanium) into spheres and leaves metal cores maintained in continuous shape (Fig. 9). Therefore,

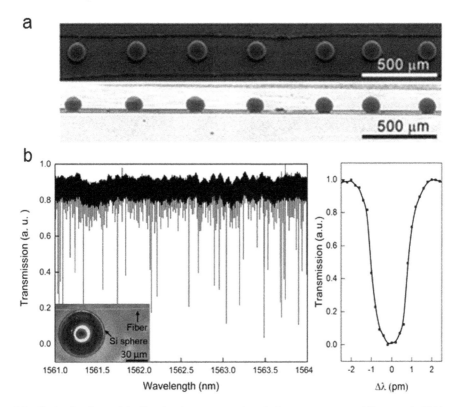

Fig. 8 Applications of in-fiber breakup generated optical resonator arrays (Aktas et al. 2014; Zhang et al. 2017). **a** Profile and top SEM micrographs of the As$_2$Se$_3$ optical resonator arrays. Copyright 2014, WileyVCH. Reproduced with permission. **b** Transmission spectra of a silicon spherical particles with a quality factor Q of 7.1 × 10^5. Copyright 2014 & 2017, WileyVCH. Reproduced with permission

these spheres can serve as optoelectrical components of a device such as pressure sensors or optics, and the metal cores here serve as electrical connection wires as well as signal-transmitting guides in the fiber-based optoelectronic and photodetecting devices. Owning to the fibers' ultra-flexible, stretchable, and made with biocompatible materials features, applications such as in brain sensing and neuron stimulation as well as wearable bio-sensing are proposed. Furthermore, as the fiber from thermal drawing can be ultra-long, large scale sensing fabrics and smart machine covers can finally be developed using the advanced multifunctional fibers and the textiles.

- Outlook of in-fiber breakup process
  Although remarkable achievement has been reached and structured particles as well as assembled functional devices have been proved with the help of the thermal treatment induced in-fiber breakup process, there is still great prospect for further expansion of the in-fiber breakup process into the multifunctional multimaterial

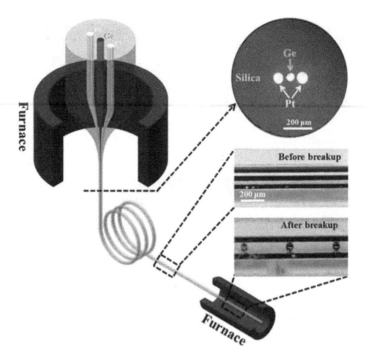

Fig. 9 Photodetecting fiber devices based on in-fiber breakup-to-contact process (Wei et al. 2017). Copyright 2017, WileyVCH. Reproduced with permission

fibers. One of the remaining issues is the combination of the iso-thermal heating method and the gradient-thermal heating method. The iso-thermal treatment is suitable for mass production of in-fiber particles but cannot avoid the generation of the satellite particles which will influence the final uniformity of the dimensions. Meanwhile, the gradient-thermal treatment is efficient in avoiding the generation of the satellite, but the production efficiency is not high. Therefore, a new combination method can meet the demand of volume-producing in-fiber particles with a precisely controlled and highly uniform diameter. Another challenging issue is the in-fiber functional device assembly. By now, the in-fiber functional unit generation and device assembly are depending on the in-fiber breakup-to-contact and selective-breakup process. To enrich the fiber internal structures as well as the functional units, more in-fiber fluid dynamic phenomena should be studied and applied to the post-drawing process.

3   Conclusion

In this chapter, the in-fiber breakup process is comprehensively discussed, including the mechanism of the in-fiber fluidic instabilities, in-fiber micro/nano-particles, wires and layered structure fabrications, and applications of in-fiber breakup technology. This in-fiber fabrication method is based on post-drawing thermal treatment induced fluidic instabilities, which can lead the continuous fiber internal structure to breakup. This technology can be used to generate in-fiber micro- to nanoscale spherical particles, rods, and wires with the precisely controlled geometry and diameter. Meanwhile, the study of the in-fiber fluidic instabilities also provides methods to avoid the breakup phenomena during the fiber thermal drawing process and to preserve the fiber internal geometry and the layered structure. By now, the thermal treatment induced in-fiber breakup process has been applied in a variety of materials with a wide range of processing temperatures from 2400 to 400 K and material viscosity ratio of 10 orders. Furthermore, the in-fiber breakup process creates new opportunities for in-fiber functional device assembly with the breakup-to-contact method and selective-breakup method. In addition, with the in-fiber thermal treatment, the study in-fiber breakup process can also contribute to the in-fiber material engineering, such as the redistribution and resolidification of elements and the built-in stress modification. In summary, the thermal treatment induced in-fiber breakup process offers a new route for many applications including but not limited to optoelectronic and photodetecting devices, electronic devices, biomedicine, drug delivery, and energy harvesting and storage.

References

O. Aktas, E. Ozgur, O. Tobail, M. Kanik, E. Huseyinoglu, M. Bayindir, Adv. Opt. Mater. **2**(7), 618–625 (2014)

S. Chandrasekhar, *Clarendon* (Oxford, 1961)

R. de Gennes, F. Brochard-Wyart, D. Quere, B. Widom, Phys. Today **57**(12), 63 (2004)

D. Deng, J.-C. Nave, X. Liang, S. Johnson, Y. Fink, Opt. Express **19**(17), 16273–16290 (2011)

D.S. Deng, N.D. Orf, A.F. Abouraddy, A.M. Stolyarov, J.D. Joannopoulos, H.A. Stone, Y. Fink, Nano Lett. **8**(12), 4265–4269 (2008)

D.S. Deng, N.D. Orf, S. Danto, A.F. Abouraddy, J.D. Joannopoulos, Y. Fink, Appl. Phys. Lett. **96**(2) (2010)

M. Fokine, A. Theodosiou, S. Song, T. Hawkins, J. Ballato, K. Kalli, U.J. Gibson, Opt. Mater. Express **7**(5), 1589–1597 (2017)

T. Frolov, W.C. Carter, M. Asta, Nano Lett. **14**(6), 3577–3581 (2014)

A. Gumennik, E.C. Levy, B. Grena, C. Hou, M. Rein, A.F. Abouraddy, J.D. Joannopoulos, Y. Fink, Proc. Natl. Acad. Sci. USA **114**(28), 7240–7245 (2017)

A. Gumennik, L. Wei, G. Lestoquoy, A.M. Stolyarov, X. Jia, P.H. Rekemeyer, M.J. Smith, X. Liang, B.J.-B. Grena, S.G. Johnson, Nat. Commun. **4**, 2216 (2013)

N. Healy, S. Mailis, N.M. Bulgakova, P.J. Sazio, T.D. Day, J.R. Sparks, H.Y. Cheng, J.V. Badding, A.C. Peacock, Nat. Mater. **13**(12), 1122–1127 (2014)

K. Kao, G.A. Hockham, Dielectric-fibre surface waveguides for optical frequencies, in Proceedings of the Institution of Electrical Engineers, IET (1966)

J.J. Kaufman, Multifunctional, Multimaterial Particle Fabrication Via an In-Fiber Fluid Instability. Electronic Theses and Dissertations, University of Central Florida, 4803 (2014)

J.J. Kaufman, R. Ottman, G. Tao, S. Shabahang, E.H. Banaei, X. Liang, S.G. Johnson, Y. Fink, R. Chakrabarti, A.F. Abouraddy, Proc. Natl. Acad. Sci. USA **110**(39), 15549–15554 (2013)

J.J. Kaufman, F. Tan, R. Ottman, R. Chakrabarti and A.F. Abouraddy, Scalable Production of Digitally Designed Multifunctional Polymeric Particles by In-Fiber Fluid Instabilities. Photonics and Fiber Technology 2016 (ACOFT, BGPP, NP), Sydney, Optical Society of America (2016)

J.J. Kaufman, G. Tao, S. Shabahang, E.-H. Banaei, D.S. Deng, X. Liang, S.G. Johnson, Y. Fink, A.F. Abouraddy, Nature **487**(7408), 463 (2012)

T. Khudiyev, O. Tobail, M. Bayindir, Sci. Rep. **4**, 4864 (2014)

X. Liang, D. Deng, J.-C. Nave, S.G. Johnson, J. Fluid Mech. **683**, 235–262 (2011)

N.N. Mansour, T.S. Lundgren, Phys. Fluids A **2**(7), 1141–1144 (1990)

J.-H. Park, L. Gu, G. Von Maltzahn, E. Ruoslahti, S.N. Bhatia, M.J. Sailor, Nat. Mater. **8**(4), 331 (2009)

L. Rayleigh, Proc. London. Math. Soc. **1**(1), 4–13 (1878)

L. Rayleigh, Proc. R. Soc. Lond. **29**(196–199), 71–97 (1879)

L. Rayleigh, Lond., Edinb., Dubl. Phil. Mag. J. Sci. **14**(87): 184–186 (1882)

S. Shabahang, J. Kaufman, D. Deng, A. Abouraddy, Appl. Phys. Lett. **99**(16), 161909 (2011)

F.H. Suhailin, N. Healy, M. Sumetsky, J. Ballato, A. Dibbs, U. Gibson, A.C. Peacock, *Kerr nonlinear switching in a core-shell microspherical resonator fabricated from the silicon fiber platform* (Science and Innovations, Optical Society of America, CLEO, 2015)

G. Tao, J.J. Kaufman, S. Shabahang, N.R. Rezvani, S.V. Sukhov, J.D. Joannopoulos, Y. Fink, A. Dogariu, A.F. Abouraddy, Proc. Natl. Acad. Sci. USA **113**(25), 6839 (2016)

S. Tomotika, Proc. R. Soc. Lond. Ser. A, Math. Phys. Sci. **150**(870), 322–337 (1935)

T. Van Buuren, L. Dinh, L. Chase, W. Siekhaus, L.J. Terminello, Phys. Rev. Lett. **80**(17), 3803 (1998)

N. Vukovic, N. Healy, F. Suhailin, P. Mehta, T. Day, J. Badding, A. Peacock, Sci. Rep. **3**, 2885 (2013)

P. Wang, T. Lee, M. Ding, A. Dhar, T. Hawkins, P. Foy, Y. Semenova, Q. Wu, J. Sahu, G. Farrell, J. Ballato, G. Brambilla, Opt. Lett. **37**(4), 728–730 (2012)

J. Ward, O. Benson, Laser Photonics Rev. **5**(4), 553–570 (2011)

A.L. Watts, N. Singh, C.G. Poulton, E.C. Magi, I.V. Kabakova, D.D. Hudson, B.J. Eggleton, JOSA B **30**(12), 3249–3253 (2013)

L. Wei, C. Hou, E. Levy, G. Lestoquoy, A. Gumennik, A.F. Abouraddy, J.D. Joannopoulos, Y. Fink, Adv. Mater. **29**(1), 1603033 (2017)

B. Xu, M. Li, F. Wang, S.G. Johnson, Y. Fink, D. Deng, Phys. Rev. Fluids **4**(7), 073902 (2019)

J. Zhang, K. Li, T. Zhang, P.J.S. Buenconsejo, M. Chen, Z. Wang, M. Zhang, Z. Wang, L. Wei, Adv. Funct. Mater. **27**(43), 1703245 (2017)

J. Zhang, Z. Wang, Z. Wang, T. Zhang, L. Wei, ACS Appl. Mater. Interfaces **11**(48), 45330–45337 (2019)

Nano- and Micro-structuring of Materials Using Polymer Cold Drawing Process

Ming Chen, Zhixun Wang, and Ke He

Abstract Cold drawing is a well-established manufacturing process for tailoring properties of polymers in polymer industries. It refers to the phenomenon that necks formed during the tensile drawing of polymer fibers, films, or any other shapes. The necking reduces the height and width of the polymer sample being stretched and elongates it in length. The cross-sectional area in the necking region becomes smaller and keeps in a constant value with traveling to the remaining non-necking regions until the non-necking regions disappeared. Fundamentally, cold drawing is the movements and orientations of entangled polymer chains under external forces. It happens both for amorphous and for semicrystalline polymers in specific conditions. Though the cold drawing process has long been used in industrial applications, it has been rediscovered as a method of nano- and micro-structuring of materials recently. Materials used for structuring are wrapped in polymer fiber to form a core-shell structure or attached to the surface of the polymer substrate, then experience ordered fragmentations during the cold drawing of the composite structure, and finally formed regular patterns in nano- and micro-scale. This robust nano- and micro-structuring process is easy to perform and can be conducted using bare hands in some situations. Further, this method can be applied to various materials and is suitable for large-scale fabrication. This chapter discusses nano- and micro-structuring of materials including crystal, glass, and polymer with the core-shell structured fiber and film-based flat geometries.

Ming Chen and Zhixun Wang—Equally contributed.

M. Chen (✉) · K. He
Center for Information Photonics and Energy Materials, Shenzhen Institutes of Advanced Technology, Chinese Academy of Sciences, Shenzhen 518055, People's Republic of China
e-mail: ming.chen2@siat.ac.cn

Z. Wang
School of Electrical and Electronic Engineering, Nanyang Technological University, 50 Nanyang Avenue, Singapore 639798, Singapore

K. He
Department of Nano Science and Technology Institute, University of Science and Technology of China, Suzhou 215123, People's Republic of China

© Springer Nature Singapore Pte Ltd. 2020
L. Wei (ed.), *Advanced Fiber Sensing Technologies*,
Progress in Optical Science and Photonics 9,
https://doi.org/10.1007/978-981-15-5507-7_12

Keywords Cold drawing · Nano- and micro-structuring · Controlled fragmentations · Shear-lag theory · Metal film · Two-dimensional materials · Micro-ribbons

1   Introduction

In the polymer industries, cold drawing is a successful technique used in the production of polymer with enhanced mechanical properties, such as high tensile strength polyester and nylon fibers (Carraher 2017; Ziabicki 1976). First discovered in the 1940s (Carothers and Hill 1932), cold drawing refers to the phenomenon that many polymers, when subjected to tensile stress, may undergo non-homogeneous deformation with one or multiple necking regions formed after the initial small homogeneous elastic deformation. The movements of polymer chains are the underlying mechanism of this phenomenon. The 'frozen' polymer chains can move under the external tensile force. The nature of this mechanical instability is high-elastic state deformation or forced high-elastic state deformation since the deformation only recovers when the specimen being annealed at around the glass transition temperature (for amorphous polymers) or the melting temperature (for semicrystalline polymers). The interesting part of this non-homogeneous deformation is, clear boundaries or necking fronts, lying normal to the drawing direction can be observed between the necking and non-necking regions. The necking regions elongated in the drawing direction and shrank in the dimension perpendicular to the drawing direction. With the cold drawing process goes on, the necking regions extend by the traveling of necking fronts until the non-necking regions disappeared. In the propagation of necking fronts, strong strain localized only in the necking fronts, while the necking and non-necking regions are stable. Eventually, the specimen becomes stronger after the necking fronts went through it, which is called strain hardening (Vincent 1960; Coleman 1985). While the forming of neck fronts is influenced by many factors, such as the drawing temperature, the drawing rate, and the molecular weight, many polymers can be cold drawn under ambient temperature.

   The studies on cold drawing focused on single material in the past decades (Argon 2013), cold drawing of a composite structure emerged later which concentrated on fiber reinforcement composites (Loos et al. 2001; Friedrich et al. 2005; Fakirov et al. 2007). Yet until recently, research works on using cold drawing to form nano- and micro-structures have been reported (Shabahang et al. 2016; Xu et al. 2017; Wu et al. 2019; Loke et al. 2019; Chen et al. 2017, 2019; Wang et al. 2010; Nairn 1997; Galiotis and Paipetis 1998; Li et al. 2018). There are two geometries have been investigated, namely the core-shell structured fiber and planar film-based composite. The designed core-shell structured fibers are consisting of a ductile polymer shell and a core of another relatively brittle material. The polymer shell experiences cold drawing under tensile stretching; consequently, the core material be affected through the stress transferred through the interfaces. When the stress in the core material exceeds the tensile strength of it, the core material breaks. Rather than random fragmentations, a sequentially ordered break-up of the core material happens and results in a periodic chain of rods separated by voids. The strong localized strain in the necking fronts leads to the controllable fragmentations. Core material only breaks after the necking fronts propagated through, and other parts located in the non-necking regions remain intact. The size of fragments depends on the diameter and Young's modulus of the core material. From this method, various multicomponents nano- and micro-structures can be produced on a large scale. In addition, the uniformly sized rods of the core material can be etched out via the selective dissolution of the polymer shell. In principle, any material which does not undergoes cold drawing can be used as the core material, including crystals, glasses, and polymers. For the planar film-based composite, a similar phenomenon can be observed. A thin layer of relatively brittle material, for example, glass or metal can be deposited on a polymer film, which can be cold drawn. After the process of cold drawing, the thin deposited layer fragments in the transverse direction sequentially and resulting in ordered ribbons array. The formed pattern has a featured size of nanometers to microns. Further, submicron ribbons of monolayer or few layers two-dimensional transition metal dichalcogenides (TMDs) and graphene can be produced by this lithography-free approach. This versatile method can be applied to other low-dimensional materials.

To understand the underlying mechanism of the ordered fragmentations, the shear-lag theory was modified to explain this phenomenon. The traditional shear-lag theory was first reported in the 1940s and well-established for the mechanical analysis of composites, especially for fiber-reinforced composites (Wang et al. 2010; Nairn 1997; Galiotis and Paipetis 1998). In the traditional shear-lag model, the strain is distributed globally in contrast to the cold drawing process where the strain is localized in the necking fronts. Also, the traditional shear-lag fragmentations typically occur at random locations simultaneously, while the target material only fragments with the necking front passing through in the cold drawing process. Besides, the traditional shear-lag model and the cold drawing process share the same interfacial stress transfer mechanism. A modified shear-lag model can be applied in the analysis of the cold drawing process.

2   Cold Drawing of Core-Shell Structured Fibers

A wire-like core material embedded in a polymer fiber matrix assembles a core-shell structured fiber composite. The core-shell structured fiber provides a good platform to study the cold drawing on composite in which the shell undergoes cold drawing and the core does not. The mismatch in mechanical properties for the core and the shell materials leads to different mechanical behaviors for them during the cold drawing process.

2.1 Fabrication of Core-Shell Structured Fibers

The 1-mm-diameter fibers used for cold drawing have a shell of thermoplastic polymer polyethersulfone (PES). It is chosen to be used as the shell material because it undergoes cold drawing at ambient temperature, using other polymers (which undergo cold drawing) can get similar results. To demonstrate the universality of this method, a wide range of materials is used as core material, including four glasses: silica (SiO_2), tellurite glass $70TeO_2–20ZnO–5K_2O–5Na_2O$ (TeG), phosphate glass (PeG), and chalcogenide glasses As_2Se_3 (ChG); five polymers: polystyrene (PS), acrylonitrile butadiene styrene (ABS), polyethylene oxide (PEO), silk, and human hair; and three crystals: silicon (Si), germanium (Ge), and ice. The fibers consisting of a 20-μm-diameter core of ChG are fabricated via fiber thermal drawing technique (PES and ChG are thermally compatible), which is a scale-down process of a macroscopic scale perform, and the process is diagramed in Fig. 1a. The same technique is applied for thermally compatible core materials. For other core materials which are thermally incompatible with PES, they are placed into hollow PES fibers (prepared from thermal drawing) and annealed to ensure the strong adhesion between the shell and the core (Fig. 1b). The fibers with ice core are fabricated by first filling water into hollow PES fibers and then frozen at -5 °C.

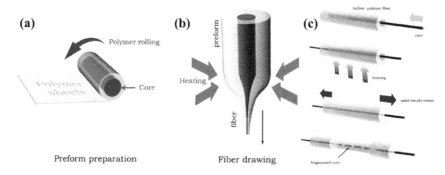

Fig. 1 Schematic illustration for the fabrication of core-shell structured fibers. **a** Schematic depiction of the thin-film-rolling process. A cylindrical rod of a material to be used as a core is rolled with thin PES sheets and then thermally consolidated under vacuum to fuse the thin films and yield a preform. **b** The preform is heated in a custom-built fiber draw tower, and axial tension is applied to elongate the preform into a fiber (along the direction of the arrow). **c** Schematic depiction of combining a thermally incompatible micro-wire and a hollow polymer fiber for cold drawing. Adapted with permission from Shabahang et al. (2016)

2.2 Ordered Fragmentation of the Core Material

The core-shell-structured fibers are subjected to cold drawing after preparation. The cold drawing is conducted by fixing the ends of the specimen on a travel stage; then, the travel stage stretches the specimen at a constant speed. Interestingly, this simple process can be also done using bare hands, as shown in Fig. 2a.

All the tested core materials fragmented into ordered rods after the cold drawing. For the characterization of the fragmentation process, optical images are taken for the different stages of the cold drawing process for the PES-cladding and ChG-core fiber. Figure 2e is the optical image taken before the neck forms, the initial stage of necking, and the propagation of the necking front, respectively. The core ChG is observed to fragment into ordered micro-rods after the cold drawing process. The enlarged image in Fig. 2f shows the details of the fragmentation. The core in neck regions has fractured into rods, while other parts located in non-necking regions remain intact. It can be seen clearly that the fragmentation of core material happens in the necking front. The extracted ChG rods shown in Fig. 2j demonstrate the potential of this method as a mass fabrication tool for micro-rods. Figure 2i illustrates the whole process. It is worth noting that this method is not limited by fiber geometries, and as shown in Fig. 2b–d, cylindrical, rectangular, and triangular fibers can be cold drawn in spite of their difference in the shape of cross section.

Some results can be found with a systematic investigation of the obtained rods. In Fig. 3a, a linearly proportional relationship between the average length L and the diameter D of the rods can be clearly observed both for ChG and TeG. The linear proportionality $f = L/D$ depends on the core material. Figure 3b plots the relationship between f and the Young's modulus of eleven different core materials. It can be found empirically that $f \propto \sqrt{E}$, where E is the Young's modulus of the core material. Considering the similarity in geometries between the core-shell structured fibers and fiber-reinforced composites, the interfacial mechanical behaviors of the core-shell structured fibers could possibly follow the classic shear-lag theory. Verified with experiments, the modified shear-lag model $\overline{L}/D = \sqrt{E/\Omega}$ (Galiotis and Paipetis 1998; Thostenson et al. 2002; Kim and Nairn 2002; Figueroa et al. 1991), where Ω is a characteristic stress (equals to 0.1 GPa as shown in Fig. 3c), is in a good agreement with experimental results (Table 1).

3 Cold Drawing of Film-Based Flat Composites

Apart from the core-shell fiber structures, cold drawing of film-based flat composites can lead to interesting results as well. There are mainly two types of film-based flat composites studied, one is the two-layer structure, and the other one is the polymer film/core film/polymer film sandwiched structure (Xu et al. 2017; Wu et al. 2019; Chen et al. 2017, 2019). The interfacial mechanical behaviors between core film

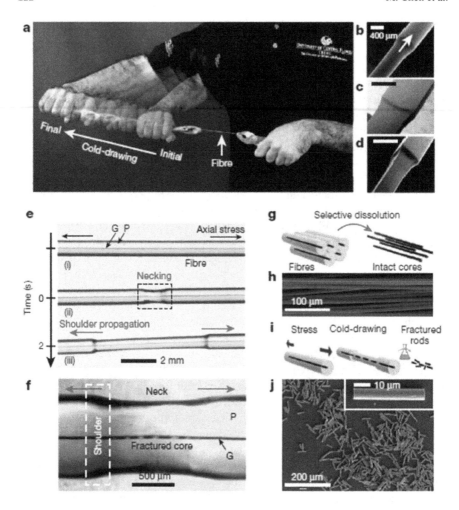

Fig. 2 Fragmentation *via* a cold drawing-induced, propagating, mechanical–geometric instability. **a** Photograph of a polymer fiber undergoing cold drawing under axial stress at a speed of approximately 5 mm s^{-1}. Multiple shots taken over 1 min are overlaid to highlight the extent of fiber elongation. **b–d** Scanning electron microscope (SEM) micro-graphs of the propagating shoulder in polymer (PES) fibers with cross sections that are circular (diameter of 0.7 mm; (**b**), rectangular (side lengths of 0.2 and 1 mm; (**c**) and equilaterally triangular (side length of 0.4 mm; (**d**). Scale bars, 400 μm. **e** Transmission optical micro-graphs of a multimaterial cylindrical fiber undergoing cold drawing at 3 mm s^{-1} captured at three different stages: (**i**) initially intact fiber; (**ii**) neck formation; and (**iii**) shoulder propagation, leaving behind a fractured core after fragmentation. The cladding is a polymer 'P' (PES), and the core is a glass 'G' (As$_2$Se$_3$). **f** A magnified transmission micro-graph of the neck region, corresponding to the dashed black rectangle in e. The dashed white rectangle highlights the propagating instability, wherein fragmentation takes place. **g** Schematic of selective dissolution of the polymer cladding to retrieve intact cores. **h** SEM micro-graph of retrieved intact glass cores from multiple fibers. **i, j** Schematic (**i**) and SEM micro-graph (**j**) of retrieved nano-fragmented micro-rods by selective dissolution from a cold-drawn fiber. Inset in **j** is an SEM micro-graph of a single micro-rod. Adapted with permission from Shabahang et al. (2016)

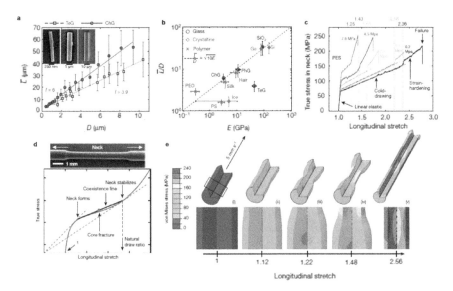

Fig. 3 Characterization of fragmentation induced by cold drawing of a PES fiber. **a** Measurements of the average length $\overline{L}$ of fragmented micro- and nano-rods of chalcogenide glass (ChG; As_2Se_2; red circles) and tellurite glass (TeG; $70TeO_2$–$20ZnO$–$5K_2O$–$5Na_2O$; blue squares) of diameter D in a PES fiber upon cold drawing. The red solid and blue dashed lines are linear fits with slopes $f = \overline{L}/D = 6$ and 3.9 for ChG and TeG, respectively. Vertical error bars represent the root-mean-squared (r.m.s.) length dispersion of rods at each value of D. Insets are SEM micro-graphs of individual rods resulting from the cold drawing-driven fragmentation of As_2Se_3 cores of diameters (from left to right) 200 nm, 1, and 10 μm. **b** Measured values of f for a host of materials embedded in a PES fiber plotted against their Young's modulus E; the dashed line corresponds to the ansatz $f \approx \sqrt{E/\Omega}$, with $\Omega = 0.1$ GPa (such that $f \approx \sqrt{10E}$ when E is in gigapascals). Vertical error bars represent the measured r.m.s. dispersion in f; horizontal error bars correspond to the uncertainty in the measured E (those for TeG and ice reflect the range of reported values). PhG, phosphate glass; PEO, polyethylene oxide; PS, polystyrene. **c** Stress–strain measurements of cylindrical PES fibers produced by thermal drawing at different prestress values (ranging from 0.2 to 7.8 MPa; see colored-coded labels) identifying the four stages of linear elasticity, cold drawing, strain-hardening, and failure. The coexistence (dashed) lines and natural draw ratios (vertical dotted lines; values given above the plot) at neck stabilization (both defined in **d**) are identified. **d** Schematic representation of Considére model. The blue curve corresponds to the true stress versus stretch in a strain-controlled experiment. The solid black 'coexistence' line indicates necked and unnecked regions coexisting at equal engineering stresses as local stretch varies from the onset of necking to the natural draw ratio at which stable neck propagation occurs. The dashed black lines serve as guides to the eye. Above the plot is an SEM micro-graph of a necked region in a PES fiber; the arrows indicate the direction of the axial stress and shoulder propagation. **e** Von Mises stress distributions from finite-element simulations of in-fiber core (As_2Se_3) fragmentation during cold drawing of a PES fiber. The five steps (**i**–**v**) correspond to increasing stretch values. Top panels depict the full fiber; bottoms panels show the regions corresponding to that highlighted by the rectangle in (**i**). Adapted with permission from Shabahang et al. (2016)

Table 1 Measured values of $\overline{L}/D$ and measured and reported values of E, for different core materials

Material	$\frac{\overline{L}}{D}$	St. Dev.	E (GPa)	Δ (GPa)	E (GPa)	Δ (GPa)
	Measured		Measured		Reported	
SiO$_2$	33	11.95	88.5	5.5	70	65–73.1
Si	32.5	9.56	152	7.5	150	130–202
Ge	30	9.26	78	6	103	102.7–103
PhG[42–44]	9.5	3.19	11	2	55	31.3–79
Silk[45–47]	8.2	2.07	9	3.2	16	3.8–17
ChG[48–50]	6	1.57	3.2	0.1	38.6	18–40
TeG[51]	3.9	1.07	N/A[1]	N/A[1]	44	37.1–50.7
Hair[52,53]	4.9	1.46	3.7	1.1	4	2.5–7.5
PEO[54,55]	2.9	0.56	0.3	0.1	5	0.2–7
Ice[20,44,56]	1.7	0.46	N/A[2]	N/A[2]	5.1	0.3–10
PS	1.6	0.25	2.7	0.7	3.2	3–3.5

Adapted with permission from Shabahang et al. (2016)

and polymer film follow the shear-lag theory for both types. For the core material, metals like silver (Ag), gold (Au), aluminum (Al), copper (Cu), platinum (Pt), and nickel (Ni) have been used, and the resulted nano-structures find a great potential in the applications of surface-enhanced Raman scattering (SERS) sensing (Wu et al. 2019), optical superlenses (Li et al. 2018), and localized surface plasmon resonance (LSPR) sensing (Xu et al. 2017). Further, two-dimensional materials, monolayer and few-layer transition metal dichalcogenides (TMDs), and graphene are used as core film both for investigating the mechanical behaviors of two-dimensional materials/polymer composites and fabricating two-dimensional materials submicron ribbons (Chen et al. 2017, 2019). This method is demonstrated as a versatile approach for large-scale two-dimensional material ribbons array production, and the fabricated ribbons array has a wide range of applications in nano-electronics.

3.1   Metal-Polymer Composites

A thin layer of metal elements can be deposited onto a polymer film via e-beam technique. A thin layer (~25 nm) of Ag was deposited onto a 20-μm-thick poly(ε-caprolactone) (PCL) film before conducting the cold drawing process at a stretching speed of ~5 mm/s, as shown in Fig. 4.

After cold drawing, the Ag layer on PCL film has fragmented into ordered ribbons perpendicular to the drawing direction. And uniformly submicron wrinkles can be observed on the whole length of the ribbons, which is lying in the direction of drawing.

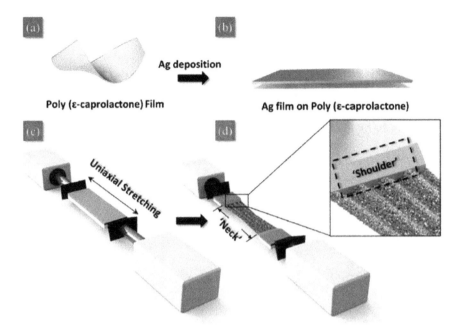

Fig. 4 Schematic diagram of stretching polymer SPR film under an external mechanic force. **a** Flexible PCL polymer film. **b** Ag film is deposited on PCL polymer film by an electron beam evaporator. **c** The polymer SPR film is fixed onto a mechanical machine. **d** The uniaxial stretching of polymer SPR film at a constant stretching speed. Two primary characteristics are observed: neck formation and shoulder propagation. The inset highlights the nano-structures' formation in the neck region. Adapted with permission from Xu et al. (2017), copyright (2020) American Chemical Society

The scanning electron microscope (SEM) images of the PCL film (without Ag layer deposited on) and Ag/PCL film after cold drawing are demonstrated in Fig. 5a, b.

While the ribbons have a feature size of tens of micrometers, the nano-grooves are observed to have a width of a hundred of nanometers between each wrinkle, shown in Fig. 5c. On each Ag wrinkle, there are uniformly distributed nano-gaps with a size of tens of nanometers on the surface. The atomic force microscope (AFM) was used to verify the surface morphology of resulted Ag ribbons with wrinkles, as illustrated in Fig. 5d. From the AFM images, the wrinkles are found to have a height of ~100 nm and a width of ~1 μm. To shed a light on the formation of wrinkles on the ribbon surface, a PCL film without Ag layer deposited on has been cold drawn (Fig. 5a), as can be seen that many wrinkles formed on the surface of PCL film, considering the nature of semicrystalline polymer for PCL film, Poisson's ratio (~0.45) of PCL, and the relatively fast drawing rate, the formation of wrinkled surface can be attributed to the deformation of PCL film. During the cold drawing process, the PCL crystals reoriented and further shrank laterally, and the distance between each adjacent lamellae increased within a crystal. For PCL film with Ag layer deposited on, the Ag particles act as blocks preventing the wrinkles from overlapping and

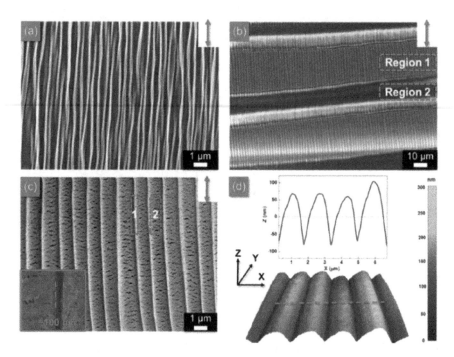

Fig. 5 Surface morphology of stretched polymer SPR film without (**a**) and with (**b, c**) 25 nm Ag film. The arrows denote the direction of externally applied stress. The polymer SPR film at a width of 4 cm is drawn from 4 cm (initial length) to 10 cm (stretched length). The inset in panel c shows the dimension of the nano-groove. The dashed boxes in panel c indicate the nano-groove (1) and nano-gaps (2), respectively. (**d**) Morphologies of four complete periods of polymer SPR film demonstrate 3D wave-shaped structures. The inset in panel **d** illustrates line scanning obtained across the dashed line. Adapted with permission from Xu et al. (2017), copyright (2020) American Chemical Society

thus forms the Ag ribbons with nano-grooves. The nano-gaps are formed during the drawing process since the deposited Ag layer is not single crystalline but consisting of many Ag particles, when subjected to large deformation, the nano-gaps formed on the boundaries between each particle. The nano-gaps function as hotspots and make the Ag/PCL film a good photonic surface for using in SERS. Compared to unstretched film, the stretched Ag/PCL film provides an enhancement (more than 10 times) in SERS, as Fig. 6 shows.

Au, Ni, and Al have been used as the deposition layer, and the surface morphology is shown in Fig. 7. For Au/PCL composites, similar surface morphology of Ag/PCL can be observed. After cold drawing, Au ribbons with wrinkles (nano-grooves and nano-gaps) are obtained. It can be used as a photonic surface for enhanced SERS as well. For Ni and Al, different surface morphology is found. The deposited metal layer fragmented into a web-like network. The different ductility and adhesion to the PCL layer lead to the distinct morphologies.

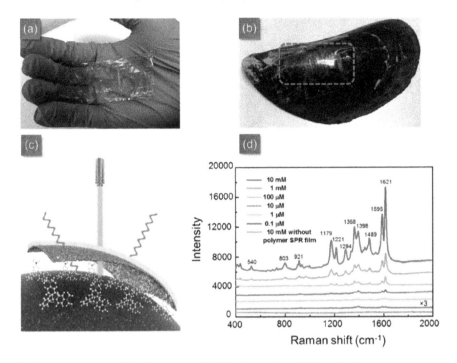

Fig. 6 Demonstration of flexible polymer SPR film for practical SERS applications. **a** A photograph image of stretched polymer SPR film (8 cm × 4 cm). **b** A photograph image of polymer SPR film attached to the green mussel surface contaminated by MG molecules. **c** Schematic diagram of contacting polymer SPR film onto the green mussel and collecting the SERS signals from the backside surface. **d** In situ detection of MG molecules on the green mussel surface at various concentrations from 10 mM to 0.1 μM. The SERS spectra are obtained at the laser excitation wavelength of 514 nm, where the power is 1.5 mW, the acquisition time is 10 s, and the accumulation time is 1 s. Adapted with permission from Xu et al. (2017), copyright (2020) American Chemical Society

Polycarbonate (PC) film can be used as a polymer film matrix as well. Different from semicrystalline PCL, PC is an amorphous polymer with a Poisson ratio of ~0.3. A layer of Au in different thickness (Fig. 8g) was deposited onto PC (125 μm thick) via e-beam. After the cold drawing process, the Au layer fractured into strips and no wrinkle observed. Au layer with different thicknesses from 10 to 50 nm has been studied, and similar surface morphology is found. The average width of resulted Au stripes is proportional to the thickness as shown in Fig. 8f. After cold drawing, the Au/PC composites are subjected to drawing again at the perpendicular direction (2D stretching). The Au strips are torn into distributed Au islands. The fragment length for Au layers with different thicknesses is shown in Fig. 8g.

The cold drawing of Cu and Pt layer (20 nm) deposited PC composites is studied for further investigations. After cold drawing, Cu and Pt fractured into strips similar to the Au strips. And the surface morphologies of Cu/PC and Pt/PC are islands-like, similar to the results in Au/PC (Fig. 9). The different mechanical properties

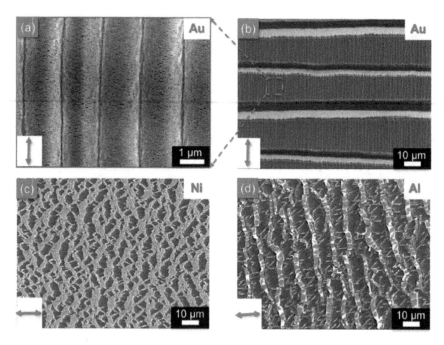

Fig. 7 Surface morphology of various metallic materials deposited on the PCL polymer films. **a, b** Au. **c** Ni. **d** Al. All the thicknesses of metallic films are 25 nm, and the composite film is stretched from 4 to 10 cm at a width of 4 cm. The red arrows denote the stretching direction. Adapted with permission from Xu et al. (2017), copyright (2020) American Chemical Society

and adhesion to the metal layer of PC and PCL films determine the distinct surface morphologies for metal layer after cold drawing.

The fabricated Au/PC film is highly conformal, flexible, and semitransparent, and it can be used on complex surfaces for on-site SERS sensing. As an example, the on-site SERS sensing on basil leaves and apples has been studied using the cold-drawn Au/PC film. The schematic process is shown in Fig. 10a. After cold drawing, the resulted films are attached to the surface of the leaves and apples with different concentrations of thiram. The detection of thiram residue on foods is crucial as the thiram can lead to serious health issues (Wang et al. 2015). The strongest Raman peak for thiram is at ~1376 cm^{-1}, and it is monitored during the test of on-site SERS sensing. The results are demonstrated in Fig. 10c and e. As can be seen, for the basil leaves, the lower limit of detection (LOD) for thiram using Au/PC film is about 48 ng/cm^2, and it is 0.48 ng/cm^2 for apples. The different LOD values may arise from the surface morphologies and tissue compositions.

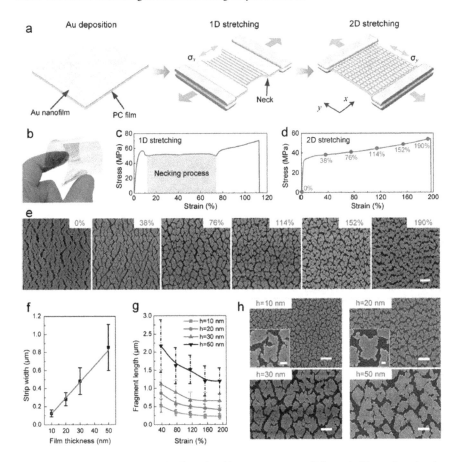

Fig. 8 Fabrication and characterization of gold nano-patterns. **a** Schematic illustration showing the fabrication process of metallic micro/nano-patterns on flexible substrates through two-step mechanical stretching. **b** Photograph of a centimeter-scale flexible photonic surface fully covered with gold nano-patterns fabricated by 2D stretching. **c, d** Stress–strain curves for the gold/PC film composite in 1D and 2D stretching processes, respectively. **e** SEM images of the gold patterns at different stages of 2D stretching (thickness of the gold film, 30 nm). **f** Average width of the gold strips fabricated by 1D stretching as a function of gold film thickness. **g** Average length of the gold patterns at different stages of 2D stretching. **h** SEM images of the 2D gold patterns with different thicknesses (strain, 152%). Scale bars: (**e**) 2 mm; (**h**) 1 mm; inset picture, 100 nm. Reproduced from Li et al. (2018) with permission from the Royal Society of Chemistry

3.2 Two-Dimensional Materials–Polymer Composites

Materials structuring using the cold drawing method have been proven to be highly effective, low cost and suitable for mass production. It is natural to think whether this method can be applied to two-dimensional materials. Different from bulk materials, two-dimensional materials have their unique properties in many aspects, and

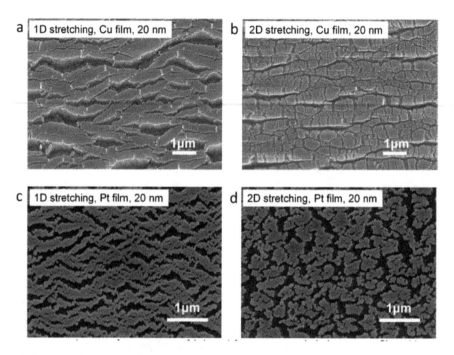

Fig. 9 SEM images of nano-patterns fabricated from copper and platinum nano-films. **a** Copper nano-stripes after 1D stretching. **b** Copper nano-patterns after 2D stretching. **c** Platinum nano-stripes after 1D stretching. **d** Platinum nano-patterns after 2D stretching. Film thicknesses, 20 nm; substrate, 125-μm-thick PC film; elongation for 2D stretching, 152%. Reproduced from Li et al. (2018) with permission from the Royal Society of Chemistry

their excellent electrical and mechanical properties make them a good candidate for flexible electronics (Zhang et al. 2016; Yang et al. 2017; Liu et al. 2014; Gong et al. 2016; Castellanos-Gomez et al. 2012; Bertolazzi et al. 2011). Many researches work on two-dimensional material-based flexible electronic devices are focused on two-dimensional materials/polymer assembly thanks to the superior flexibility of polymers. Thus, the investigation of cold drawing on two-dimensional materials/polymer composites not only develops a potential fabrication tool for two-dimensional ribbons but also gaining an understanding of the mechanical behaviors in two-dimensional materials/polymer assembly, which is essential for the evaluation of device stability for related flexible electronics. In the family of two-dimensional materials, the TMDs and graphene attract many interests. They are transferred onto a 125 μm PC film via the wet transfer method and then subjected to cold drawing.

Figure 11 shows the experimental process for the cold drawing of CVD-grown single crystal monolayer WS_2 on PC film. The resulted WS_2 submicron ribbons can be observed clearly using a fluorescence microscope (Fig. 11e). The ordered fragmentation of WS_2 only happens in the necking front, as Fig. 11b–d shows. Figure 11f–j gives a clearer comparative view of three WS_2 crystals before and after the cold drawing process.

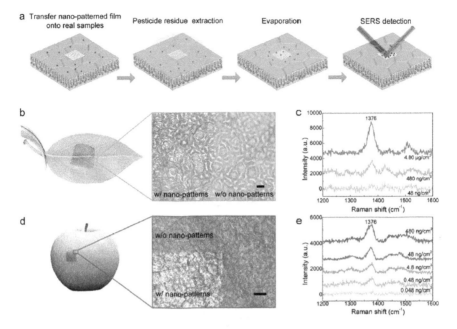

Fig. 10 Ultra-flexible and conformal gold nano-patterned film for on-site SERS sensing. **a** Schematic illustration showing the process of on-site SERS sensing. **b** Photograph of a basil leaf coated with an ultra-flexible gold nano-patterned film and micro-graph of the surface morphology. **c** On-site SERS spectra acquired using ultra-flexible gold nano-patterned films from basil leaves with different concentrations of the thiram residue (background subtracted). **d** Photograph of an apple covered with an ultra-flexible gold nano-patterned film and micro-graph of the surface morphology. **e** On-site SERS spectra acquired using ultra-flexible gold nano-patterned films from apple peels with different concentrations of the thiram residue. Scale bar: (**b**) 50 mm; (**d**) 100 mm. Reproduced from Li et al. (2018) with permission from the Royal Society of Chemistry

The ordered fragmentation also occurs on few-layer WS_2 after the cold drawing, as shown in Fig. 12d, e. To study the versatility of this method, monolayer MoS_2, $MoSe_2$, and WSe_2 are also investigated under different drawing speeds. Again, the drawing speed is proven to have neglectable effects on the fragmentations. And similar to WS_2, other TMDs also fractured into ordered ribbons array. The resulted ribbons have many freshly produced edges and have a better performance on hydrogen evolution reaction (HER). Figure 13 shows the comparison of HER performance over MoS_2 ribbons array-based and pristine WS_2-based bottom-gated field-effect transistors (FETs). The MoS_2 ribbons array-based FET has enhanced performance on HER.

As another representative two-dimensional material, graphene is the thinnest and strongest material ever found. Also, large-scale polycrystalline graphene can be grown using CVD. The results for cold drawing of graphene/PC film are shown in Fig. 14.

After cold drawing, the graphene sheet was torn into submicron ribbons array. This versatile method can be applied to small single crystals and large-scale polycrystals.

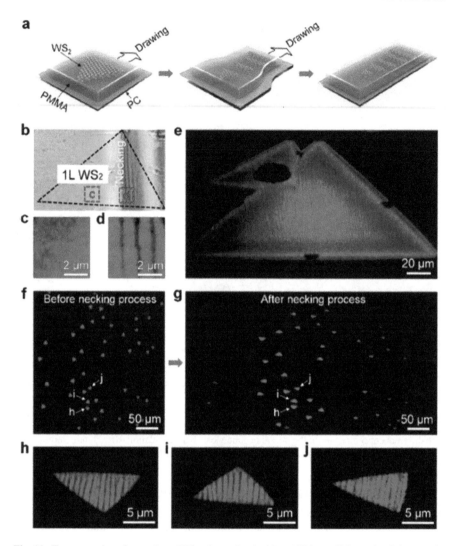

Fig. 11 Fragmentation of monolayer WS_2 via mechanical instabilities. **a** Schematic of the experimental steps to fragment monolayer WS_2. **b** Optical image of a monolayer WS_2 during the necking process, highlighting the propagating mechanical instabilities, wherein fragmentation occurs. The triangle represents 1L WS_2. **c, d** Fluorescence images of monolayer WS_2 during the necking process, corresponding to the dashed rectangles c and d in (**b**). (**d**) After necking propagation, WS_2 monolayers are fragmented into well-ordered nano-ribbons. (**c**) Before necking process, WS_2 monolayers are maintaining integrity. **e** Fluorescence image of a monolayer WS_2 with sample size of ~100 μm after necking process. **f** Fluorescence images of WS_2 monolayers before (located on SiO_2/Si surface) and **g** after necking process (drawing speed: 0.1 mm/s). **h–j** Magnified fluorescence images of monolayer WS_2 nano-ribbons with different orientations, corresponding to (**h**), (**i**) and (**j**) in (**f**) and (**g**). Adapted with permission from Chen et al. (2017), copyright (2020) American Chemical Society

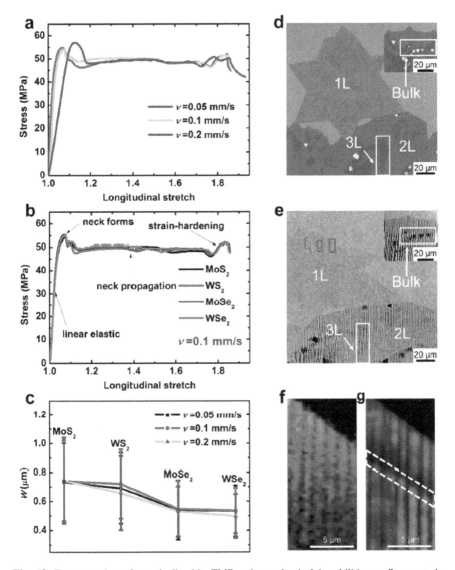

Fig. 12 Fragmentation of atomically thin TMDs via mechanical instabilities. **a** Stress–strain measurements of PC/WS$_2$ monolayers/PMMA at different drawing speeds. **b** Stress–strain measurements of PC/TMDs monolayers/PMMA at drawing speed of 0.1 mm/s. **c** Width of fragmented TMDs monolayers at different drawing speeds. **d, e** Optical images of mono- to multilayer thick WSe$_2$ film before and after the necking process, respectively. **f, g** Raman and PL mapping images (mapping at the $A_1' + E'$ mode and the strongest PL peak from the A exciton) of the section in (**e**) enclosed in a red solid rectangle. The dashed white parallelogram in (**g**) shows the grain boundary in the monolayer WSe$_2$. Adapted with permission from Chen et al. (2017), copyright (2020) American Chemical Society

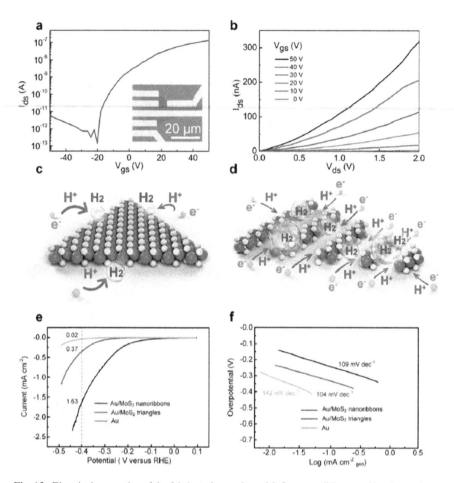

Fig. 13 Electrical properties of the fabricated monolayer MoS₂ nano-ribbons and its electrochemical performance for hydrogen evolution reaction (HER). **a** Transfer characteristic ($I_{ds} - V_{gs}$) of the FET at $V_{ds} = 1$ V, showing an on/off ratio of 106. Inset shows the fabricated FET device based on MoS₂ nano-ribbon arrays. The Si/SiO₂ substrate with the SiO₂ thickness of 270 nm is used as bake-gate. Source/drain (S/D) metal contacts are defined by electron beam lithography, and Cr/Au (5/50 nm) is deposited by thermal evaporation followed by the lift-off process. The channel width $W \sim 10$ μm and channel length $L \sim 2$ μm. **b** Output characteristics ($I_{ds} - V_{ds}$) of the FET under different gate voltages ranging from 0 to 50 V. **c, d** Schematic representations illustrate the function of the edges of the monolayer MoS₂ triangle and monolayer MoS₂ nano-ribbons as the active sites for HER. **e, f** HER polarizing curves and Tafel plots for supporting Au (cyan), monolayer MoS₂ triangles (red), and monolayer MoS₂ nano-ribbons (black). Adapted with permission from Chen et al. (2017), copyright (2020) American Chemical Society

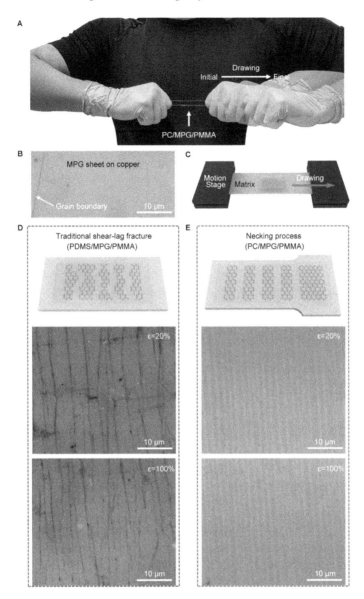

Fig. 14 Fragmentation of Monolayer Polycrystalline Graphene Sheet on a Polydimethylsiloxane Substrate versus on a Polycarbonate Substrate. **a** Photograph of a PC/MPG/PMMA sample undergoing necking process under axial stress. **b** An optical image of the initial MPG sheet on copper. **c** Schematics of MPG sheet under necking processes. **d** Schematics of MPG sheet under the traditional shear-lag fracture (PDMS/MPG/PMMA) and optical morphologies of MPG sheet under the traditional shear-lag fracture with an applied strain of 20% and 100%, respectively. Cracks are random, and the sizes of the fragment pieces decrease as the strain increases. **e** Schematics of MPG sheet under the necking process (PC/MPG/PMMA) and optical morphology of MPG sheet under the necking process with an applied strain of 20% and 100%, respectively. MPG sheet is fragmented into well-ordered ribbons with similar sizes at different strains. Adapted with permission from Chen et al. (2019)

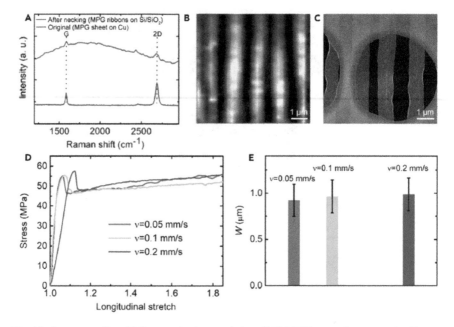

Fig. 15 Spectroscopic and Microscopic characteristics of MPG Ribbons. **a** Representative Raman spectra of MPG in different process steps. **b, c** Representative Raman mapping (**b**) and SEM (**c**) images of the resulting MPG ribbons. **d** Stress–strain measurements of PC/MPG/PMMA at different drawing speeds. **e** Width of the resulting MPG ribbons at different drawing speeds. The error bar is the SD. Adapted with permission from Chen et al. (2019)

The scale of two-dimensional materials has no limit on the occurrence of ordered fragmentation.

The Raman spectra in Fig. 15a demonstrate the high quality of resulted graphene ribbons. The morphology of graphene ribbons is further investigated using AFM and tunneling electron microscopy (TEM). The AFM and TEM images are shown in Fig. 15b, c, respectively.

To demonstrate the potential of resulted graphene ribbons on flexible electronics, the graphene sheet-based and ribbons-based FETs are fabricated and tested on nitrogen-doping and pH sensing. Compared to graphene sheet and sheet-based FETS, the graphene ribbons have an enhanced nitrogen-doping effect and the ribbons-based FET shows a higher sensitivity as the ribbons array owns a larger number of active edges, as shown in Fig. 16.

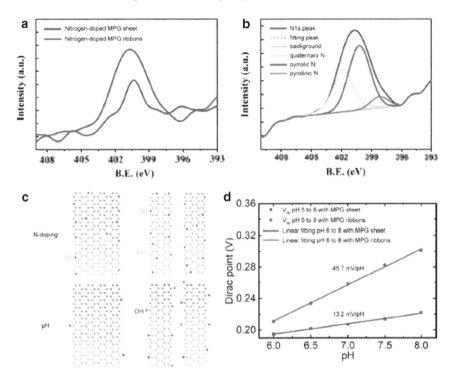

Fig. 16 Enhanced nitrogen-doping effect and pH response performance based on the fabricated MPG ribbons. **a** High-resolution XPS spectra of N1s region of nitrogen-doped MPG sheet and MPG ribbons. **b** Spectrum of N1s region of nitrogen-doped MPG ribbons showing contributions from pyrrolic, pyridinic, and quaternary nitrogen species. **c** Bonding configurations for nitrogen atom (N1, quaternary nitrogen; N2, pyridinic nitrogen; N3, pyrrolic nitrogen) and hydroxyl groups in nitrogen-doped MPG sheet and MPG ribbons. **d** V_{Dir} position versus pH for MPG sheet (blue) and MPG ribbons (red), including linear fittings to the data points (solid lines). Adapted with permission from Chen et al. (2019)

4 Conclusions

Cold drawing, a classic manufacturing process, has been rediscovered as a low-cost robust fabrication tool for materials nano- and micro-structuring. There are mainly two different geometries using in the study, core-shell structured fiber and film-based flat composites. After cold drawing, ordered fragmentation occurred. Using the core-shell fiber structures, micron rods can be produced, and micro-ribbons can be obtained by the film-base flat structures, both on a large scale. The fragmentation process can be analyzed by the shear-lag theory, which explains the mechanical behaviors on the perspective of composites and predicts the resulted morphologies of materials. This versatile method can be applied to a wide range of materials, including metals, glasses, polymers, and even two-dimensional materials. Depends on the mechanical properties of the target material, different results are obtained. For metal/polymer film-based flat composites, a network of metal strips or islands can be produced, which finds its great potential in SERS. For core materials embedded in a polymer shell, micron rods are fabricated and can be etched out for usage. The shear-lag theory is also applicable to explain the mechanical behaviors of two-dimensional materials/polymer composites. In other words, the materials structuring using cold drawing are versatile on two-dimensional materials, which enable a new platform for the study of two-dimensional material-based flexible electronics.

References

A.S. Argon, *The Physics of Deformation and Fracture of Polymers* (New York, Cambridge, 2013)

S. Bertolazzi, J. Brivio, A. Kis, Stretching and breaking of ultrathin MoS2. ACS Nano **5**, 9703–9709 (2011)

W.H. Carothers, J.W. Hill, Studies of polymerization and ring formation. XV. Artificial fibers from synthetic linear condensation superpolymers. J. Am. Chem. Soc. **54**, 1579–1587 (1932)

C.E. Carraher Jr., *Introduction to Polymer Chemistry* (CRC Press, 2017)

A. Castellanos-Gomez et al., Elastic properties of freely suspended MoS2 nanosheets. Adv. Mater. **24**, 772–775 (2012)

M. Chen et al., Ordered and atomically perfect fragmentation of layered transition metal dichalcogenides via mechanical instabilities. ACS Nano **11**, 9191–9199 (2017). https://doi.org/10.1021/acsnano.7b04158

M. Chen et al., Controlled fragmentation of single-atom-thick polycrystalline graphene. Matter (2019). https://doi.org/10.1016/j.matt.2019.11.004

B.D. Coleman, On the cold drawing of polymers. Comput. Math Appl. **11**, 35–65 (1985)

S. Fakirov, D. Bhattacharyya, R.J.T. Lin, C. Fuchs, K. Friedrich, Contribution of coalescence to microfibril formation in polymer blends during cold drawing. J. Macromol. Sci. Part B-Phys. **46**, 183–193 (2007). https://doi.org/10.1080/00222340601044375

J.C. Figueroa, T.E. Carney, L.S. Schadler, C. Laird, Micromechanics of single filament composites. Compos. Sci. Technol. **42**, 77–101 (1991). https://doi.org/10.1016/0266-3538(91)90013-f

K. Friedrich et al., Microfibrillar reinforced composites from PET/PP blends: processing, morphology and mechanical properties. Compos. Sci. Technol. **65**, 107–116 (2005). https://doi.org/10.1016/j.compscitech.2004.06.008

C. Galiotis, A. Paipetis, Definition and measurement of the shear-lag parameter, beta, as an index of the stress transfer efficiency in polymer composites. J. Mater. Sci. **33**, 1137–1143 (1998). https://doi.org/10.1023/a:1004357121802

Y. Gong, V. Carozo, H. Li, M. Terrones, T.N. Jackson, High flex cycle testing of CVD monolayer WS2 TFTs on thin flexible polyimide. 2D Mater. **3**, 021008 (2016)

B.W. Kim, J.A. Nairn, Observations of fiber fracture and interfacial debonding phenomena using the fragmentation test in single fiber composites. J. Compos. Mater. **36**, 1825–1858 (2002). https://doi.org/10.1177/0021998302036015243

K. Li et al., Formation of ultra-flexible, conformal, and nano-patterned photonic surfaces via polymer cold-drawing. J. Mater. Chem. C **6**, 4649–4657 (2018). https://doi.org/10.1039/c8tc00884a

K. Liu et al., Elastic properties of chemical-vapor-deposited monolayer MoS2, WS2, and their bilayer heterostructures. Nano Lett. **14**, 5097–5103 (2014)

G. Loke, W. Yan, T. Khudiyev, G. Noel, Y. Fink, Recent progress and perspectives of thermally drawn multimaterial fiber electronics. Adv. Mater. e1904911 (2019). https://doi.org/10.1002/adma.201904911

J. Loos, T. Schimanski, J. Hofman, T. Peijs, P.J. Lemstra, Morphological investigations of polypropylene single-fibre reinforced polypropylene model composites. Polymer **42**, 3827–3834 (2001). https://doi.org/10.1016/s0032-3861(00)00660-1

J.A. Nairn, On the use of shear-lag methods for analysis of stress transfer unidirectional composites. Mech. Mater. **26**, 63–80 (1997). https://doi.org/10.1016/s0167-6636(97)00023-9

S. Shabahang et al., Controlled fragmentation of multimaterial fibres and films via polymer cold-drawing. Nature **534**, 529–533 (2016). https://doi.org/10.1038/nature17980

E.T. Thostenson, W.Z. Li, D.Z. Wang, Z.F. Ren, T.W. Chou, Carbon nanotube/carbon fiber hybrid multiscale composites. J. Appl. Phys. **91**, 6034–6037 (2002). https://doi.org/10.1063/1.1466880

P. Vincent, The necking and cold-drawing of rigid plastics. Polymer **1**, 7–19 (1960)

X. Wang, B. Zhang, S. Du, Y. Wu, X. Sun, Numerical simulation of the fiber fragmentation process in single-fiber composites. Mater. Des. **31**, 2464–2470 (2010). https://doi.org/10.1016/j.matdes.2009.11.050

Q. Wang, D. Wu, Z. Chen, Ag dendritic nanostructures for rapid detection of thiram based on surface-enhanced Raman scattering. Rsc Adv. **5**, 70553–70557 (2015)

T. Wu et al., Ultrawideband surface enhanced Raman scattering in hybrid graphene fragmented-gold substrates via cold-etching. Adv. Opt. Mater. **7** (2019). https://doi.org/10.1002/adom.201900905

K. Xu et al., Uniaxially stretched flexible surface plasmon resonance film for versatile surface enhanced Raman scattering diagnostics. ACS Appl. Mater. Interfaces. **9**, 26341–26349 (2017). https://doi.org/10.1021/acsami.7b06669

Y. Yang et al., Brittle fracture of 2D $MoSe_2$. Adv. Mater. **29**, 1604201 (2017)

R. Zhang, V. Koutsos, R. Cheung, Elastic properties of suspended multilayer WSe2. Appl. Phys. Lett. **108**, 042104 (2016)

A. Ziabicki, *Fundamentals of Fibre Formation* (Wiley, 1976)

Fiber-Based Triboelectric Nanogenerators

Mengxiao Chen, Xun Han, Xiandi Wang, and Lei Wei

Abstract With the fast-growing demand for flexible and wearable electronics, fibers show great application prospects in this multidisciplinary area due to its breathability, washability, flexibility, and lightweight. Combining its advantages with the newly developed triboelectric nanogenerator technology, wearable biomotion energy harvesting, and multifunctional self-power sensors play an increasingly important role in this coming intelligent era. This technology is to utilize low frequency mechanical energy from dailylife mechanical motions to large-scale seawave energy. Based on the common triboelectric effect, electricity would be generated out of physically manipulating two materials with opposite surface charges. Simply by walking or running, the output current is enough to power sensors, microcontrollers, memories, arithmetic logic units, displays, and even wireless transmitters. Fiber TENGs are expected to both work as convenient energy harvesters to supply wearable electronics and directly work as self-powered real-time wearable sensors to monitor personal healthcare and bio-motions. Through textile method, fibers are also fabricated to into fabric and textile TENGs which makes the harvesting energy a higher amount level with more efficient area and could perform as a multifunctional cloth itself. In this chapter, we will introduce this emerging technique from the basic working principles, and the thoughts of device designs, to the versatile applications as energy harvesters and self-powered sensors. For remaining challenges to face both

M. Chen · L. Wei (✉)
School of Electrical and Electronic Engineering, Nanyang Technological University,
Jurong West 639798, Singapore
e-mail: wei.lei@ntu.edu.sg

M. Chen
e-mail: chenmx@ntu.edu.sg

X. Han
College of Mechatronics and Control Engineering, Shenzhen University, 518060 Shenzhen, China
e-mail: hanxun@szu.edu.cn

X. Wang
Biosensor National Special Laboratory, Key Laboratory for Biomedical Engineering of Education Ministry, Department of Biomedical Engineering, Zhejiang University, 310027 Hangzhou, China
e-mail: wxd_sy@126.com

© Springer Nature Singapore Pte Ltd. 2020
L. Wei (ed.), *Advanced Fiber Sensing Technologies*,
Progress in Optical Science and Photonics 9,
https://doi.org/10.1007/978-981-15-5507-7_13

for energy harvesting textile TENG and TENG sensors, we will discuss and provide some practical insights and viewpoints.

Keywords Multifunctional fiber sensor · Wearable electronics · Triboelectric nanogenerator · Self-powered textile · Fiber TENG

1 Introduction

The triboelectric nanogenerator (TENG) is a newly developed energy-harvesting technology that converts mechanical energy into electrical power through a coupling effect of contact-electrification and electrostatic induction (Dong et al. 1902). Based on the common triboelectric effect, it would generate electricity out of physically manipulating two materials with opposite surface charges. It has drawn attention from all over the world to utilize low frequency daily-life mechanical energy and seawave energy. Simply by walking or running, the output current is enough to power micro-controllers, memories, displays, sensors, arithmetic logic units and even wireless transmitters. Among the demonstrated device form factors, fiber-based TENGs are able to fully utilize fiber electronics' advances of large specific surface area while being lightweight, as well as its feasibility to achieve scalable fabrication.

So far, great progress has been made in this area. TENG fibers have been achieved through dip coating, spray coating or deposition of functional layers onto textiles fibers (Huang et al. 2015; Zhu et al. 2016), wet- or electro-spinning (Yu et al. 2017), and thermal drawing. The fabricated TENG fibers perform good stability, easy pack-aging, waterproof, and most importantly, they are born to be woven into textiles, which means it can lead to wearable and large area applications. Single core shell structured fiber TENG was fabricated and was used to harvest hand motions to power the digital watch and calculator. Textile TENG was designed through weaving two surface material fibers and was intergraded with fiber DSSC component to harvest ambient mechanical energy and solar energy (Wang et al. 2015), then store into a fiber shaped super capacitor (Dong et al. 2017). A V_{OC} of 0.74 V and a J_{SC} of 11.92 mA/cm^2 were achieved, corresponding to an overall power conversion efficiency of 5.64% (Wen et al. 2016).

In this chapter, we will introduce this emerging technique starting from the basic working principle and structure designs (e.g., how current generated between fibers; what kind of working modes could be adopted to make fiber TENGs work, and what kind of structures could be designed to fully utilize the fiber surface.) to the integrating methods and multiple feasible applications (e.g., how to design an out circuit to store the instantaneous electricity and how to develop the fiber device into a large area sensor). The remaining challenges and potential solutions toward fiber-based TENGs are discussed in the end to provide some practical insights and viewpoints.

2 Working Principles and Structure Designs of Fiber-Based TENGs

The triboelectric nanogenerator (TENG) is a newly developed energy-harvesting technology that converts mechanical energy into electrical power based on a combination effect of the electrification effect/triboelectric effect and electrostatic induction. Triboelectric effect describes the phenomenon that the surface of one material becomes electrically charged after it contacts with another material through friction. The primary models are mostly based on plate structures, and then on this basis, fiber TENGs were developed.

2.1 Basic Operating Principle and Working Modes of Fiber TENGs

Here we simply take the general operating condition to illustrate how TENG fibers work in a contact-separation process. As shown in Fig. 1, in one cycle of a contact–separation, charge migration occurs between the upper fiber and lower fiber, both

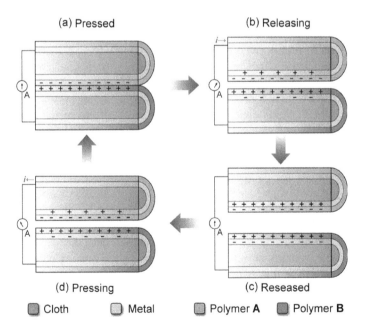

Fig. 1 Basic operating principle of the fiber-based triboelectric generator on the short-circuit condition. Schematic of the electricity-generation process in a full cycle under the vertical contact-separation mode. Two fibers with different polymer shell. **a** Original state. The two fibers are in full contact under pressed. **b** Releasing from each other. **c** Fully released. **d** Get close to each other

of which contain a supporting core, a conductive layer and a polymer shell. In the beginning, there is no charge migration, as there is no electrical potential difference. When the upper and lower fiber contact with each other, electrification occurs at the interfaces, generating opposite polarized charges on fiber surfaces with the same amount (Fig. 1a). The polarities of the two sides are up to their natural ability to keep and lose electrons (triboelectric sequence). Since the opposite triboelectric charges on the two fiber surfaces coincide at the same amount, there is actually no electrical potential difference between them. Once the upper fiber gradually moving away from the lower one, negative and positive charges are induced in their electrodes correspondingly, according to the electrostatic induction effect (Fig. 1b). As the two fibers further separate, the accumulated electrical potential difference drives electrons to flow between the electrodes, as a result, an instantaneous current is generated. When two fibers completely separated, the negative and positive charges on fiber surfaces are fully equilibrated by the electrostatic induced charges on their electrodes (Fig. 1c). At this time, it is electrostatic balanced, so there is no electrical signal in the circuit. In the reverse process, when the two fibers are approaching back to each other, to compensate the electrical potential difference, the previously electrostatic induced charges flow back through the external load (Fig. 1d). While the system returns to the original state (Fig. 1a), the positive and negative triboelectric charges are fully offset again. In this process, a contact–separation circle between the upper and lower fibers generates an instantaneous alternating electrical signal.

Depend on the above basic operating principle, several working modes could be designed and adopted in fiber-based TENG devices. Three most common used working modes are introduced in Fig. 2. The first one is the vertical contact-separation or lateral-sliding mode. There are two kinds of fibers with inner electrodes, and the two electrodes are connected to form a circuit, the contact-separation process can be achieved both in a vertical contact-separation and lateral-sliding process as illustrated in Fig. 1. Figure 2a shows the simple working process of this mode, and based on this mode, not only two single fibers work, but also two kinds of fibers could work together by weaving them into cloth-like devices. Another working mode is the single-electrode one, through which, only one working fiber needs electrode inside. As shown in Fig. 2b, the circuit formed between the fiber electrode and the ground. The triboelectric effect and electrostatic induction still work in the same way, but the currents go throughout loads between the fiber electrode and the ground. This working mode does not require a connection between two working fibers, so it expands a large amount of applications. For example, only the sensing part with electrodes fixed on target objects (e.g., human body), the moving part is free from connecting wires. The third working mode is the free stranding triboelectric-layer mode as show in Fig. 2c. The inner electrodes of the same kind fibers are jointed to form the circuit, so the currents go throughout loads between the two connected fibers. In this mode, also only one working component needs to be fixed, and the other one is free. When woven into fabrics, odd numbered fibers connected together, even numbered fibers connected together, then out load is put in the middle of the two. These three working modes cover most working conditions of fiber and fabric-based TENGs, and device designs are conducted on this basis.

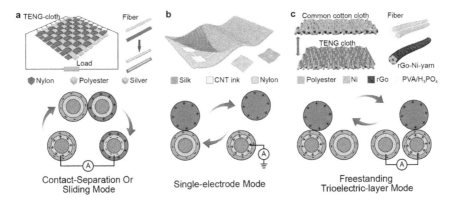

Fig. 2 Three working modes when fibers developed into textile TENGs. a Vertical contact-separation or lateral-sliding mode. Two inner fiber electrodes are attached together. Reproduced with permission (Zhou et al. 2014). Copyright 2014, American Chemical Society. **b** Single-electrode working mode. The electrical signals are generated through the contact-separation process between the grounding fibers and dielectric fabrics. Reproduced with permission (Cao et al. 2018). Copyright 2018, American Chemical Society. **c** Freestanding triboelectric-layer mode. The inner electrodes of the same kind fibers are jointed together as an interdigital electrode. Reproduced with permission (Pu et al. 2016). Copyright 2016, Wiley-VCH

2.2 Structure Designs for Fiber TENG Devices

Fiber TENGs can be divided into two kinds: single fiber device and woven fabric and textile TENGs. For single fiber TENG devices, they usually designed as core shell structure, and work in a single-electrode mode. Three typic device designs are listed in Fig. 3. Figure 3a shows a soft single fiber TENG device, triboelectric effect happens at the interface of the silicone rubber surface and the copper wire surface, the fiber electrode layer is connected with the out load and then to the ground. Figure 3b, c are another two core shell structured single fiber devices. Figure 3b reports a stretchable triboelectric fiber (STEF) formed a multilayered core shell and wrinkle structure. The key working principle for electrical signal generation is the reversible distance change induced by the Poisson's ratio difference between the core fiber (silver-coated nylon/ polyurethane) and the shell (wrinkled polyvinylidene fluoride-co-trifluoroethylene/carbon nanotube layer) during tensile deformations. In Fig. 3c, the fiber device includes a polydimethylsiloxane (PDMS) substrate supporting fiber, two multilayer aligned carbon nanotube (CNT) sheets served as inner and outer electrodes, and PDMS and polymethyl methacrylate (PMMA) were used as contact materials for triboelectrification. In addition, an air gap was implemented for effective contacts and separations of the triboelectric materials. To summarize, single fiber TENG device, core shell (multilayer core shell) structures are utilized to form the functional triboelectrification polymer layer and conductive electrode layer, and a gap between layers is necessary to induce the charge variation process.

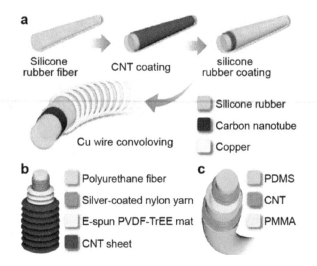

Fig. 3 Structure design of single fiber TENG device. a–c Typical preparation process of a single fiber and different fiber structures for single fiber-based TENGs. Reproduced with permission (Yu et al. 2017; He et al. 2017; Sim et al. 2016). Copyright 2017, Wiley-VCH. Copyright 2016, Springer Nature. Copyright 2017, Royal Society of Chemistry

As investigated above, besides gap distance change between coaxial layers, motions and movements among fibers can also be utilized to generate electrical outputs. Fabric and textile TENGs are developed by weaving fibers together to obtain a larger area energy harvester, and fibers with different functions (e.g., solar cell fibers and Li-battery fibers) can be easily integrated in the same time. In Fig. 4a, a wearable power unit integrated by a whole-textile TENG cloth and a flexible lithium-ion battery (LIB) belt is demonstrated. With the help of a conformal Ni film, common flexible but insulating polyester fabrics were transformed into conductive. Ni film works both as electrodes in the TENG cloth and as current collectors in the LIB belt. When being worn on the human body, the TENG cloth is an energy collector of various human motions, converting the mechanical energy into electricity; LIB was charged by the TENG cloth. Another textile TENG was fabricated by direct weaving of Cu-coated polyethylene terephthalate (Cu-PET) warp yarns and polyimide (PI)-coated Cu-PET (PI-Cu-PET) weft yarns through an industrial sample weaving loom as shown Fig. 4b. As illustrated in Fig. 4b, the textile TENG consists of 2-ply Cu-PET yarns and PI-Cu-PET yarns, and they were woven as warp (vertical) and weft (horizontal), respectively. With Cu and PI serving as the triboelectric materials, the 2-ply Cu-PET warp yarn acts as both electrode and frictional surface, while the PI-Cu-PET yarn provides another frictional surface with its inner 2-ply Cu-PET yarn acting as its electrode. It is worth to mention that this textile TENG does not require an additional substrate when works, every yarn crisscross intersection of this single-layer TENG works as an individual TENG. Furtherly, a double-faced 3D interlock textile was proposed in Fig. 4c. Cotton yarn was chosen as the positive triboelectric material, a four-ply twisted polyamide (PA) yarn coated with Ag was used as the

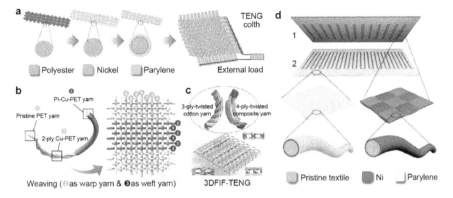

Fig. 4 **Schematic illustration of the fabrication of TENG textile based on the freestanding triboelectric-layer mode. a** Fabrication of a two-dimensional TENG cloth. Reproduced with permission (Pu et al. 2015). Copyright 2015, Wiley-VCH. **b** Textile TENG cloth based on Cu-PET yarn and 2-ply Cu-PET yarn. Reproduced with permission (Zhao et al. 2016). Copyright 2016, Wiley-VCH. **c** A three-dimensional double-faced interlock fabric TENG cloth. Reproduced with permission. (Chen, et al. 2019) Copyright 2019, Elsevier. **d** The scheme of a TENG fabrics consisting of a slider fabric (1) and a stator fabric (2). Reproduced with permission (Pu et al. 2016). Copyright 2016, Wiley-VCH

conductive electrode. The interlock fabric is knitted with both cotton yarn and PA composite yarn. Two rib-stitch structures are knitted with each other to obtain the interlock structure. This interlock stitch is also a double-faced effect interlock structure. Cotton loop and PA composite yarn loop are back to back, that is, each cotton loop is right behind PA composite yarn loop of another face and so as to each PA composite yarn loop. This work presents good electrical output performances during contact-separation operation, bend-stretch mode and only-stretch mode in transverse direction. Another wearable TENG fabric is depicted in Fig. 4d with a pair structure. The upper fabric (1) has a series of parallel grating electrode segments, and its number, size, and spacing are identical to the below part (2). All of the metal electrodes/segments are composed of woven fibers with conformal Ni coatings. For the below fabrics (2), an additional parylene layer was applied on the top of Ni by chemical vapor deposition (CVD), serving as a friction layer when contacts with the parallel Ni on part (1). These two fabrics could be placed on the sleeve and underneath the arm functioning as a pair of sliding-mode TENGs. They also have a good performance after machine washing.

3 Practical Applications of Fiber-Based TENGs

Fiber-based TENGs can be used to harvest mechanical energy both in exploiting small mechanical energy and large-scale energy generation, and to be used as a multifunctional sensor to monitor human movement and health. Various structures have been designed to harvest different types of energies such as the waterwaves, wind, raindrops, and human motion among which it is most promising in wearable related devices. So in this section we mainly introduce small and wearable applications.

3.1 Fiber TENGs Working as Energy Harvesters

By weaving the fiber TENGs into clothes, carpet and other fabrics, energies from human motion can be easily harvested to construct the self-powered system. Six fiber TENGs were integrated in parallel to enhance the output current as a power supply. A digital watch/calculator can be lit up by the integrated fiber TENGs with hand motions (Fig. 5a). To support the calculator continuously work, the pulsed output of fiber TENGs was rectified to continuous direct output through a designed circuit, including a rectifier and a commercial capacitor (Fig. 5b). In addition, the LED arrays can also be lighted up by the fiber TENGs. As shown in Fig. 5c, d, the word "HELP" consisting of 35 green LEDs was sufficiently lighted up by the 3DFIF-TENG wrapping up through bare hand tapping. In another reported work, 27 green LEDs can be lit up by a fabric structured TENGs through a human foot stepping (Fig. 5e, f). It is worth note that the fabric structured TENG demonstrates satisfactory efficiency for energy harvesting even though with half-covered active area by a human's foot, which shows potential application in square and subway.

Lightweight and self-charging power modulus with high efficiency of energy harvesting and storage are urgently desired in wearable electronics. One straightforward approach is to integrate the current energy-harvesting devices, such as TENG and DSSC, with conventional rechargeable energy storage technologies, e.g., Li-ions battery and super capacitor, on the fabric. Chen et al. reported a micro-cable power textile based on the integration of triboelectric textiles and photovoltaic for harvesting mechanical and solar energy simultaneously. Under ambient sunlight and mechanical excitation, the hybrid power textile can charge a 2mF commercial capacitor up to 2 V in 1 min, demonstrating strong capability of energy harvesting (Fig. 6a). In addition, the textile can also serve as a wearable power supply to charge a cell phone (Fig. 6b) and drive an electronic watch continuously (Fig. 6c). As shown in Fig. 6d, the output power from each component demonstrates similar trend with load resistance increasing but in different range. Thus, considering the various operational resistance of electronic devices, the optimization of the two components hybridization can not only improve output power but also expand the range of load resistances. To sustainably operate the wearable devices, the supercapacitor and Li-ions battery are utilized to store the harvested energy. Figure 6e demonstrates a self-charging power textile integrated with fabric supercapacitors to power wearable electronics. It was attached on outwears t harvest energy from ambient environment and subsequently store into three yarned supercapacitors. The equivalent circuit of the textile is shown in Fig. 6f, which uses a bridge rectifier to convert AC to DC and charge the connected supercapacitors. Pu and co-workers adopted the Li-ions battery belt (LIB) to store the delivered power from a TENG cloth for the self-charging power unit.

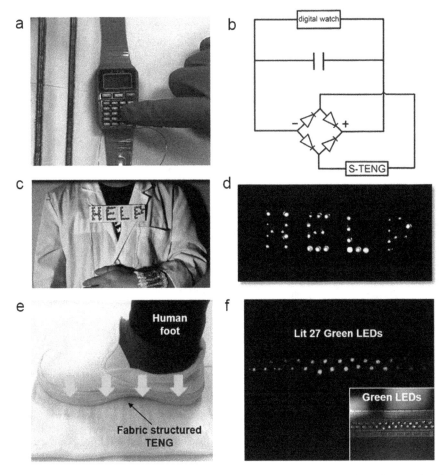

Fig. 5 Primary fiber-based TENG energy harvesters with simple external circuits. **a** Photography of a working digital watch/calculator lit up by the integrated fiber TENG and **b** corresponding managing circuit of this self-powered system for continuous direct outputs. Reproduced with permission (He et al. 2017). Copyright 2017, Wiley-VCH. **c** Photography of a double-faced interlock fabric TENG wrapped on the wrist and **d** LEDs marked as alphabets "HELP" lit up by tapping the TENG. Reproduced with permission (Chen et al. 2019). Copyright 2019, Elsevier. **e** Photography of a footstep on a fabric structured TENG composed of 6 × 6 fibers and **f** 27 commercial green LED lit up upon walking step. Reproduced with permission (Kim et al. 2015). Copyright 2015, American Chemical Society

A heartbeat meter was then powered by this LIB and corresponding signals were transmitted to a smart phone through wireless technologies, to realize self-powered healthcare monitoring (Fig. 6g, h). As shown in Fig. 6i, the self-charging power unit possessed simple equivalent circuit, which can be easily modified for other wearable electronic devices.

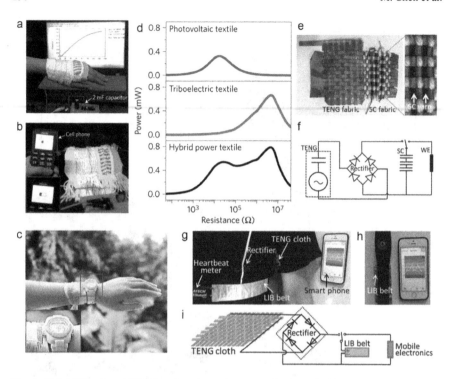

Fig. 6 Hybrid fiber-based TENGwork as energy harvesters integrated with energy storage components. Demonstration of the hybrid power textile to charge **a** a 2mF commercial capacitor, **b** a cell phone and **c** continuously drive an electronic watch under natural daylight and mechanical excitation. **d** Output power of individual components and the hybrid power textile as a function of the load resistances. Reproduced with permission (Chen et al. 2016). Copyright 2016, Wiley-VCH. **e** Photograph of the self-charging power textile yarned with supercapacitor fabric to be used as wearable electronics. **f** Circuit diagram of a wearable electronics device driven by the self-charging powered textile. Reproduced with permission (Pu et al. 2016). Copyright 2016, Wiley-VCH. **g** Photography of the integrated self-charging power system with energies been harvested by TENG cloth, stored in LIB belt, and then to power a heartbeat meterstrap, which is in remote communication with a smart phone. **h** Backside of the heartbeat meter showing that it is driven by the LIB belt. **i** Circuit diagram of the self-charging power system. Reproduced with permission (Pu et al. 2015). Copyright 2015, Wiley-VCH

3.2 Fiber TENG Sensors

TENG can harvest not only energy from ambient environment, but also be used as self-powered multifunctional sensors. Kanik and co-workers reported a fiber-based microfluidic TENG for energy harvesting from droplets as well as biochemical sensing. As shown in Fig. 7a, b, the system consists of two components: microfluidic chip and fiber-based TENG. The microfluidic chip has two inlets to generate droplets connected with DI water and air supplies. There is one outlet coupled with a PVDF hollow fiber to allow droplets transfer inside, by tuning relative pressure between the

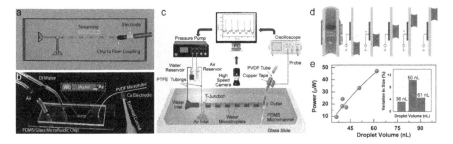

Fig. 7 A TENG microfluidics chip fabricated by a thermal drawing PVDF hollow fiber to sense droplets. Schematic diagram (**a**) and photography (**b**) of the droplets sensing system consisting a PVDF hollow fiber coupling to the outlet of a droplet generator microfluidic chip. Inset is the droplet with biconcave shape generated in microfluidic channel. Scale bar is 300 μm. **c** Experimental setup for droplets sensing, including droplet generation by tuning the pressure between water and air inlets and droplet sensing through PVDF hollow fiber and Cu electrodes. **d** Schematic diagram of the mechanism of induced charges transport during droplet went through the PVDF tube. **e** Dependence of generated power on the droplet volume. Inset is the size variation from different droplet volume. Reproduced with permission (Kanik et al. 2016). Copyright 2016, Wiley-VCH

two inlets. The fiber-based TENG comprises of a PVDF hollow fiber wrapped up by a copper tape electrode connecting with the external analyzing circuit. The system was characterized by experimental setups shown in Fig. 7c through measuring the output voltage and current which are closely related to droplet size. The working mechanism is shown in Fig. 7d, it is a single-electrode working mode as demonstrated in Fig. 2b previously. Droplets surfaces keep positive charged. Electrons transfer between coper tape electrode and out circuit play the role to realize electrostatic equilibrium when there's potential change as the droplet moves near and moves away. Detail charge transfer process could be referred to Sect. 2.1 contents. From the working mechanism, the output signal value is related to the overlapping area of droplet on Cu electrode, which is decided by the droplet volume. So the output power increases with the droplet volume, and a maximum power of 47 μW in sensing region is achieved through the longest droplet (Fig. 7e). In addition, the output power variations are also observed which is caused by the droplets size variation.

As pressure sensors have been developed through various methods, monitoring sleep behavior with posture detecting and real-time physiological analysis is highly desired. The TENGs with superior pressure sensitivity, low-cost, and simple fabrication process have emerged as a potential candidate for sleep monitoring. Based on a 9 × 11 TENG array Lin and co-workers reported a large-scale and washable self-powered smart textile to monitor sleep behavior and give out warnings. The smart textile which mainly consists of three layers can serve as a bedsheet for sleep monitor (Fig. 8a). The top layer and bottom layer bedsheets are both comprised of conductive fiber arrays in perpendicular column and row. The middle layer is a wavy-shaped PET film, which is sandwiched between the top and bottom conductive fiber arrays. The SEM image of the conductive fiber is presented in Fig. 8c, demonstrating uniform Ag coating on the fiber rendering the smooth surface. The photography of

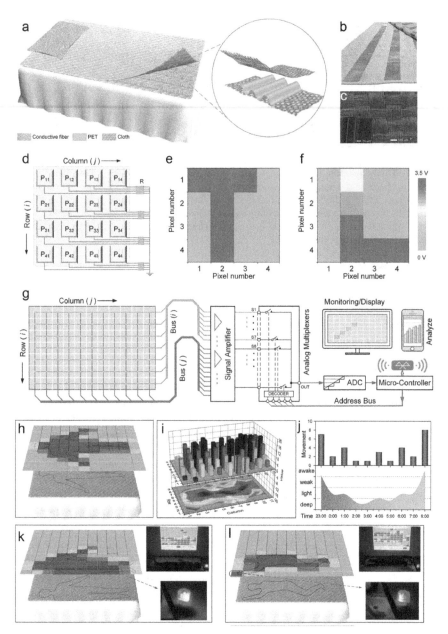

Conductive fiber   PET   Cloth

◄**Fig. 8** A fiber-based TENG work as self-powered smart textile to monitor sleeping behavior. **a** Schematic of the smart textile based on a 9 × 11 TENGs array. Inset to show an enlarged view of a single unit TENG textile structure. **b** Photograph of a self-powered smart textile. **c** SEM image to show the conductive fiber in the smart textile. **d** Schematic illustration of the customized multichannel data acquisition system to monitor sleeping behavior and record in real-time. The output voltage from a 4 × 4 sensors array by applying external pressure through a T-shaped mass with **e** uniform and **f** gradient cross section height. **g** Schematic diagram to illustrate the working principle of the fiber TENG based self-powered smart textile. **h** 2D pressure mapping results showing the position and posture of human body. **i** Distribution of the press number for a sleeper over an entire night. **j** Statistic results of the active number for a sleeping period from 23:00 p.m. to 08:00 a.m. and the corresponding sleep quality report. **k** Self-powered falling warning and (l) help-seeking system based on the smart textile. Reproduced with permission (Lin et al. 2018). Copyright 2018, Wiley-VCH

the as-fabricated smart textile bedsheet is shown in Fig. 8b, with the dimension of 2 m × 1.5 m. According to the mechanism of TENG, external pressure applied on a TENG unit will lead to an electrical output.

To characterize the pressure sensitivity of the smart textile, a customized multi-channel acquisition system was designed to connect a 4 × 4 TENG array with each pixel addressable (Fig. 8d). The external pressure was applied on the device through a T-shaped mass with the same cross section height and gradient cross section height, respectively. As shown in Fig. 8e, f, both the pressure distribution and pressure variations can be clearly mapped through the output voltage from all pixels with different colors corresponding to the pressure variation. In addition, a comprehensive activity recognition system was developed based on the smart textile to integrate posture detecting, real-time monitoring and falling warning (Fig. 8g). The pressure signals were detected through the smart textile, then were transferred to the analog-to-digital converter and the micro-controller through row/column data bus to generate a pressure map, and subsequently a data stream consisted of pressure images was sent the external devices through Bluetooth module, thus realizing real-time monitoring. The real-time sleep behavior was monitored and summarized through this integrated system. The pressure distribution of a tester lying on the smart textile is shown in Fig. 8h, which also exhibits the posture and position of the tester. The distribution of sensor pressing number of a tester over the entire night was recorded, which present that the tester lay in the middle of the textile for most of the time while the activity of his head and legs occupied the wider range (Fig. 8i). In addition, sleep quality analysis can also be realized through recording the active number of the textile. Figure 8j shows the statistic results of the active number in a sleep period from 11:00 p.m. to 8:00 a.m. and corresponding sleep quality analysis. The activities were mostly concentrated in the fall asleep and wake up time, which is accord with a normal sleep circle. Besides, a deep sleep of 3 h around 2:30 a.m. was observed, which indicates a good sleep quality of the tester.

The smart textile can be further developed for falling warning and help-seeking system, not only limited in sleep behavior monitoring. As shown in Fig. 8k, when sleeper unconsciously moved to the bed edge and press the five corresponding

TENGs, alerting message would be sent to the external devices and the alarm lamp was lit up. Additionally, a TENG unit at a corner on the smart bedsheet was defined as the help-seeking button, through which sleepers can press to inform family members and/or their doctors (Fig. 8l). These novel functions of the smart textile have provided new prospects for real-time monitoring and other related healthcare service.

Textile-based TENGs have demonstrated remarkable superiority in self-powered active sensing for wearable electronics. A critical challenge of the large-scale manufacturing of the textile TENGs hampers its practical applications. Zhao et al. reported a machine washable textile TENG through loom weaving metallic yarns for respiratory monitoring. The textile TENG is consisted of the 2-ply Cu-PET yarns and PI-Cu-PET yarns woven as warp and weft, respectively. Researchers developed a customized signal process system to deal with the relative low frequency respiratory rate and filter the false signal caused by chest movement and noise from environment when collect the respiratory data (Fig. 9a). During inhalation, chest cavity and abdominal cavity will expand to stretch the textile TENG, generating the output voltage, while the exhalation can induce contraction of chest cavity and abdominal cavity to release the textile TENG, thus leading to an opposite output signal. A typical raw respiratory signal is shown in Fig. 9b to reveal a complete breathing cycle including inhalation and exhalation. After signal processing, a clear sinusoidal wave is observed, defining the datum line of K to recognize the effective respiratory signal (Fig. 9c). Four states of respiratory including deep and shallow, rapid and slow, were investigated through the textile TENG (Fig. 9d). Corresponding respiratory rate and tidal volume representing the respiratory depth were calculated and summarized in Fig. 9e and f, respectively. The four respiratory states can be clearly recognized from the results, demonstrating good sensitivity as a respiratory sensor.

4   Conclusion

This newly developed combination technology of fiber fabrication and TENG devices provides versatile possibilities to achieve various fiber, fabric and textile devices to power wearable and flexible electronics and directly work as self-powered multifunctional sensors. It brings a viable path to achieve soft electronics. However, fiber/textile-based TENG, there still remains some challenges to face both for energy-harvesting textile TENG and TENG sensors. One is that most fabric-based TENGs require two or more fabrics to do contact-separation and contact sliding motion which limits the application. Another one is that: as the original 1D structure, most applications are to be woven into 2D structures, but actually it could even be further developed into 3D structures which reveal unlimited potential unique applications. So further directions could be on the fiber inside structure design and 3D applications (e.g., 3D weaving to irregular shaped surfaces).

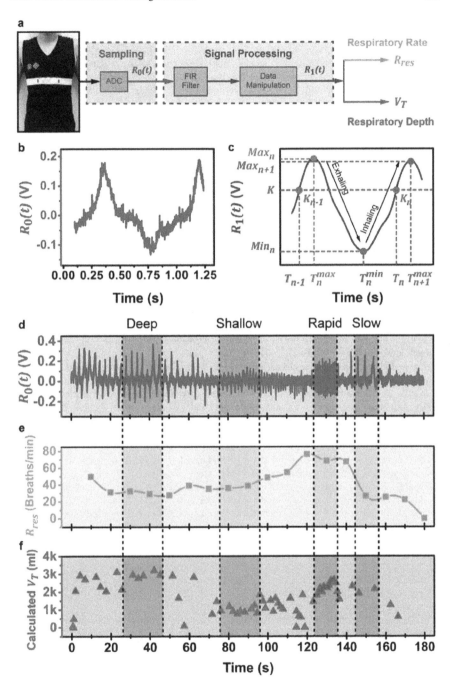

◀**Fig. 9** A textile TENG as a self-powered wearable respiratory monitor. **a** Customized data acquisition system for respiratory monitoring. **b** Raw respiratory signal from a complete breath cycle. **c** Processed respiratory signal with parameters marked for the respiratory data analysis. **d** Raw respiratory signal from four different respiratory states including deep, shallow, rapid, and slow. **e** Corresponding respiratory rate and **f** the tidal volume from the four respiratory states. Reproduced with permission (Zhao et al. 2016). Copyright 2016, Wiley-VCH

References

C. Chen, et al., 3D double-faced interlock fabric triboelectric nanogenerator for bio-motion energy harvesting and as self-powered stretching and 3D tactile sensors. Mater. Today (2019)

R. Cao et al., Screen-printed washable electronic textiles as self-powered touch/gesture tribo-sensors for intelligent human–machine interaction. ACS Nano **12**, 5190–5196 (2018)

J. Chen et al., Micro-cable structured textile for simultaneously harvesting solar and mechanical energy. Nature Energy **1**, 1–8 (2016)

K. Dong et al., A highly stretchable and washable all-yarn-based self-charging knitting power textile composed of fiber triboelectric nanogenerators and supercapacitors. ACS Nano **11**, 9490–9499 (2017)

K. Dong, X. Peng, Z.L. Wang, Fiber/fabric-based piezoelectric and triboelectric nanogenerators for flexible/stretchable and wearable electronics and artificial intelligence. Adv. Mater. 1902549 (2019)

X. He et al., A highly stretchable fiber-based triboelectric nanogenerator for self-powered wearable electronics. Adv. Func. Mater. **27**, 1604378 (2017)

T. Huang et al., Human walking-driven wearable all-fiber triboelectric nanogenerator containing electrospun polyvinylidene fluoride piezoelectric nanofibers. Nano Energy **14**, 226–235 (2015)

M. Kanik, M. Marcali, M. Yunusa, C. Elbuken, M. Bayindir, Continuous triboelectric power harvesting and biochemical sensing inside poly (vinylidene fluoride) hollow fibers using microfluidic droplet generation. Adv. Mater. Technol. **1**, 1600190 (2016)

K.N. Kim et al., Highly stretchable 2D fabrics for wearable triboelectric nanogenerator under harsh environments. ACS Nano **9**, 6394–6400 (2015)

Z. Lin et al., Large-scale and washable smart textiles based on triboelectric nanogenerator arrays for self-powered sleeping monitoring. Adv. Func. Mater. **28**, 1704112 (2018)

X. Pu et al., A self-charging power unit by integration of a textile triboelectric nanogenerator and a flexible lithium-ion battery for wearable electronics. Adv. Mater. **27**, 2472–2478 (2015)

X. Pu et al., Wearable self-charging power textile based on flexible yarn supercapacitors and fabric nanogenerators. Adv. Mater. **28**, 98–105 (2016a)

X. Pu et al., Wearable power-textiles by integrating fabric triboelectric nanogenerators and fiber-shaped dye-sensitized solar cells. Adv. Energy Mater. **6**, 1601048 (2016b)

H.J. Sim et al., Stretchable triboelectric fiber for self-powered kinematic sensing textile. Scient. Rep. **6**, 1–7 (2016)

J. Wang et al., A flexible fiber-based supercapacitor–triboelectric-nanogenerator power system for wearable electronics. Adv. Mater. **27**, 4830–4836 (2015)

Z. Wen et al., Self-powered textile for wearable electronics by hybridizing fiber-shaped nanogenerators, solar cells, and supercapacitors. Sci. Adv. **2**, e1600097 (2016)

X. Yu et al., A coaxial triboelectric nanogenerator fiber for energy harvesting and sensing under deformation. J. Mater. Chem.A **5**, 6032–6037 (2017)

Z. Zhao et al., Machine-washable textile triboelectric nanogenerators for effective human respiratory monitoring through loom weaving of metallic yarns. Adv. Mater. **28**, 10267–10274 (2016)

T. Zhou et al., Woven structured triboelectric nanogenerator for wearable devices. ACS Appl. Mater. Interfaces. **6**, 14695–14701 (2014)

M. Zhu et al., 3D spacer fabric based multifunctional triboelectric nanogenerator with great feasibility for mechanized large-scale production. Nano energy **27**, 439–446 (2016)

Fiber-Shaped Energy-Storage Devices

Qichong Zhang, Bing He, Jiao Yang, Ping Man, and Lei Wei

Abstract With the rapid development of science and technology, portable and wearable electronic devices presented a prominent technological trend for future lifestyles and they have brought enormous convenience to our daily life. In order to realize the wearability of the whole equipment, it is necessary to develop matching high flexible, lightweight and small volume energy supply devices. The volume of planar flexible energy-storage device is too large to be integrated into the fabric, so it is difficult to give full play to the advantages of energy storage device. The fiber-shaped energy storage devices with their unique advantages of tiny volume, high flexibility and remarkable wearability have triggered wide attention. Thus, developing high-performance fiber-shaped energy storage devices is recognized as a promising strategy to address the above issues. This chapter discusses the design principles and device performance of fiber-shaped energy storage devices. In the first section, design principles of fiber-shaped energy storage devices with fiber electrode, electrolyte and device configurations are presented. In the next section, the development of fiber-shaped energy storage devices, including supercapacitors, nonaqueous and aqueous batteries, are comprehensively summarized, with particular emphasis on electrochemical and mechanical properties. The existing challenges and future directions are finally discussed to provide some useful insights from the viewpoint of practical applications.

Keywords Fiber-shaped energy storage device · Supercapacitor · Lithium-ion battery · Lithium-sulfur battery · Lithium-air battery · Zinc-ion battery · Aqueous battery · Alkaline battery

Q. Zhang (✉) · J. Yang · L. Wei (✉)
School of Electrical and Electronic Engineering,
Nanyang Technological University, Singapore 639798, Singapore
e-mail: zhangqc@ntu.edu.sg

L. Wei
e-mail: weilei@ntu.edu.sg

B. He · P. Man
Division of Advanced Nanomaterials, Suzhou Institute of Nano-Tech
and Nano-Bionics, Chinese Academy of Sciences, 215123 Suzhou, China

© Springer Nature Singapore Pte Ltd. 2020   259
L. Wei (ed.), *Advanced Fiber Sensing Technologies*,
Progress in Optical Science and Photonics 9,
https://doi.org/10.1007/978-981-15-5507-7_14

1 Introduction

Early attempts in wearable energy storage include mounting existing components on clothes or other accessories, such as batteries and supercapacitors that are rigid and unwashable, and have hence limited the broad uptake of wearable technologies (Lu et al. 2013; Liu et al. 2012; Guan et al. 2016). To improve device flexibility, flexible functional components (e.g., electrodes, electrolytes, and current collectors) have been integrated into flexible films. In addition, the film-shaped substrates are often impermeable to air and/or moisture, vulnerable to twisting or wrapping and of low wear comfort. Fiber electronics represent one of the emerging technologies with broad potential impact on wearable, flexible and conformable electronic systems (Xu, et al. 2019; Wang et al. 1971; Weng et al. 2016). Flexible textile electronic devices require flexible textile/fiber energy storage devices as compatible power suppliers. To match flexible textile electronic devices, the energy storage devices should have similar textile/fiber shapes with excellent flexibility, mechanical stability, light weight and can also bear deformations in all dimensions. Thus, intense effort has recently been directed toward fiber-shaped energy storage devices, including fiber-shaped supercapacitors, nonaqueous and aqueous batteries, by incorporating active materials into fibers, and then weaving or knitting these functional fibers into fabrics or textiles to make the devices air/moisture permeable (Zhai et al. 2019; Chen et al. 1806; Zhu et al. 1961; Mo et al. 2019). Their 1D structures offer various advantages over their conventionally rigid and bulky counterparts: (1) they show a higher degree of mechanical flexibility; (2) allow easy scaling up by self-integration; (3) are easily packed into small spaces with diverse shapes; and (4) have shape advantage for integrating with other fiber-shaped devices (which is preferable for the fabrication of multifunctional wearable systems).

Over the past decades, researches on fiber-shaped energy storage devices expanded very fast and great progresses have been achieved. The key contents related to the fiber-shaped energy storage devices, including the synthesis of fiber electrode, the design of device configurations, electrochemical and mechanical properties are carefully discussed for supercapacitors, nonaqueous and aqueous batteries. The remaining challenges and potential solutions toward fiber-shaped energy storage devices with improved performances for next generation wearable electronics are finally discussed to provide insights for the future development.

2 Design Principles of Fiber-Shaped Energy Storage Device

2.1 Fiber Electrode

Fiber electrode, as the key component to realize the high-performing fiber-shaped energy storage device, usually consists of two parts: current collector and active substrate. The desired fiber current collector is required to possess superior conductivity and excellent mechanical flexibility. Generally, metal wires with unexceptionable electrical conductivity are heavy and rigid, while the lightweight polymeric fibers with good flexibility are less conductive. By contrast, carbon-based fibers (carbon fiber, carbon nanotubes fibers and graphene fibers) have great potential as flexible and lightweight current collectors in the fiber-shaped energy storage device owing to their favourable conductivity, relatively low mass density, high tensile strength and high surface area (Yu et al. 2016; Chen et al. 1806; Zhu et al. 1961; Mo et al. 2019). Their scanning electron microscopy (SEM) images of abovementioned carbon-based fibers are presented in Fig. 1a–c. Besides, the interfacial adhesion, mass density, specific surface area and production cost should also be taken into

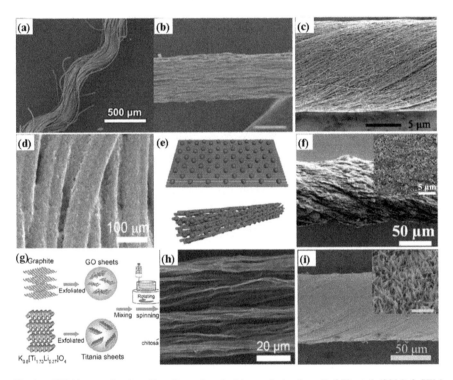

Fig. 1 a SEM image of carbon fiber. Reproduced with permission from Ref. Yu et al. (2016). **b** SEM image of graphene. Reproduced with permission from Ref. Zheng et al. (2014). **c** SEM image of carbon nanotube fiber. Reproduced with permission from Ref. Chen et al. (2013). **d** SEM image of the carbon cloth fiber after coating LiMn$_2$O$_4$ nanoparticles. Reproduced with permission from Ref. Hoshide et al. (2017). **e** Schematic illustration of the synthesis of high-performance CNT-LMO composite fiber. **f** SEM images of CNT-LM2017O composite fiber at increasing magnification. **e–f** Reproduced with permission from Ref. Weng et al. (2014). **g** Schematic illustration of the fabrication process for the hybrid fiber of titania/rGO. **h** SEM image of as-spun dry titania/GO fiber. **g–h** Reproduced with permission from Ref. Hoshide et al. (2017). **i** SEM images of vanadium nitride nanowire arrays on carbon fiber with increasing magnifications. Reproduced with permission from Ref. Zhang et al. (2017)

account for fiber electrode for fiber electrode. Electrode fabrication is key step in designing high-performance fiber-shaped energy-storage device. Up to now, there are three basic strategies for manufacturing fiber electrodes in current research. (1) Directly coating active materials onto the fiber substrates. The surface-coating method is a widespread cost-efficient method for large-scale fiber electrode production. However, the major drawback of this surface-coating method is the limited adhesion between the active layer and the curved surface of the fiber substrate, which may lead to unsatisfactory electrochemical performance and abscission of active substance during physical deformation. As shown in Fig. 1d, the fiber lithium-ion cathode was made by coating slurry of $LiMn_2O_4$ (LMO) nanoparticles on a carbon cloth wire (Hoshide et al. 2017). (2) Incorporating active materials into fiber substrates by utilizing spinning techniques(dry-spinning and wet-spinning) to achieve hybrid fiber electrodes. Figure 1e shows that LMO particles are directly deposited onto the aligned CNT sheet, followed by scrolling into a CNT-LMO composite fiber. Figure 1f shows a typical SEM image of the resulting CNT-LMO fiber with a uniform diameter of around 100 μm (Weng et al. 2014). As schematically illustrated in Fig. 1g, the hybrid fiber was achieved by a scalable approach of wet-spinning the mixture dispersion of two-dimensional titanium oxide sheets and graphene oxide (GO) (Hoshide et al. 2017). Notably, the hybrid fiber shows ordered stacking structure, as observed by SEM characterizations (Fig. 1h). (3) In situ growing/synthesizing active materials on the surfaces of fiber substrates. Developing binder-free electrodes would not only simplify the fabrication process and eliminate the undesirable interfaces produced by additional binders/additives, but also provide large surface area and enhanced charge transfer efficiency. Electrodeposition and hydrothermal/solvothermal processes are facile method suitable for the in situ growth of active materials. Vanadium nitride nanowire arrays are densely grown on the carbon nanotube fiber after the hydrothermal synthesis and annealing process, as shown in Fig. 1i (Zhang et al. 2017a, b, c).

2.2 Electrolyte

Electrolytes are another critical component of the fiber-shaped energy-storage device, as they are associated with the electrochemically stable potential window as well as the ionic transport when electrochemical reactions occur. Currently, electrolytes used for energy storage devices are generally classified into two categories: aqueous and nonaqueous electrolytes (liquid electrolytes and organic liquid/aprotic electrolyte). Aqueous energy-storage device possess their own advantages and uniqueness including the ease of fabrication, safe operation, and low cost, whereas nonaqueous batteries normally suffer from safety issues due to the toxic constituent elements and the flammability risk, which require more stringent and complicated fabrication processes. Compared with high voltage of nonaqueous batteries, the working voltage of aqueous electrolyte is usually limited by the intrinsic characteristic voltage of water splitting. The electrolytes of fiber-shaped energy-storage device can also be classified into three categories: liquid, gel, and solid according to their existence states. In the fabrication of fiber-shaped energy-storage device, directly packing liquid electrolytes along a fiber electrode is difficult, and the leakage and short circuiting concerns become more prominent when these battery fibers undergo various deformations. In this case, mechanically robust and flexible separators infiltrated with the liquid electrolyte, as well as effective encapsulation, are highly required. In addition, the use of gel and solid polymer as mechanical frameworks incorporated with ionic conductive additives has been a mainstream strategy to satisfy the requirements of electrolytes for fiber-shaped energy-storage device. These quasi-solid-state and solid-state electrolytes can lead to good mechanical flexibility and stability; thus, they can simplify the coating and hermetic sealing process for fiber-shaped energy-storage device assembly and reduce the device volume.

2.3 Device Configurations

Three types of device configurations including parallel, twisted and coaxial have been widely studied for fiber-shaped energy-storage device, as illustrated in Fig. 2a–c (D. 17). Figure 2a shows that the parallel device is assembled by placing two parallel fibers on a planar substrate. However, the introduction of flexible planar substrate may lead to have a large volume, which is not suitable for large-scale integration in flexible fabrics. The twisted device can be produced by twisting two fiber electrodes with precoating a solid state electrolyte together through a rotation-translation setup. This twisted configuration displays a structure similar to that of fabric filaments; thus, the twisted-type device appear to be highly suitable for wearable applications, and can easily be woven into scale-up energy textiles. However, two fiber electrodes that comprise twisted structures are easily separated during violent bending. A coaxial energy-storage device has a multilayered structure, where a core fiber electrode and an outer layer electrode with a polymer gel sandwiched between them. To certain extent, this configuration is similar to the sandwiched structure in a planar energy-storage device. Compared with parallel and twisted configurations, the coaxial-structured device possess enhanced interfacial areas between electrodes that can favor high

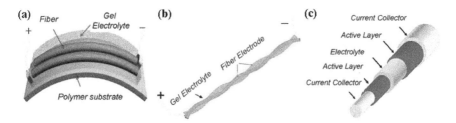

Fig. 2 Schematic illustrations of three fiber-shaped energy-storage device configurations: **a** parallel, **b** twisted, **c** coaxial. **a–c** Reproduced with permission from Ref. Yu et al. (2015)

active material mass loading, as well as a more stable structure to withstand repeatable deformations. However, precisely controlling the assembling process on such small diameter of the fibers and on long fibers remains challenging, which is accompanied by the limited flexibility, thus seriously restricting the manufacturing scalability.

3   Recent Advances of Fiber-Shaped Energy-Storage Device

3.1   Fiber-Shaped Supercapacitors

Supercapacitors are recognized as one of the most robust power supplies owing to their high power density, short charging time, long cycle life, wide working temperature range, safety and no pollution. With the rapid development of portable and wearable electronic devices, the traditional planar flexible supercapacitors can't effectively meet the need of light, miniaturization and integration of wearable devices. The fiber-shaped supercapacitors have attracted tremendous attention as a result of the integration of the tiny volume, high flexibility and remarkable wearability. Recent advances of fiber-shaped supercapacitors consisting of (symmetrical supercapacitors and Asymmetric Supercapacitors) are introduced detailedly from the viewpoint of device configuration, electrode materials and functionality.

As shown in Fig. 3a, an all-solid-state fiber-shaped supercapacitor was assembled by adopting two parallel fiber electrodes on a flexible polymeric substrate with polyvinyl alcohol (PVA)/H_3PO_4 as the gelled electrolyte, without binder, current collector, separator or any other packaging material. The hierarchically structured carbon microfiber fabricated made of an interconnected network of aligned single-walled carbon nanotubes with interposed nitrogen-doped reduced graphene oxide sheets via silica capillary column as a linear hydrothermal microreactor (D. 18). Figure 3b demonstrates that the aligned SWNTs interposed between the rGO layers not only reduce the stacking of the rGO, but also provide well developed porosity in the hybrid fiber. Its galvanostatic charge/discharge (GCD) curves (Fig. 3c) have a triangular shape with a coulombic efficiency of 98%, indicating excellent reversibility and good charge propagation between the two fiber electrodes. The twisted fiber-shaped supercapacitors were prepared by tightly intertwining two coaxial rGO + carbon nanotubes@sodium carboxymethyl cellulose (rGO + CNT@CMC) fibers with precoating a layer of H_3PO_4-PVA gel electrolyte (Fig. 3d) (L. 19). Figure 3e illustrates that the coaxial fibers contained the electrically conductive core assembled homogeneously by rGO and CNTs. Evidently, the rGO + CNT@CMC fiber-shaped supercapacitors shows regular triangular shaped GCD curves (Fig. 3f). The novel coaxial fiber-shaped supercapacitor were successfully fabricated from the aligned CNT fiber and sheet, which functioned as two electrodes with a polymer gel sandwiched between them (Fig. 3g) (X. 13). Obviously, the coaxial supercapacitors show

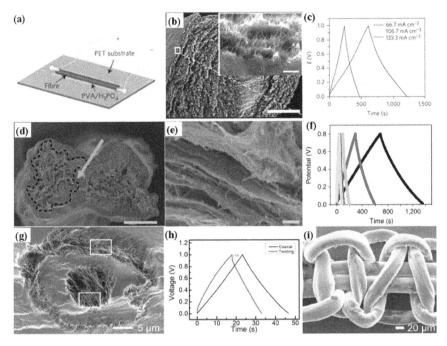

Fig. 3 **a** Schematic of a parallel fiber-shaped supercapacitor constructed using two fiber-3 electrodes on a polyester substrate. SEM images of the cross-section of fiber. **c** GCD curves at various current densities. **a–c** Reproduced with permission from Ref. Yu et al. (2014). **d** SEM images of cross-sectional twisted fiber-shaped supercapacitor. **e** SEM images of coaxial rGO + CNT@CMC fiber. **f** GCD curves at current density increased from 0.1 to 0.2, 0.5, 0.8 and 1 mA cm^{-2}. **d–f** Reproduced with permission from Ref. Kou et al. (2014). **g** SEM images the cross-section of the coaxial fiber-shaped supercapacitor. **h** GCD curves for coaxial (solid line) and twisted (dotted line) EDLCs at the same current of 1 μA. **i** SEM image of several EDLC fibers being woven into a textile structure. **g–i** Reproduced with permission from Ref. Chen et al. (2013)

longer discharge time than twisted ones, which indicates the superior electrochemical performance of novel coaxial structure (Fig. 3h). As shown in Fig. 3i, the coaxial supercapacitors with high integrity could be easily woven.

Asymmetric supercapacitors can increase the working voltage through combining two materials with different potential windows in the same electrolyte. MIL–88–Fe metal-organic framework (MOF) derived spindle-like α–Fe$_2$O$_3$ was directly grown on oxidized CNTF (S–α–Fe$_2$O$_3$/OCNTF) as advanced anodes for wearable asymmetric supercapacitors (Fig. 4a) (Z. 20). As shown in Fig. 4b, the operating potential windows of S–α–Fe$_2$O$_3$/OCNTF and Na–MnO$_2$ NSs/CNTF are -1 to 0 V and 0 to 1.2 V, respectively, from which we can infer that the maximum operating voltage for the as-assembled device can achieve 2.2 V. As expected, the GCD curve of the as-assembled S–α–Fe$_2$O$_3$/OCTNF//Na–MnO$_2$/CNTF device collected at 4 mA cm^{-2} given in Fig. 4c still maintains a symmetrical triangle shape even at a high potential window up to 2.2 V. A novel asymmetric coaxial fiber-shaped supercapacitors

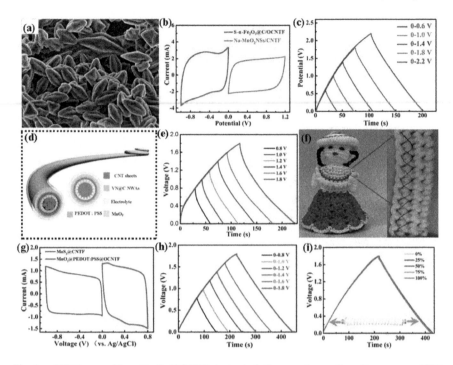

Fig. 4 **a** SEM image of S–α–Fe$_2$O$_3$/OCTNF electrode. **b** Comparative cyclic voltammogram (CV) curves of the S–α–Fe$_2$O$_3$/OCNTF and Na–MnO$_2$/CNTF at a scan rate of 25 mV s^{-1}. **c** GCD curves of the as-fabricated FASC device collected over different voltages from 0.6 to 2.2 V at a current density of 4 mA cm^{-2}. **a–c** Reproduced with permission from Ref. Zhou et al. (2018). **d** Schematic illustrations of the ACFSS. **e** GCD curves of the VN@C NWA/CNTF//MnO$_2$/PEDOT:PSS/CNT sheet composite device collected over different voltages from 0.8 to 1.8 V at a current density of 2 mA cm^{-2}. **f** Photograph of the as-prepared ACFSS woven into flexible textiles. **d–f** Reproduced with permission from Ref. Zhang et al. (2017). **g** Comparative CV curves of the MoS$_2$@CNTF and MnO$_2$@PEDOT:PSS@OCNTF at a scan rate of 25 mV s^{-1} in a three-electrode configuration. **h** GCD curves of the as-fabricated stretchable FASC device collected over different voltages from 0.8 to 1.8 V at a current density of 2 mA cm^{-2}. **i** GCD curves of the as-prepared stretchable FASCs with increasing strain from 0% to 100%. **g–i** Reproduced with permission from Ref. Zhang et al. (2017)

was successfully fabricated was schematically illustrated in Fig. 4d by wrapping aligned CNT composite sheets around aligned vanadium nitride (VN) nanowire arrays (NWAs) on CNTF (VN@C NWA/CNTF) (Zhang et al. 2017a, b, c). Figure 4e illustrates that the GCD curves of the asymmetric coaxial fiber-shaped supercapacitors (ACFSS) at a current density of 2 mA cm^{-2} were still nearly symmetric at an operating potential as high as 1.8 V, suggesting that the device exhibited excellent capacitive characteristics. To further demonstrate the potential applications for the newly developed FSSs as efficient energy-storage components for wearable devices, these ACFSS devices were woven into the flexible powering textile shown in Fig. 4f to demonstrate the favorable stitchability of ACFSS devices. High-performance stretchable fiber-shaped asymmetric supercapacitors (FASCs) were successfully fabricated by adopting hierarchically-structured MnO_2@PEDOT:PSS@OCNTF as the positive electrode, MoS_2@CNTF as the negative electrode and elastic fiber as substrate. As shown in Fig. 6g, the operating potential windows for MoS_2@CNTF and MnO_2@PEDOT:PSS@OCNTF are -1–0 V and 0–0.8 V, respectively (Zhang et al. 2017a, b, c). Thus, it is expected that the maximum operating voltage will reach 1.8 V for the as-assembled stretchable FASC. Furthermore, the GCD curves of the stretchable FASC at a current density of 2 mA cm^{-2} are still nearly symmetric with operating potentials as high as 1.8 V. As seen in Fig. 4i, there are negligible changes in the GCD curves at a current density of 2 mA cm^{-2} with increasing strains from 0% to 100%, indicating the great flexibility of as-prepared FASCs.

3.2 Fiber-Shaped Nonaqueous Lithium-Based Batteries

3.2.1 Fiber-Shaped Lithium-Ion Batteries (LIBs)

With attributes of a high energy density, a long cycle lifespan, and a high operation voltage, aprotic LIBs have been utilized as one of the most common commercial power sources for various portable electronic products. The realization and development of high-performance fiber-shaped LIBs depends on the effective exploitation of novel electrode materials. A fiber-shaped LIB with better performance was developed from a CNT/LMO hybrid fiber cathode and a CNT/$Li_4Ti_5O_{12}$ (LTO) hybrid fiber anode in a parallel structure (Fig. 5a) (J. 22). This fiber-shaped LIB was highly flexible and could be bent into various formats (Fig. 5b). As shown in Fig. 5c, the resulting fiber-shaped LIB exhibited a discharge plateau voltage of 2.5 V at 0.1 mA, and the specific capacity reached 138 mAh g^{-1}, which could be retained at 85% after 100 cycles. An elastic LIB was produced from two springlike fibers bearing LMO and LTO nanoparticles as positive and negative electrodes. Figure 5d shows the CNT/LTO fiber where CNTs acted as conducting scaffolds and the LTO nanoparticles were well dispersed among them (Y. 23). Figure 5e reveals a typical fiber overtwisted from multiple CNT/LTO fibers with coiled loops aligning along the fiber axis. The typical charge-discharge profile of the fiber-shaped CNT/LTO//CNT/LMO battery at 0.1 mA cm^{-1} shows average discharge voltage plateau of 2.5 V (Fig. 5f). 3D printing technology was used as a fast, scalable, and universal method to prepare fiber-shape cathode and anode electrodes. Electrode materials of $LiFePO_4$ (LFP) and LTO, were incorporated into the poly (vinylidene fluoride) (PVDF) solution with CNT conductive additive to make the cathode and anode inks, respectively, for printing LFP (cathode) and LTO (anode) fibers (Fig. 5g) (Weng et al. 2014). PVDF-co-hexafluoropropylene (PVDF-co-HFP) was coated both on the surfaces of LFP and LTO fibers and soaked with limited LiPF6 in EC/DEC solution to serve as quasi-solid-state electrolyte and electronic insulating separator layer. The coated LFP and LTO fibers were then twisted together into one integrated fiber with the protection of an outer heat shrink tube. A twisted yarn structure is shown in Fig. 5h

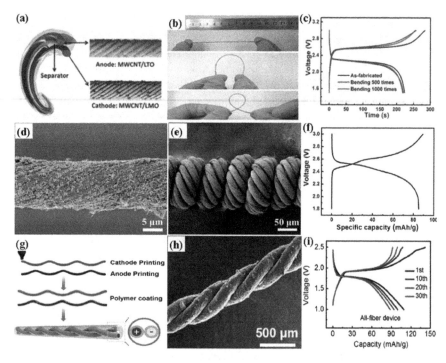

Fig. 5 **a** Schematic illustration of the parallel structure of a fiber-shaped LIB from the CNT/LTO and CNT/LMO electrode. **b** Photographs of the parallel fiber-shaped LIB being deformed into different shapes. **c** Voltage profiles of the parallel fiber-shaped LIB before and after bending. **a–c** Reproduced with permission from Ref. Ren et al. (2014). **d** SEM image of the CNT/LTO fiber electrode. **e** SEM image of springlike CNT/LTO fibers. **f** Charge and discharge profiles of the LIB battery at a 0.1 mAcm^{-1}. **d–f** Reproduced with permission from Ref. Zhang et al. (2014). **g** 3D printing fabrication process of all-fiber flexible LIBs. **h** SEM image of yarn composed of three LFP fibers. **i** Charge and discharge profiles of the all-fiber LIB device. **g–i** Reproduced with permission from Ref. Wang et al. (2017)

with a tight connection. All fiber LIB devices with an average discharge plateau of around 3.37 V had a tiny capacity drop of less than 3% after 30 cycles (Fig. 5i).

3.2.2 Fiber-Shaped Lithium-Sulfur (Li–S) Batteries

Li–S batteries have arousing interest and are regarded as a promising successor to conventional LIBs because of their their high theoretical capacity (1675 mAh g^{-1}) and energy density (2600 Wh g^{-1}). 1D cable-shaped Li–S battery was firstly fabricated using a carbon nanostructured hybrid fiber as the sulfur cathode. The obtained CMK-3@S hybrid fiber can be wrapped around a metal wire and the resulting fiber electrode has a diameter of 254 μm (Fig. 6a) (X. 25). The magnified image in Fig. 6b displays that S-imbibed particles were apt to be tangled by conductive CNTs due

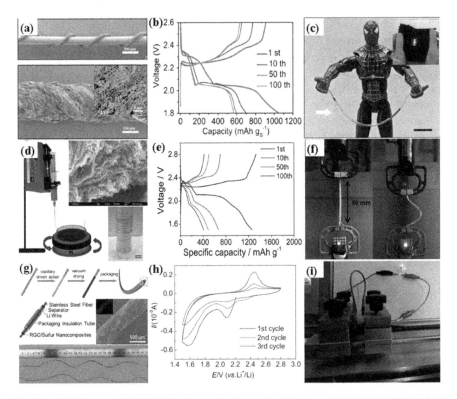

Fig. 6 **a** SEM image of a hybrid fiber intertwined around a Ti wire and CMK-3@S-CNT hybrid fiber and SEM images of CMK-3@S-CNT hybrid fiber. **b** GCD profiles at increasing cycle numbers at 0.1 C. **c** A bent cable-shaped lithium sulfur battery lighting up a red LED (Inset: three white LEDs powered by the fabric under stretching deformations). **a–c** Reproduced with permission from Ref. Fang et al. (2016). **d** Schematic illustration of wet-spinning process of rGO/CNT/S gel fibers and structure characterization of CMK-3@S-CNT hybrid fiber. **e** Charge/discharge profiles determined at 0.1 C. **f** Cable battery powering a LED subjected to bending in compression. **d–f** Reproduced with permission from Ref. Chong et al. (2017). **g** Scheme of synthesis for wire-shaped lithium sulfur electrode by the capillary action and structure characterization of S-containing hybrid fiber cathode. **h** rGO/S/SSF hybrid fiber electrode at scanning rate of 0.2 mV s^{-1} within 1.5–2.8 V versus Li/Li$^+$. **i** The photographs display a red LED lit up by the wire-shaped Li–S battery bent at 90° under water without any waterproof measures. **g–i** Reproduced with permission from Ref. Liu et al. (2017)

to the ultrahigh number density of CNTs in the sheet. The GCD curves in Fig. 6b shows two characteristic voltage plateaus of Li–S batteries. The cable-shaped battery is able to light up a 1.8 V red light-emitting diode (Fig. 6c). Ultralight composite fibers consisting of reduced graphene oxide/carbon nanotube filled with a large amount of sulfur (rGO/CNT/S) are prepared by a facile, one-pot wetspinning method (Fig. 6d) (W. G. 26). The GCD profiles of the rGO/CNT/S fiber electrode exhibited two discharge plateaus of typical LSBs at 2.24–2.37 and 2.0–2.11 V (vs. Li/Li$^+$) (Fig. 6e). The robustness and stability of the battery's discharge characteristics were evaluated in a cyclic bending test of a 50 mm long cable battery to a minimum bending radius of ≈6.5 mm (equivalent to a maximum 50% linear displacement) for 30 cycles at a constant crosshead speed (Fig. 6f). As illustrated in Fig. 6g, a new and general strategy is proposed to produce freestanding sulfur-containing fibrous electrodes using industrially weavable stainless steel fibers (SSF) as supports and current collectors (R. 27). A coaxial fiber-shaped Li–S battery was subsequently fabricated by coupling the hybrid fiber electrode with a lithium wire in parallel with an outer encapsulation of a heat-shrink tube. The CV curves (Fig. 6h) of rGO/S/SSF hybrid fibrous cathode are similar to those of the coin cell, only the current density is smaller. The resulting fiber-shaped Li–S battery was able to light up a 1.8 V red light-emitting diode (LED) under bending deformation (Fig. 6i).

3.2.3 Fiber-Shaped Lithium-Air Batteries

The lithium-air (Li-air) battery exhibits a high theoretical specific energy density of 3500 Whk^{-1}, 5–10 times higher than the commercial LIB, and thus has attracted increasing interest and is proposed as a promising energy storage candidate. As schematically shown in Fig. 7a, a coaxial structure all-solid-state Li-air battery was successfully assembled by adopting Li wire as the inner anode, aligned CNT sheet air electrode as the outer cathode and gel polymer electrolyte sandwiched in between them (Zhang et al. 2017a, b, c). Figure 7b shows typical charge and discharge curves at a high current density of 1400 mA g^{-1} with a cutoff capacity of 500 mAh g^{-1}. The fiber-shaped Li-air batteries can be stably operated even under a dynamic bending and releasing process at a speed of 10 degrees per second (Fig. 7c). A novel silicon-oxygen battery (SOB) fiber with high energy density and ultra-high flexibility by designing a solid-state coaxial architecture with the lithiated silicon/CNT hybrid fiber as inner anode, a polymer gel as middle electrolyte and a bare CNT sheet as outer cathode (Fig. 7d) (Zhang et al. 2017a, b, c). The hybrid fiber showed a uniform diameter of approximately 160 μm and Si nanoparticles with diameters of 50 nm were uniformly dispersed in the hybrid fiber (Fig. 7e). Under such deformations, the charge and discharge behaviors of the SOB fiber showed no significant changes (Fig. 7f). A quasi-solid-state flexible fiber-shaped Li–CO$_2$ battery was successfully assembled by employing ultrafine Mo$_2$C nanoparticles anchored on a CNT cloth freestanding hybrid film as the cathode (Fig. 7g) (Zhou et al. 2017). As can be seen from the SEM images of CC@Mo$_2$C nanoparticles, numerous CNTs loaded with Mo$_2$C nanoparticles entangle with each other, forming a highly interconnected 3D porous network with sub-micrometer-sized macropores (Fig. 7h). This freestanding CNT/Mo$_2$C hybrid cathode shows obvious charge-discharge plateaus and good cycle stability (Fig. 7i).

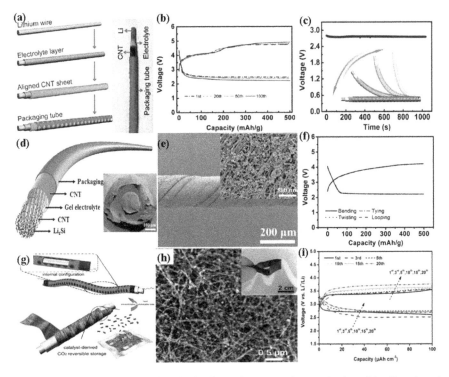

Fig. 7 **a** Schematic illustration of the fabrication and structure characterization of the fiber-shaped Li-air battery and Photograph of a fiber-shaped Li-air battery. **b** Charge and discharge curves at current density of 1400 mAg^{-1} in air. **c** Charge and discharge curves under a dynamic bending and releasing process at a speed of 10 degree per second. **a–c** Reproduced with permission from Ref. Zhang et al. (2016). **d** SEM images of silicon/CNT hybrid fiber at low and high magnifications. **d** Schematic illustration to the structure of an SOB fiber and cross-sectional SEM image of the SOB fiber. **f** Charge and discharge curves of the SOB fiber before and after different bending cycles. **d–f** Reproduced with permission from Ref. Zhang et al. (2017). **g** Schematics illustrating the quasi-solid-state flexible fiber-shaped Li–CO$_2$ battery. **h** SEM images of CC@Mo$_2$C nanoparticles with an optical photo(Inset: an optical photo CC@Mo$_2$C nanoparticles freestanding film). **i** Cycling behavior of CC@Mo$_2$C nanoparticles electrode for the selected cycles under CO$_2$ at a current of 20 μA cm^{-2} and curtailing capacity of 100 μAh cm^{-2}. **g–i** Reproduced with permission from Ref. Zhou et al. (2019)

3.3 Fiber-Shaped Aqueous Batteries

3.3.1 Fiber-Shaped Aqueous Lithium-Ion and Sodium-Ion Batteries

Aqueous rechargeable Li/Na ion batteries featured with low manufacturing cost, good safety, environmental friendliness and high rate capability are particularly noteworthy. The ever-growing demand for wearable and portable electronic stimulates the exploitation of safe and high-performance wearable power sources. To meet this target, fiber-shaped aqueous rechargeable Li/Na ion batteries have stimulated ever-increasing attentions by reason of their remarkable advantages of tiny volume, extraordinary flexibility and remarkable wearability. A new fiber-shaped aqueous LIB (FALIB) is developed using a polyimide/carbon nanotube hybrid fiber as the anode and $LiMn_2O_4$/carbon nanotube hybrid fiber as the cathode (Fig. 8a) (Zhang et al. 2016a, b). Remarkably, the charge and discharge curves were well maintained under increasing current rates, and the specific capacity was 101 mAh g^{-1} even at a current rate of 100 C (Fig. 8b). As shown in Fig. 8c, these fiber-shaped aqueous LIBs can be woven into various flexible structures, such as textiles, that can be bent, folded and twisted into various architectures. A novel fiber-shaped aqueous SIB (FASIB) was successfully assembled by putting the fiber-shaped $CNT/Na_{0.44}MnO_2$ and $CNT/NaTi_2(PO_4)_3$@C electrodes directly in a tube containing 1 M Na_2SO_4 aqueous electrolyte (Fig. 8d) (Z. 32). The as-assembled full cell delivered a discharge specific capacity of 46 mAh g^{-1} at a current density of 0.1 A g^{-1} with three voltage plateaus at 1.3, 1.1, and 0.9 V (Fig. 8e). Even after bending at 180° for 100 times,

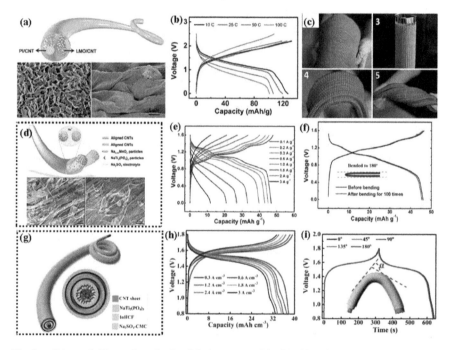

Fig. 8 **a** Schematic illustration of a simplified structure of the FALIB and structure characterization of fiber electrodes. **b** Charge and discharge curves of the FALIB under increasing current rates. **c** Energy textile woven with FALIBs under bending, folding and twisting. **a–c** Reproduced with permission from Ref. Zhang et al. (2016). **d** Schematic illustration of a simplified structure of the FASIB and structure characterization of fiber electrodes. **e** GCD curves of the FASIB at different current densities. **f** GCD curves before and after bending of the flexible FASIB at a current density of 0.2 A g^{-1}. **d–f** Reproduced with permission from Ref. Guo et al. (2017). **g** Schematics illustrating the CFASIB. **h** GCD curves of as-assembled CFASIB. **i** GCD curves of the CFASIB at different bending angles. **g–i** Reproduced with permission from Ref. Zhang et al. (2019)

the GCD curves of the fiber-shaped aqueous SIBs remained almost unchanged, indicating a high flexibility of the fiber-shaped aqueous SIB (Fig. 8f). As displayed in Fig. 8g (Q. 35), an advanced quasi-solid-state coaxial-fiber aqueous SIB (CFASIB) are successfully constructed by adopting $NaTi_2(PO_4)_3$@CNTF as the inner anode, InHCF composition as the outer cathode and gel electrolyte sandwiched in between them. The GCD curves of as-fabricated CFASIB in Fig. 8h show distinct charge-discharge voltage plateaus with minimal voltage hysteresis and excellent symmetry. Notably, the CFARSIB device could be bent from $0°$ to $180°$ without deteriorating its electrochemical performance, indicating its exceptional flexibility (Fig. 8i).

3.3.2 Fiber-Shaped Aqueous Zinc-Ion Batteries (ZIBs)

ZIBs have attracted tremendous attention since metallic zinc exhibits their unique merits of high theoretical capacity (820 mA h g^{-1}), low electrochemical potential (-0.763 V vs. standard hydrogen electrode), good rate performance, low cost and intrinsic safety. A high-performance waterproof and stretchable fiber-shaped ZIB was developed by double-helix yarn electrodes and a cross-linked PAM-based polymer electrolyte, as schematically illustrated in Fig. 9a (H. 34). The fiber-shaped aqueous ZIB (FAZIB) maintained a high capacity retention of 94.8% after cycling 100 times when a strain of 300% was applied, which is close to the value of 97.2% under normal conditions (Fig. 9b). It can be seen that the FAZIB has a capacity retention of 96.5% after being fully immersed in water for 12 h, demonstrating good waterproof capability and high durability (Figs. 9c). The first coaxial-fiber AZIB (CFAZIB) was successfully developed by adopting Zn nanosheet arrays (NSAs) on carbon CNTF as the core electrode and ZnHCF composite on aligned CNT Sheets as the outer electrode with $ZnSO_4$-carboxymethyl cellulose sodium as the gel electrolyte (Q. 35). The electrochemical performance of our CAZIB was further investigated at different current densities from 0.1 to 1 A cm^{-3}, and the corresponding charge-discharge curves in Fig. 9e presented remarkable symmetry and high discharge plateaus, indicating its desirable battery behavior. To further demonstrate the favorable weavability of the newly developed CFAZIBs, two long CFAZIBs devices were woven into the flexible textile in series (Figs. 9f), and the resulting energy textile could illuminate a 3.3 V blue LED. As shown in Fig. 9g, binder-free V-MOF as high-performance cathode was firstly used to fabricated all-solid-state flexible FAZIB (B. 36). 3D freestanding V-MOFs were uniformly grown on the surface of the CNTF via a simple solvothermal reaction (Fig. 9h). Encouragingly, the fiber-shaped V-MOF//Zn battery still maintained a high volumetric capacity of 67.98 mAh cm^{-3} at the discharge current density of 0.1 A cm^{-3} after rapid charging at the high current density of 2.0 A cm^{-3} (Fig. 9i).

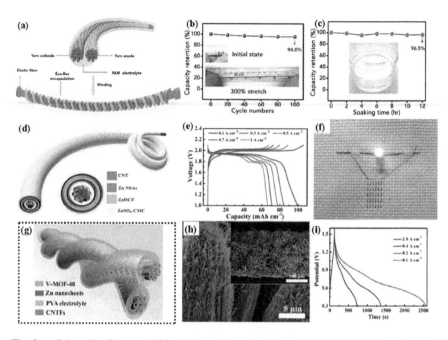

Fig. 9 **a** Schematic diagram of fabrication and encapsulation of the FAZIB. **b** Dependence of capacity retention on cycle numbers with a strain of 300% **c** Capacity retention test of the FAZIB for 12 h continuous underwater immersion in deionized water at 24 °C. **a–c** Reproduced with permission from Ref. Li et al. (2018). **d** Schematic illustrations of the CFAZIB. **e** Charge-discharge curves collected at different current densities. **f** Photograph of a blue LED illuminated by the charged energy textile consisting of our CFAZIBs. **d–f** Reproduced with permission from Ref. Zhang et al. (2019). **g** Schematic illustration of the all-solid-state fiber-shaped V-MOF//Zn battery. **h** SEM images of V-MOF/CNTF at different levels of magnification. **i** GCD curves of the assembled battery during charging at 2.0 A cm^{-3} and discharging at 0.1, 0.2, and 0.4 A cm^{-3}. **g–i** Reproduced with permission from Ref. He et al. (2019)

3.3.3 Fiber-Shaped Aqueous Alkaline Batteries

Aqueous alkaline batteries are considered promising energy storage sources for wearable and portable devices. The fabrication process for the stretchable fiber-shaped Ag–Zn battery is presented in Fig. 10a (Zamarayeva et al. 2017). Figure 10b shows schematics of the self-similar serpentine and optical images of the full battery assembled around such a currentcollector in relaxed and stretched configurations. The GCD curves for the 2nd (flat configuration), 12th (stretched configuration), and 22nd (flat configuration) electrochemical cycles are presented in Fig. 10c. The battery exhibits stable performance in both flat and 100% stretched configurations with minor fluctuations in the capacity in both cases. Figure 10d illustrates the schematic diagram of the fiber-shaped Ni–Zn battery, where a Zn fiber anode is twined tightly by Ni–NiO fiber cathode (Zeng et al. 2017). Such structure was further verified by the SEM image

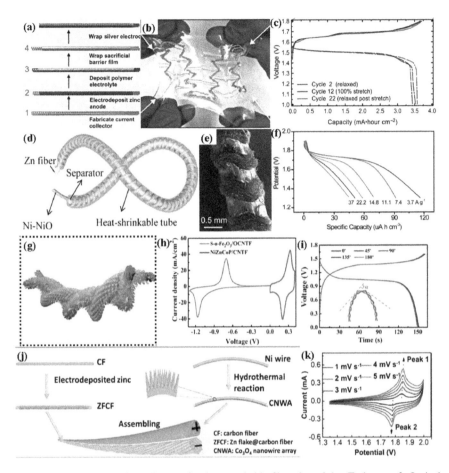

Fig. 10 **a** The assembly flow diagram for the stretchable fiber-shaped Ag–Zn battery. **b** Optical images of the stretchable fiber-shaped Ag–Zn battery. **c** GCD curves for the 2nd (flat configuration), 12th (stretched configuration), and 22 nd (flat configuration) electrochemical cycles of stretchable fiber-shaped Ag–Zn battery. **a–c** Reproduced with permission from Ref. Zamarayeva et al. (2017). **d** Schematic diagram of the flexible quasi-solid-state fiber-shaped Ni–Zn battery. **e** SEM image of the quasi-solid-state fiber-shaped Ni–Zn battery. **f** Discharge curves of the quasi-solid-state fiber-shaped Ni–Zn battery at various current densities. **d–f** Reproduced with permission from Ref. Zeng et al. (2017). **g** Schematic fabrication process of the twisted fiber-shaped Ni–Fe battery. **h** Comparison CV curves of S–α–Fe$_2$O$_3$/OCNTF anode and NiZnCoP@CNTF cathode at a scan rate of 10 mV s^{-1}. **i** GCD curves of the as-assembled FARBs bent at various angles at a current density of 2 mA cm^{-2}. **g–i** Reproduced with permission from Ref. Zhang et al. (2018). **j** Schematic illustration of fabrication of aqueous fiber-shaped Co–Zn battery. **k** CV curves of fiber-shaped Co–Zn battery at different scan rates. **j–k** Reproduced with permission from Ref. Guan et al. (2019)

of the fiber-shaped Ni–NiO//Zn battery device (Fig. 10e). The as-fabricated fiber-shaped Ni–Zn battery possesses a high discharge profiles witness a plateau around 1.7–1.8 V (Fig. 10f). The fabrication process of the twisted fiber-shaped Ni-Fe battery is schematically depicted in Fig. 10g and the device was prepared successfully by twisting NiZnCoP/CNTF and S–α–Fe$_2$O$_3$/OCNTF electrodes together (Zhang et al. 2018). Figure 10h comparatively demonstrates the CV curves of the as obtained S–α–Fe$_2$O$_3$/OCNTF and NiZnCoP@CNTF electrodes at 10 mV s^{-1} and it should be noted that both the electrodes consist of a couple of prominent redox peaks. The GCD curves of the assembled fiber-shaped Ni–Fe battery, shown in Fig. 6f, nearly overlap completely at different bending angles ranging from 0° to 180°, indicating that the device possesses excellent mechanical stability. The fabrication process of the fiber-shaped Co–Zn battery is schematically illustrated in Fig. 10j (Q. 40). In a typical synthesis, zinc flakes were deposited on the carbon fibers as the anode and Co$_3$O$_4$ nanowires arrays were homogeneously grown on Ni wire as the cathode via a hydrothermal method. Cyclic voltammetry measurements were tested at different scan rates ranging from 1 to 5 mV s^{-1} (Fig. 10k) and two obvious redox peaks were presented in Fig. 10k.

4 Conclusion

In conclusion, to meet the ever growing requirements for powering portable and wearable electronics, increased efforts have been dedicated to developing high-performance fiber-shaped energy storage devices for better adapting to the various aggressive deformations. In this chapter, the state-of-the-art fiber-shaped energy storage devices have been critically looked into from the viewpoint of new design principles, materials and device fabrication. Although fiber-shaped supercapacitors feature the advantages of high power density, rapid charge/discharge capability and superior cyclability, their unstable discharge platform and low energy density are still serious hindrances for powering wearable energy-consuming devices. Meanwhile, fiber-shaped nonaqueous batteries are unable to wearable due to the use of the toxicity and flammability of organic electrolytes. The construction of fiber-shaped aqueous rechargeable batteries is thus an extremely effective strategy to achieve satisfactory safety, stable output voltage and high energy density. To date, numerous energy storage mechanisms have been successfully realized in the fiber-shaped aqueous rechargeable batteries with acceptable performances. Significant attempt will be continuing to pursuing better electrochemical performances, mechanical flexibility and robustness, and at the same time to make the fiber-shaped aqueous rechargeable batteries playing an important part in the coming age of internet of things, where there will be rapid advances in the development of new materials and device fabrication with the overall low manufacturing cost.

References

D. Chen, K. Jiang, T. Huang, G. Shen, Adv. Mater. **190**, 2019 (1806)

X. Chen, L. Qiu, J. Ren, G. Guan, H. Lin, Z. Zhang, P. Chen, Y. Wang, H. Peng, Adv. Mater. **6436**, 25 (2013)

W.G. Chong, J.-Q. Huang, Z.-L. Xu, X. Qin, X. Wang, J.-K. Kim, Adv. Funct. Mater. **1604815**, 27 (2017)

X. Fang, W. Weng, J. Ren, H. Peng, Adv. Mater. **491**, 28 (2016)

C. Guan, W. Zhao, Y. Hu, Q. Ke, X. Li, H. Zhang, J. Wang, Adv. Energy Mater. **1601034**, 6 (2016)

Q. Guan, Y. Li, X. Bi, J. Yang, J. Zhou, X. Li, J. Cheng, Z. Wang, B. Wang, J. Lu, Adv. Energy Mater. **1901434**, 9 (2019)

Z. Guo, Y. Zhao, Y. Ding, X. Dong, L. Chen, J. Cao, C. Wang, Y. Xia, H. Peng, Y. Wang, Chem **348**, 3 (2017)

B. He, Q. Zhang, P. Man, Z. Zhou, C. Li, Q. Li, L. Xie, X. Wang, H. Pang, Y. Yao, Nano Energy **103935**, 64 (2019)

T. Hoshide, Y. Zheng, J. Hou, Z. Wang, Q. Li, Z. Zhao, R. Ma, T. Sasaki, F. Geng, Nano Lett. **3543**, 17 (2017)

L. Kou, T. Huang, B. Zheng, Y. Han, X. Zhao, K. Gopalsamy, H. Sun, C. Gao, Nat. Commun. **3754**, 5 (2014)

H. Li, Z. Liu, G. Liang, Y. Huang, Y. Huang, M. Zhu, Z. Pei, Q. Xue, Z. Tang, Y. Wang, B. Li, C. Zhi, ACS Nano **3140**, 12 (2018)

B. Liu, J. Zhang, X. Wang, G. Chen, D. Chen, C. Zhou, G. Shen, Nano Lett. **3005**, 12 (2012)

R. Liu, Y. Liu, J. Chen, Q. Kang, L. Wang, W. Zhou, Z. Huang, X. Lin, Y. Li, P. Li, X. Feng, G. Wu, Y. Ma, W. Huang, Nano Energy **325**, 33 (2017)

X. Lu, M. Yu, G. Wang, T. Zhai, S. Xie, Y. Ling, Y. Tong, Y. Li, Adv. Mater. **267**, 25 (2013)

F. Mo, G. Liang, Z. Huang, H. Li, D. Wang, C. Zhi, Adv Mater. 1902151 (2019)

J. Ren, Y. Zhang, W. Bai, X. Chen, Z. Zhang, X. Fang, W. Weng, Y. Wang, H. Peng, Angew. Chem. Int. Ed. **7864**, 53 (2014)

L. Wang, X. Fu, J. He, X. Shi, T. Chen, P. Chen, B. Wang, H. Peng, Adv. Mater. **190**, 2019 (1971)

Y. Wang, C. Chen, H. Xie, T. Gao, Y. Yao, G. Pastel, X. Han, Y. Li, J. Zhao, K.K. Fu, L. Hu, Adv. Funct. Mater. **1703140**, 27 (2017)

W. Weng, Q. Sun, Y. Zhang, H. Lin, J. Ren, X. Lu, M. Wang, H. Peng, Nano Lett. **3432**, 14 (2014)

W. Weng, P. Chen, S. He, X. Sun, H. Peng, Angew. Chem. Int. Ed. **6140**, 55 (2016)

X. Xu, S. Xie, Y. Zhang, H. Peng, Angew. Chem. Int. Ed. **2**, 58 (2019)

D. Yu, K. Goh, H. Wang, L. Wei, W. Jiang, Q. Zhang, L. Dai, Y. Chen, Nat. Nanotech. **555**, 9 (2014)

D. Yu, Q. Qian, L. Wei, W. Jiang, K. Goh, J. Wei, J. Zhang, Y. Chen, Chem. Soc. Rev. **647**, 44 (2015)

N. Yu, H. Yin, W. Zhang, Y. Liu, Z. Tang, M.-Q. Zhu, Adv. Energy Mater. **1501458**, 6 (2016)

A. M. Zamarayeva, A. E. Ostfeld, M. Wang, J. K. Duey, I. Deckman, B. P. Lechêne, G. Davies, D. A. Steingart, A. C. Arias, Sci. Adv. **e1602051**, 3 (2017)

Y. Zeng, Y. Meng, Z. Lai, X. Zhang, M. Yu, P. Fang, M. Wu, Y. Tong, X. Lu, Adv. Mater. **1702698**, 29 (2017)

S. Zhai, H. E. Karahan, C. Wang, Z. Pei, L. Wei, Y. Chen, Adv Mater. 1902387 (2019)

Q. Zhang, P. Man, B. He, C. Li, Q. Li, Z. Pan, Z. Wang, J. Yang, Z. Wang, Z. Zhou, X. Lu, Z. Niu, Y. Yao, L. Wei, Nano Energy 104212 (2019)

Y. Zhang, W. Bai, X. Cheng, J. Ren, W. Weng, P. Chen, X. Fang, Z. Zhang, H. Peng, Angew. Chem. Int. Ed. **14564**, 53 (2014)

Y. Zhang, L. Wang, Z. Guo, Y. Xu, Y. Wang, H. Peng, Angew. Chem. Int. Ed. **4487**, 55 (2016a)

Y. Zhang, Y. Wang, L. Wang, C.-M. Lo, Y. Zhao, Y. Jiao, G. Zheng, H. Peng, J. Mater. Chem. A **9002**, 4 (2016b)

Q. Zhang, X. Wang, Z. Pan, J. Sun, J. Zhao, J. Zhang, C. Zhang, L. Tang, J. Luo, B. Song, Z. Zhang, W. Lu, Q. Li, Y. Zhang, Y. Yao, Nano Lett. **2719**, 17 (2017a)

Q. Zhang, J. Sun, Z. Pan, J. Zhang, J. Zhao, X. Wang, C. Zhang, Y. Yao, W. Lu, Q. Li, Y. Zhang, Z. Zhang, Nano Energy **219**, 39 (2017b)

Y. Zhang, Y. Jiao, L. Lu, L. Wang, T. Chen, H. Peng, Angew. Chem. Int. Ed. **13741**, 56 (2017c)

Q. Zhang, Z. Zhou, Z. Pan, J. Sun, B. He, Q. Li, T. Zhang, J. Zhao, L. Tang, Z. Zhang, L. Wei, Y. Yao, Adv Sci. **1801462**, 5 (2018)

Q. Zhang, C. Li, Q. Li, Z. Pan, J. Sun, Z. Zhou, B. He, P. Man, L. Xie, L. Kang, X. Wang, J. Yang, T. Zhang, P.P. Shum, Q. Li, Y. Yao, L. Wei, Nano Lett. **4035**, 19 (2019)

B. Zheng, T. Huang, L. Kou, X. Zhao, K. Gopalsamy, C. Gao, J. Mater. Chem. A **9736**, 2 (2014)

Z. Zhou, Q. Zhang, J. Sun, B. He, J. Guo, Q. Li, C. Li, L. Xie, Y. Yao, ACS Nano **9333**, 12 (2018)

J. Zhou, X. Li, C. Yang, Y. Li, K. Guo, J. Cheng, D. Yuan, C. Song, J. Lu, B. Wang, Adv. Mater. **1804439**, 31 (2019)

Y.H. Zhu, X.Y. Yang, T. Liu, X.B. Zhang, Adv. Mater. **190**, 2019 (1961)

Brillouin Fiber Laser Sensors

Yi Liu, Zhaomin Tong, Yao Shang, Bingchen Han, Qing Bai, Rongrong Guo, and Pengfei Chen

Abstract A single longitudinal mode (SLM) Brillouin fiber laser (BFL) with cascaded ring (CR) Fabry–Pérot resonator, a SLM triple ring (TR) BFL with a saturable absorber ring (SAR) resonator and a stable multiwavelength (MW) SLM dual ring BFL (MW-SLM-DRBFL) are proposed and demonstrated. By optimizing

Y. Liu (✉) · Q. Bai · R. Guo · P. Chen
Key Laboratory of Advanced Transducers and Intelligent Control System, Ministry of Education and Shanxi Province, 030024 Taiyuan, People's Republic of China
e-mail: liuyi@tyut.edu.cn

Q. Bai
e-mail: baiqing@tyut.edu.cn

R. Guo
e-mail: guorongrong0831@link.tyut.edu.cn

P. Chen
e-mail: chenpengfei0823@link.tyut.edu.cn

College of Physics and Optoelectronics, Institute of Optoelectronic Engineering, Taiyuan University of Technology, 030024 Taiyuan, People's Republic of China

Z. Tong
State Key Laboratory of Quantum Optics and Quantum Optics Devices, Institute of Laser Spectroscopy, Shanxi University, 030006 Taiyuan, Shanxi, China
e-mail: zhaomin.tong@sxu.edu.cn

Collaborative Innovation Center of Extreme Optics, Shanxi University, 030006 Taiyuan, Shanxi, People's Republic of China

Y. Shang
College of Software, Taiyuan University of Technology, 030024 Taiyuan, People's Republic of China
e-mail: liunian2274224@163.com

Polytechnic Institute Taiyuan University of Technology, 030024 Taiyuan, People's Republic of China

B. Han
Department of Physics, Taiyuan Normal University, 030619 Shanxi, People's Republic of China
e-mail: han_bchen@126.com

© Springer Nature Singapore Pte Ltd. 2020
L. Wei (ed.), *Advanced Fiber Sensing Technologies*,
Progress in Optical Science and Photonics 9,
https://doi.org/10.1007/978-981-15-5507-7_15

the CR length of the single-mode fiber cavity at 100 m (or 50 m) and 10 m, stable SLM operation is obtained with 0.41 kHz (or 3.23 kHz). TR-BFL with approximately 65-Hz linewidth and 185 linewidth-reduction ratio is composed of a 1-km-long single-mode fiber (SMF) ring, a 100-m-long SMF ring, and an SAR with 8-m-long unpumped Erbium-doped fiber (UP-EDF), respectively. 7 stable SLM lasing wavelengths with DR configuration of 100 and 10 m length SMF are obtained with 0.084 nm wavelength spacing and 15 dB average optical signal-to-noise ratio (OSNR) through the cascaded stimulated Brillouin scattering (cSBS) and four-wave mixing (FWM). A MW SLM Brillouin–Erbium fiber laser (BEFL) sensor with ultrahigh resolution is proposed and demonstrated and Iezzi et al. proposed and investigated experimentally a distributed higher order Stokes SBS temperature fiber sensor. The one short common cavity of MW-SLM-BEFL with 50 m of SMF as the fiber under test (FUT) and 100 m of SMF as the reference realize 3.104 MHz/°C sensitivity and approximately 10^{-6} °C ultrahigh resolution in the short term of the third-order Stokes wavelength. While maintaining a fairly normal spatial resolution over a few kilometers of sensing length using time gating technology, sensitivity is increased by several folds to over 4 MHz/°C.

Keywords Stimulated Brillouin Scattering · Cascaded stimulated Brillouin Scattering · Brillouin fiber laser · Single longitudinal mode · Saturable absorber ring · Multiwavelength · Dual ring · Optical signal to noise ratio · Four wave mixing · Brillouin–Erbium fiber laser · Cascaded ring Fabry–Pérot resonator · Triple ring · High-order stokes wavelength · Brillouin fiber laser sensor · Temperature sensor · Distributed temperature fiber sensor

1 Introduction

Brillouin fiber laser (BFL) with narrow linewidth and high signal noise ratio have broad applications in optical communication system (Spirin et al. 2012), radio over fiber (Li et al. 2013), optical bistability (Zhang et al. 2015a, b), distributed/point fiber sensing (Liu et al. 2015, b), and superluminal light propagation (Zhang et al. 2015a, b) because of the signal advancement which is inversely proportional to the linewidth. Meanwhile, extra attention has been attracted to single longitudinal mode (SLM) BFL with high quality, low loss and cost. When the cavity mode spacing of about 10 or 20 m cavity (Gross et al. 2010) length is comparable to the Brillouin gain, the BFL easily operates in SLM status.

Over the past few decades, fiber laser (FL) sensors have demonstrated high signal-to-noise ratios (SNRs) and narrow linewidths; such characteristics indicate that these sensors have an inherent high resolution (Kersey et al. 1997). Such sensors have been applied in accelerometers (Ames and Maguire 2007), acoustic microphones (Wooler et al. 2007), strain sensors (Yang et al. 2007), and temperature sensors (Yang et al. 2007; Liu et al. 2010). In particular, more attention has been given to the measurement of temperature, which is one of the seven fundamental physical quantities. Currently, there are three main types of FL temperature sensors, including a SLM-FL sensor (Yang et al. 2007), a multilongitudinal mode (MLM) FL sensor (Liu et al. 2010), and a multiwavelength (MW) FL sensor. On the one hand, for the SLM-FL and MW-FL sensors (Mandal et al. 2006). Wavelength matched fiber Bragg gratings (FBGs) are used as wavelength selectors and sensor elements.

Distributed BS fiber sensors are called Brillouin Optical Domain Reflectometry (BOTDR) (Culverhouse et al. 1989) and Brillouin Optical Time Domain Analysis (BOTDA) (Kurashima et al. 1990). Nowadays, other techniques such as BOCDA (Hotate and Hasegawa 2000) also exist, usually exhibiting a very high spatial resolution (mm scale), but have limited sensing reach (usually in the order of hundreds of meters). The spatial resolution of BOCDA system is determined by the modulation parameters of the light source such as the amplitude and frequency compared to BOTDR and BOTDA systems in which the resolution is based on the decay time of an acoustic wave. Since its discovery, many groups have worked on increasing the sensing distance achieving up to 150 km (Alahbabi et al. 2005) as well as maximizing the spatial resolution down to 2 cm in standard optical fiber (Dong et al. 2012) using conventional BOTDR/A systems. However, the sensitivity of ~1.1 MHz/°C in temperature (Nikles et al. 1997) and of ~0.05 MHz/$\mu\varepsilon$ in strain (Bao and Chen 2011) has remained mostly unchanged using the first spontaneous Stokes shift.

2 Main Body Text

- Optical fiber Scattering

  - Scattering category

When the light passes through the non-homogeneous medium, the propagation characteristics of the light will change because of the defects and non-homogeneous of the medium itself. A part of light will change the direction of transmission and spread to different directions, so the light intensity can be observed in all directions of the medium, which is the scattering effect of light. There are three types of light scattering: Rayleigh scattering, Brillouin scattering (BS), and Raman scattering, as shown in Fig. 1.

Rayleigh scattering phenomenon in optical fiber is caused by the local fluctuation of refractive index caused by the micro impurity mixed in the manufacturing process or the random fluctuation of density in the manufacturing process, which makes the light scattering in all directions.

Raman scattering was first discovered by C.V. Raman when he studied light scattering (Raman 1982). It refers to the different physical effects between the frequency of partially scattered light and that of incident light in light scattering.

- Spontaneous and SBS

Fig. 1 Scattering of light

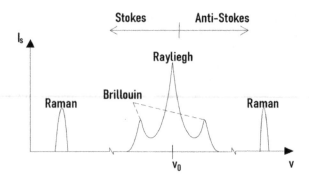

In the quantum mechanics formula, the excited BS process can be regarded as the annihilation and scattering of incident photons and the production of photons and acoustic phonons. The energy and momentum of the interacting particles must be maintained, which leads to pumping, the relationship between the frequency and wave vector of photons, Stokes photons, and acoustic phonons is as follows:

$$\omega_S = \omega_P - \Omega$$

$$\beta_S = \beta_P - B \tag{1}$$

Among them, the ω_s, ω_p, and Ω, respectively, represent pump photons, Stokes photon, and the frequency of the acoustic phonon; β_s, β_p, and B are the wave vectors of the three, respectively. In the formula, it is applied to the approximate relation $\beta_s \approx \beta_p$

$$|\beta| = 2\pi n_p / \lambda_p \tag{2}$$

BS is a kind of scattering caused by inelastic collision between incident photon and scattering particle in optical fiber. According to the different sound fields, BS can be divided into spontaneous and stimulated BS (SBS).

Spontaneous BS can be explained by quantum physics as follows: a pumped photon is converted into a new Stokes photon with a lower frequency and a new phonon is generated at the same time. Similarly, a pumped photon absorbs the energy of a phonon and converts it into a new, higher-frequency anti-Stokes photon.

When the pump light power reaches the threshold, SBS occurs, and the generated Stokes light propagates in the opposite direction to the incident light. The induced BS can be caused by the electrostrictive effect in the optical fiber (Kelley 2008), which is manifested by the change of medium density caused by the action of light. The backscattering Stokes light and the incident pump light act to produce sound waves through the electrostrictive effect. The propagating light wave produces a moving density grating, and the reflected Stokes light scatters back. So the frequency shift

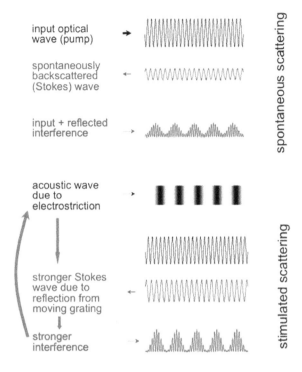

Fig. 2 Mechanism of spontaneous and SBS

down of Stokes light can also be explained by the Doppler Effect. The mechanism of spontaneous and SBS is shown in the following Fig. 2 (Kobyakov et al. 2009).

- BFL

SBS with narrowband and low noise has broad applications in frequency shifter (Kalli et al. 1991), microwave frequency generation (Shen et al. 2005), narrow bandwidth low-noise amplification (Xing et al. 2008), and FL (Yong et al. 2003; Geng et al. 2006). Meanwhile, extra attention has been attracted to SLM-BFL with high quality, low loss and cost. The investigation is mainly concentrated on the following aspects. On the one hand, in order to separate the pump and Brillouin laser and reduce the threshold power, single ring cavity (RC) with one 2×2 coupler (Stokes et al. 1982), an circulator (Wu et al. 2011), or an unbalanced Mach–Zehnder interferometer (UMZI) (Nicati et al. 1995) is configured. Circulator based single RC has the advantage that the Brillouin pump frequency does not need to match the cavity mode. On the other hand, when the cavity mode spacing of about 10 m or 20 m cavity length (Wu et al. 2011; Wang 2013) is comparable to the Brillouin gain, the BFL easily operates in SLM status. However, because of length limitation of SLM operation, it is difficult to improve the threshold power and linewidth. Then, BFL incorporating long fiber length (500 m) high nonlinear fiber (HNLF) and an unpumped Erbium-doped fiber (UP-EDF) loop (10 m) (Chen et al. 2012) which is regarded as a self-induced FBG filter with an ultra-narrow bandwidth is proved. Li et al. propose a Brillouin erbium

FLs (BEFLs) in which EDF amplifier (EDFA) can compensate the loss imported by the additional element, but the 166 high linewidth-reduction ratio is realized at cost of low optical signal-to-noise ratio (OSNR). And Liu et al. propose a SLM-BFL without additional EDFA in cavity and with cascaded ring Fabry–Perot (CR-FP) resonator, which has not plus cavity loss except elements' intrinsic loss and is scarcely to generate high-order Stokes due to low nonlinear coefficient of single-mode fiber (SMF). However, the cavity length is shorter than other optimized methods. Though Ou et al. propose a 25-km-long cavity BFL with 30 Hz linewidth without phase-locking loop, only 100 linewidth-reduction ratio is obtained due to larger loss caused by FP interferometer (FPI) and a part of coreless fiber (CLF). Lastly, besides of the temperature control, a servo loop is realized by autotracking the pump light frequency or adjusting the cavity length with piezoelectric transducer (PZT) (Yong et al. 2003) or Pound–Drever–Hall (PDH) frequency-locking scheme (Geng et al. 2006) resulting from environmental perturbations.

– Principle

Assuming that a signal light with a frequency of ω_s and a continuous pump light with a frequency of ω_p propagate together in the fiber, as long as the frequency difference $\omega_s = \omega_p$ lies in the Brillouin gain bandwidth, the signal light will be amplified due to Brillouin gain. Brillouin laser oscillations are formed by adding mirrors with appropriate reflectivity to both ends of the fiber to provide feedback to Stokes light generated by SBS effect in the fiber.

The Brillouin frequency shift v_B versus the pump is given by $v_B = (2v_A/c)v_P$ (Liu et al. 2014), where v_A is the acoustic velocity in the medium, c is the velocity of light in vacuum, and v_P is the optical frequency of the pump. v_B is approximately 10 GHz in 1550 nm wavelength region. The free spectral range (FSR) in accordance with the Vernier effect is

$$FSR = n_1 FSR_1 = n_2 FSR_2 = \cdots = n_m FSR_m \tag{3}$$

where $FSR_m = c/nL_m (m = 1, 2, 3, \ldots)$ and $L_m (m = 1, 2, 3, \ldots)$ are FSRs and the length of each RC, respectively, $n_m (m = 1, 2, 3 \ldots)$ is the integer, and $n = 1.4682$ is the fiber effective index. According to (3), effective FSR is 20 MHz. The laser mode v_L only oscillates at a frequency that experiences the highest Brillouin gain and satisfies the resonance conditions synchronously which is shown in Fig. 3. Though operational mechanism of saturable absorber ring (SAR) is similar to SMF ring resonator, bandwidth of the former $\Delta f'$ is different from that of the latter Δf. According to the previously reported, we can deduce as

$$\Delta f' = \Delta f \cdot \sqrt{\frac{2t_0}{t_1} - 1} \approx \frac{FSR}{F} \cdot \sqrt{\frac{2t_0}{t_1} - 1} \tag{4}$$

where $F = \pi\sqrt{R}/(1 - R)$ is the cavity finesse, R is the rate of laser power back fed after each roundtrip, t_0 and t_1 are transmission ratio at different input power of SAR,

Fig. 3 The principle of
SLM BFL operation

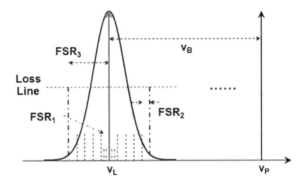

the power at t_0 is greater than that at t_1 and $\sqrt{2t_0/t_1 - 1}$ is the enhancement factor
of bandwidth.

At the same time, when the EDFA power is large enough, higher orders Stokes
$\nu_{Lx}(x = 2, 3, \ldots)$ and anti-Stokes waves are generated through the cSBS and
degenerate four-wave mixing (FWM) process which also operate in SLM status.
$\nu_{Lx}(x = 2, 3, \ldots)$ can be expressed by

$$\nu_{Lx} = \nu_{L(x-1)} - \nu_B = \nu_P - (x-1)\nu_B \, (x = 2, 3, \ldots) \tag{5}$$

– SLM Operation

To illustrate the SLM operation of BFL with CR-FP in detail, three groups (group
A with FP1-230 m, FP2-50 m. group B with FP1-50 m, FP 2-10 m. group C with FP
1-100 m, FP 2-10 m) with different fiber lengths are contrastly analyzed (Liu et al.
2013). The lengths of 230 m, 100 m, 50 m, and 10 m are corresponding to the FSRs
of nearly 870 kHz, 2 MHz, 4 MHz, and 20 MHz. Detected homodyne frequency
spectrum is shown in Fig. 4a and b for group A, (c) and (d) for group B and (e)
and (f) for group C. Figure 4a, c, and e and b, d, and f are single RC and CR-FP,
respectively.

In order to clarify the SLM operation of triple ring (TR) BFL detailedly (Liu and
Zhang 2017), different combination of RC are analyzed contrastly, Fig. 5a shows
heterodyne frequency spectrum of TR-BFL with only R1 at 20 MHz. Approximately
200 kHz frequency spacing is corresponding to FSR of R1 with 1 km. Simultaneously,
due to mode-locked of all longitudinal modes of the resonator, laser output is observed
in the time domain with a spacing of 5 μs which is shown in Fig. 5b. When R2
with 100 m is added in the main cavity, heterodyne frequency spectrum is shown
in Fig. 5c. One can observe that several side modes are suppressed with 2 MHz
frequency spacing which is FSR of R2. Once R2 and R3 with 8 m EDF are both
configured in the cavity, it becomes obvious that SLM operation status of TR-BFL
is observed in Fig. 5d.

And the linewidth result of 50 m FP1 and 10 m FP2 is same as that of the self-
heterodyne method. 0.41 kHz linewidth of 100 m FP1 and 10 m FP2 is obtained
which is in good agreement with theoretical analysis (shown in Fig. 6).

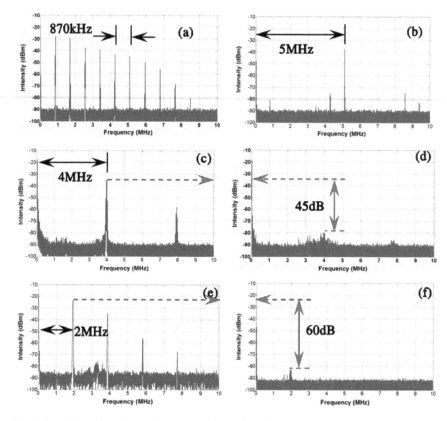

Fig. 4 Experimental results. **a** and **b** Detected homodyne frequency spectrum of the proposed laser with group A. **c** and **d** Detected homodyne frequency spectrum of the proposed laser with group B. **e** and **f** Detected homodyne frequency spectrum of the proposed laser with group C. **a, c,** and **e** with single RC and **b, d,** and **f** with CR-FP

An analytical relation linking pump laser linewidth $\Delta\upsilon_P$ and Brillouin laser linewidth $\Delta\upsilon_{BFL}$ is expressed by

$$\Delta\upsilon_{BFL} = \Delta\upsilon_P/(1 + \gamma_A/\Gamma_C)^2 \tag{6}$$

where $\gamma_A = \pi \Delta\upsilon_B$ and $\Gamma_C = -c \ln R'/nL_1$ are the decay rate of the acoustic wave and the cavity loss ratio. A TR-BFL has a coupling ratio $R' = 0.225$, and thus, the linewidth is approximately 210 times of magnitude narrower than that of the pump. Linewidth measurement is realized by delayed self-heterodyne method with a 100 km optical delay fiber line. Due to the 1 kHz measurement precision of 100 km fiber, the linewidth 1.3 kHz is measured at the −20 dB power points, which is connected with the laser real linewidth by $2\sqrt{99}$ times. The result is shown in Fig. 7 at 20 MHz center, 10 kHz Span, 500 Hz VBW and RBW. Approximately 185 linewidth-reduction ratio and 65 Hz real linewidth are obtained which are good agreement with 210

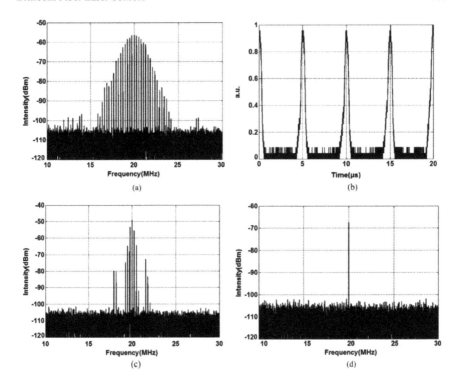

Fig. 5 Detected homodyne frequency spectra of TR-BFL **a** with only single RC of R1 and **b** corresponding time domain of laser output, **c** with double-ring cavities of R1 and R2 and **d** with TR cavities at 20-MHz center, 20-MHz span, and 3-kHzVBW and RBW

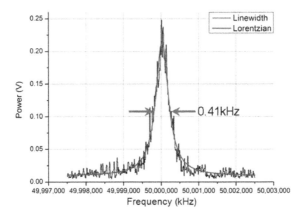

Fig. 6 Linewidth measurement of BFL with 100 m of FP1 using heterodyne beat technique with two independent pump lasers at Center 50 MHz, Span 5 kHz

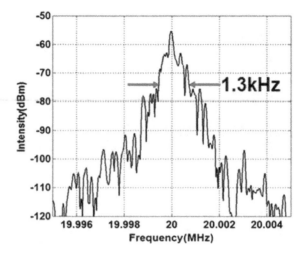

Fig. 7 Linewidth measurement of TR-BFL using a delayed self-heterodyne method at 20-MHz center, 10-kHz span, 500-Hz VBW and RBW

theoretical linewidth-reduction ratio and 48 Hz theoretical linewidth by considering the polarization between the pump and the Brillouin gain. Because 65 Hz linewidth corresponds to an optical delay length of 3143.57 km, it is difficult to obtain the completely incoherent self-mixing for the TR-BFL with the 100 km delay fiber.

– MW Operation

A MW SLM dual ring (DR) BFL is experimentally demonstrated (Liu et al. 2015a, b). The pump light from a tunable laser source (TLS) with 100 kHz linewidth and maximum output power of 3.7 dBm travels through a 50/50 optical coupler (OC1) and then is amplified by the EDFA. To achieve maximum gain in BFL, the polarization of the pump is kept parallel to that of the Stokes wave by a polarization controller (PC1). Then, the pump is injected to the single-pass cavity Ring-1 (R1) clockwise by optical circulator (Cir) and passes through the cavity with only one roundtrip. PC2 is used to align the state of polarization of the R1 with 100 m SMF. R2 is constructed with a 50/50 polarization-insensitive OC3 and a SMF length (10 m) of traditional SLM-BFL. When EDFA power is higher than the threshold of SBS, 1% of Stokes which is back to OC1 and amplified by the EDFA is injected to R1 to excite next order Stokes wave and the rest circulates in the R1 cavity with multiple roundtrips anticlockwise. The output of the laser was monitored by an optical spectrum analyzer (OSA) with a spectral resolution of 0.05 nm. For the stability of the MW-DRBFL, it is enclosed in a temperature controller system with better than 0.4 °C resolution and 3–80 °C range (Fig. 8).

Due to DR configuration, each Stokes and anti-Stokes waves work on SLM operation. When pump power is enough, 7 stable SLM lasing wavelengths including the pump with a 5 dB bandwidth of 0.5 nm are generated from 1535 nm to 1565 nm. Figure 9 shows MW generation of the MW-DRBFL as a function of the different

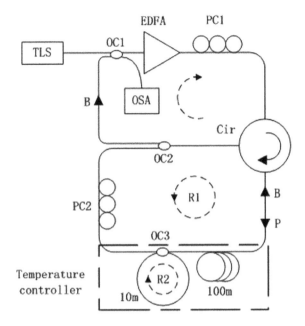

Fig. 8 Experiment setup of MW-DRBFL

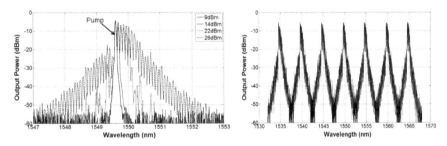

Fig. 9 MW generation in MW-DRBFL with different EDFA power and MW-DRBFL from 1535 nm to 1565 nm

EDFA power. At first, TLS is fixed at 1549.8 nm wavelength and 0 dBm power and EDFA power is fixed at 9 dBm. For an EDFA power of 14 dBm, 1st-order Stokes line with −8.05 dBm is generated. When EDFA power is increased to 22dBm, five Stokes lines and three anti-Stokes lines are observed. By increasing the EDFA power further to 26 dBm, higher order Stokes and anti-Stokes lines are generated because of the degenerate FWM process.

The relation connecting the nth-order Stokes linewidth $\Delta \upsilon_{Lm}$ of MW-DRBFL and the pump laser linewidth is given by (Smith et al. 1991)

$$\Delta \upsilon_{Lx} = \frac{\Delta \upsilon_{L(x-1)}}{\left(1 + \frac{\gamma_A}{\Gamma_c}\right)^2} = \cdots = \frac{\Delta \upsilon_{L1}}{\left(1 + \frac{\gamma_A}{\Gamma_c}\right)^{2x-1}} = \frac{\Delta \upsilon_P}{\left(1 + \frac{\gamma_A}{\Gamma_c}\right)^{2x}} \qquad (7)$$

where $\gamma_A = \pi \Delta \upsilon_B$ (Brillouin gain bandwidth $\Delta \upsilon_B = 20$ MHz) is the damping rate of the acoustic wave, $\Gamma_c = -c \ln R / n L_t$ is the cavity loss rate, and $L_t = L_1 + L_2$ is the total cavity length. Thus, the linewidth of the first Stokes line is about three orders of magnitude narrower than that of the pump. Similarly, the linewidth of high-order Stokes $\upsilon_{Lx}(x = 2, 3, \ldots)$ is the same relation with the former order Stokes $\upsilon_{L(x-1)}(x = 2, 3, \ldots)$ from (7). In experiment, the first Stokes are chose by disconnecting fiber between OC1 and OC2. And the linewidth is measured by the self-heterodyne method with a 25 km optical fiber delay line. Figure 10a (in the absence to R2) and (b) show the first Stokes homodyne frequency spectrum with 60 dB sidemode suppression value, and Fig. 10c shows 0.67 kHz linewidth (better than 0.91 kHz). However, because of the 4 kHz measurement accuracy with 25 km fiber, the linewidth is below 4 kHz. Observing from the heterodyne spectrum of MW-DRBFL (shown in Fig. 10d), there is no other the sidemode except Brillouin frequency shift (10.66 GHz) under Brillouin gain (20 MHz). Thus, each Stokes and anti-Stokes waves work on SLM operation.

- BFL sensor

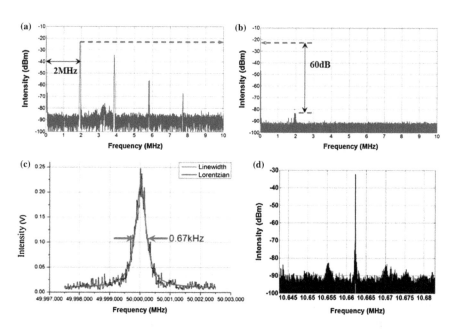

Fig. 10 **a** In absence of R2 and **b** homodyne frequency spectrum, **c** delay self-heterodyne spectrum of linewidth measurement with the first Stokes, **d** heterodyne spectrum of MW-DRBFL with 40 MHz span, 5.1 kHz RBW and VBW

In this case, the temperature information is achieved by measuring the wavelength shift under different temperatures in the optical domain or the power variation at a specific wavelength on the grating's slope. In addition, this type of FL sensor is limited by the tuning speed of the filter, the sensitivity of the detector, and the complexity of the optical interferometer, and the signal is detected by mixing with an additional TLS. On the other hand, for the MLM-FL sensor, the temperature difference can be easily detected by measuring the beat frequency generated by any two longitudinal modes. Because the disturbance can be reduced, the beat frequency can be dynamically stable. Thus, the higher the detected beat frequency is, the better the sensitivity is. However, although the 74th-order mode (1581.7 MHz) is detected, only sensitivity at the scale of 1 kHz/°C is obtained.

Recently, based on the BS effect in optical fibers. Proposed a novel DF-BFL for microwave generation (Yang et al. 2013), temperature sensing was realized based on this configuration, which has two cavities with different types of fiber. According to the linear relationship between the Brillouin frequency shift and temperature, the sensing information can be obtained by measuring the beat frequency shift as the temperature of the DF-BFL. This method may provide a stable FL sensor with 1.015 MHz/°C sensitivity. Compared to the DF-BFL, the use of the MW-BEFL based on the cascade BS effect and the linear gain from EDF could be an advanced method. Iezzi and Zhang realized temperature measurements with sensitivities of 7 MHz/°C and 13.08 MHz/°C by using the 6th-order and the 12th-order Stokes shifts with BS of two MW-BEFLs, respectively. However, due to the kilometer-scale laser cavity length (10 and 27 km), the external perturbations became the main effect factor for the temperature error, and the MLM operation status limited the improvement of the system resolution. And Liu et al. proposed a SLM MW-BEFL with short cavity as temperature sensor, and carried out experimental research.

– Principle

Figure 11 shows the high-order Stokes wavelength sensing principle based on the cascade BS effect for temperature measurement. The frequency shift υ_B with respect to the pump is given by $\upsilon_B = (2\upsilon_A/c)\lambda_P$ (Song et al. 2004; Shee et al. 2011), where υ_A is the acoustic velocity in the medium, c is the vacuum-light velocity, and υ_P is the pump wavelength. υ_B is approximately 10 GHz at a 1550 nm wavelength.

We can obtain the relationship between υ_{m1} at room temperature and $\upsilon_{m(m+1)}$ for the ΔT temperature difference of the mth-order Stokes wave (Xu and Zhang 2015).

$$\upsilon_{m1} = \upsilon_{(m-1)1} + \upsilon_B = \cdots = \upsilon_P + m\upsilon_B \tag{8}$$

$$\upsilon_{m(m+1)} = \upsilon_{(m-1)m} + \upsilon_B + \Delta\upsilon_{\Delta T} = \cdots = \upsilon_P + m\upsilon_B + \Delta\upsilon_{m\Delta T} \tag{9}$$

$$\Delta\upsilon_{m\Delta T} = c_{Bm}\Delta T = \cdots = mc_{B1}\Delta T \tag{10}$$

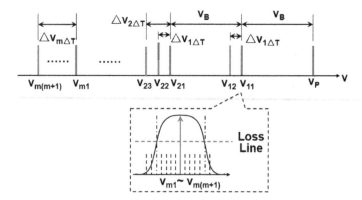

Fig. 11 Principle of the high-order Stokes wavelength for temperature sensing with a ΔT temperature difference

where v_p is the optical frequency of the pump beam, m is the order number of the Stokes wave ($m = 1, 2, 3...$), and $\Delta v_{m\Delta T}$ and c_{Bm} are the Brillouin frequency difference and the temperature coefficient of the mth-order Stokes wave, respectively; i.e., the frequency shift with temperature and the temperature coefficient of the mth-order Stokes are m-times greater than those of the first-order Stokes.

– Point Sensing Experiments

A SLM MW-BEFL with a short cavity used as the temperature sensor is proposed and investigated experimentally (Yu et al. 2006). As the configuration of the DF-BFL in Fig. 12, the pump light from a Dense Light laser (DLL) with a 10 kHz linewidth and a maximum output power of 10 dBm travels through a 50/50 OC1. Next, the light is first amplified by the EDFA. To achieve maximum Brillouin gain, the polarization of the pump is kept parallel to the Stokes polarization by a PC. The pump is injected into the RC clockwise by an optical Cir and then passes through the cavity with only one roundtrip. The RC is constructed with one 50-m fiber under test (FUT), another

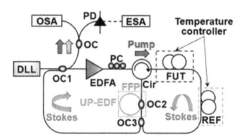

Fig. 12 Experimental Setup. DLL: Dense Light laser; OC: optical coupler; PC: polarization controller; Cir: circulator; FUT: fiber under test; REF: reference fiber; FFP: fiber Fabry–Pérot; UP-EDF: unpumped erbium-doped fiber; PD: photodetector; OSA: optical spectrum analyzer; ESA: electric spectrum analyzer

100-m reference fiber (REF), an OC3, and a FFP filter. The FFP filter is composed of a 50/50 polarization-insensitive OC2 and a 10 m UP-EDF, which acts as a saturable absorber (SA). The fiber type of the FUT and REF is standard G.652 SMF, and the FFP filter is used to keep every Stokes wavelength in SLM status. When the EDFA power is higher than the threshold of BS, 10% (OC3) of the Stokes returns back to OC1 and is amplified likewise by the EDFA clockwise. Next, the light is injected into the RC to excite the next order Stokes, and the rest circulates (90%) in the RC with multiple roundtrips anticlockwise. The output of the laser was monitored by an electric spectrum analyzer (ESA) with a 1 Hz frequency resolution and a photodetector (PD) with a 26.5 GHz bandwidth. For the convenience of temperature measurement, the FUT and REF are enclosed in two similar temperature controller systems with the following characteristics: 0.2 °C resolution, a 5–60 °C range, and no strain. In this experiment, the temperature of the REF is fixed at 20 °C, and that of the FUT is adjusted from 20–50 °C.

A 50-m-long SMF is used as the FUT, and a 100-m-long SMF is used as the REF when a different temperature is configured. With the configuration that uses a FFP filter, every Stokes wavelength of the MW-BEFL maintains SLM status. In contrast, with beat frequency linewidths of different order Stokes wavelengths, approximately 10^{-6} °C ultrahigh resolution in the short-term (linewidth) is acquired at the third-order Stokes wavelength.

Figure 13 shows the relationships between the temperature difference, and the different Stokes wavelengths are also measured for different EDFA power levels. The circular and square dots represent real frequency values, including $v_{12} - v_p, v_{22} - v_p,$ $v_{23} - v_p,$ and $v_{34} - v_{31},$ measured at a FUT temperature from 20 to 50 °C every 5 °C and at a REF temperature of 20 °C under 250, 500, and 600 mW. The x-axis is $\Delta T = T_{\text{FUT}} - T_{\text{REF}},$ and the black solid lines are the results of linear fits using a least squares fitting method. The fitting lines show the temperature dependences of the frequency shift of the FUT and the REF. As seen from the solid line, all the beat frequencies are linear with respect to temperature, while the frequency of $v_{22} - v_p$ is used for comparison.

To exactly measure the temperature resolution, we must take the linewidth of the pump into consideration. The temperature resolution depends on the short-term (linewidth) and the long-term stability of the high-order Stokes wavelength. On the one hand, the relation connecting the mth-order Stokes linewidth of Δv_m the MW-BEFL and the pump laser linewidth is given by (Debut et al. 2000)

$$\Delta v_m = \frac{\Delta v_{m-1}}{\left(1 + \frac{\gamma_A}{\Gamma_c}\right)^2} = \cdots = \frac{\Delta v_1}{\left(1 + \frac{\gamma_A}{\Gamma_c}\right)^{2m-1}} = \frac{\Delta v_p}{\left(1 + \frac{\gamma_A}{\Gamma_c}\right)^{2m}} \qquad (11)$$

where $\gamma_A = \pi \Delta v_B$ is the damping rate of the acoustic wave, $\Gamma_c = -c \ln R/n L_t$ is the cavity loss rate, and L_t is the total cavity length. In this case, the relationship between the temperature resolution T_{Rm} of the m-order Stokes wavelength and the temperature resolution T_{R1} of the 1st-order Stokes wavelength can be written as

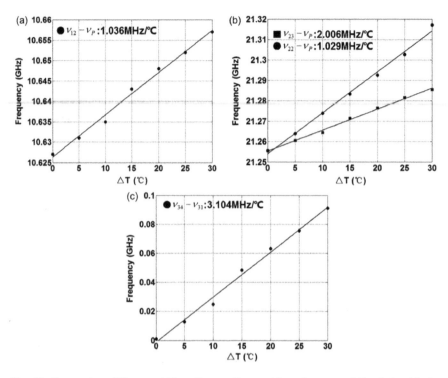

Fig. 13 Temperature difference $\Delta T = T_{\mathrm{FUT}} - T_{\mathrm{REF}}$ and beat frequency shift relationship for **a** $v_{12} - v_p$ **b** $v_{22} - v_p, v_{23} - v_p$ **c** $v_{34} - v_{31}$

$$T_{Rm} = \frac{\Delta v_m}{c_{Bm}} = 1/m\left(1 + \frac{\gamma_A}{\Gamma_c}\right)^{2^{m-1}} = T_{R1}/m\left(1 + \frac{\gamma_A}{\Gamma_c}\right)^{2^{m-1}} \tag{12}$$

Next, the beat frequency linewidths of f_1, f_2, and $v_{34} - v_{31}$ are measured using the ESA with a 1 Hz frequency resolution for a REF temperature of 20 °C and a FUT temperature of 50 °C (shown in Fig. 14). According to Eq. (11) and the configuration of the MW-BEFL, the linewidth of the 1st-order Stokes wavelength is approximately two orders of magnitude narrower than that of the pump. The linewidth is observed at the −20 dB points, which is related to the laser real linewidth by $2\sqrt{99}$ times (Bao et al. 1995). The −20 dB linewidth points of the frequencies f_1, f_2, and $v_{34} - v_{31}$, shown in Fig. 14a, c, are 12 kHz, 0.7 kHz and 0.08 kHz, respectively. Thus, 0.6 kHz, 0.035 kHz, and 0.004 kHz real linewidths are obtained for f_1, f_2, and $v_{34} - v_{31}$, respectively. The results are in good agreement with our analysis. Therefore, according to Eq. (12), the temperature resolution corresponds to approximately 6×10^{-4} °C, 1.8×10^{-5} °C, and 1.3×10^{-6} °C for f_1, f_2, and $v_{34} - v_{31}$, respectively, depending on the temperature sensitivity of the different order Stokes wavelength.

– Distributed Sensing Experiments

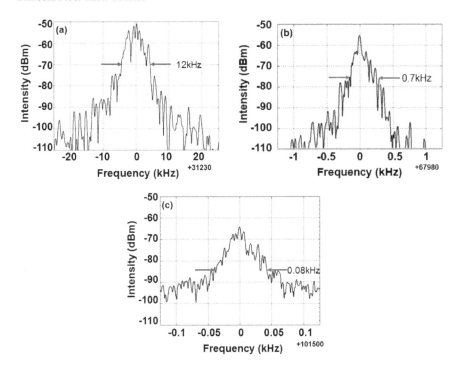

Fig. 14 Beat frequency linewidth measurement of **a** $f_1 = 31.23$ MHz, **b** $f_2 = 67.98$ MHz, and **c** $v_{34} - v_{31} = 101.5$ MHz

Since 1989, D. Culverhouse used BS in optical fibers to measure temperature, researchers in Japan, Canada, the United States and other countries have made a lot of theoretical and experimental studies, and obtained rich research results in two main distributed fiber sensing technologies based on BS, namely BOTDA (Soto et al. 2011; Li et al. 2008) and BOTDR (Shimizu et al. 1994).

Once light from the BP is separated in a 3 dB-coupler, it enters each cavity through another 3 dB coupler. As the BP power is increased, the 1st Stokes line is generated in the opposite direction and oscillates in the bottom branch of the oscillator through optical circulators as shown in Fig. 15 (Iezzi et al. 2017). A cascaded process then occurs, as more BP power is injected, light is scattered into the 1st Stokes wave. At a certain point, this Stokes wave is intense enough and reaches the threshold for the 2nd Stokes wave, which oscillates in the upper part of the cavity, e.g., same direction as the initial BP. Odd or even Stokes waves can be collected at the output with a 99:1 output coupler. The in-cavity EDFAs are used to compensate for the loss and keep the Stokes wave lasing above threshold. An optical bandpass filter is used in this configuration to decrease the ASE bandwidth of the EDFAs to only a few nanometers (~1 nm for the sensing cavity and 5 nm for the reference oscillator) reducing the potential of free-running modes competing for gain in such a configuration (Mohd Nasir et al. 2008). A 2.5 km length of fiber on a reel was placed in a temperature controlled

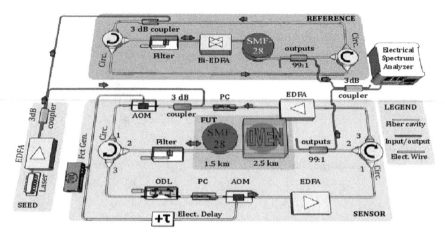

Fig. 15 Two near identical SBS ring resonators are used; one as a reference and the other as the sensor, both sharing a common seed laser through a 3 dB coupler. The signals are recombined at their respective output by a second 3 dB coupler connected to the electrical spectrum analyzer. AOMs are used as temporal gates which provide the spatial resolution of the sensor. One AOM is electrically controlled to vary the time of the overlap with the other AOM to allow a scan over the entire length of the fiber spool. The in-cavity EDFAs are used to compensate for the cavity loss

chamber, while the total length of FUT of the cavity was slightly above 4 km, giving a FSR of 49.707 kHz.

By observing the beat note generated by the nth-order Stokes waves, $S_{n sens}$ with $S_{n, ref}$, enables the detection of temperature in different regions along the 4 km fiber bundle. Figure 16 shows the variation of the beat frequency for the 2nd Stokes (blue

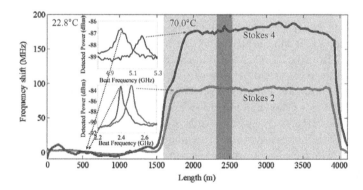

Fig. 16 A 2.5 km (area shown by the pale grey rectangle) fiber bundle is kept at 70.0 °C while the rest of the fiber (1.5 km) is maintained at room temperature of 22.8 °C. The sensor has a resolution of ~225 m (shown by the darker grey rectangle). The temperature sensing signal generated by the 4th Stokes wave is compared with the 2nd Stokes wave shown by the red and blue curves, respectively. The insets show the beat frequencies for the 2nd (bottom inset) and 4th Stokes (top inset), both for a temperature of 22.8 °C (reference oscillator) and 70.0 °C (sensing coil)

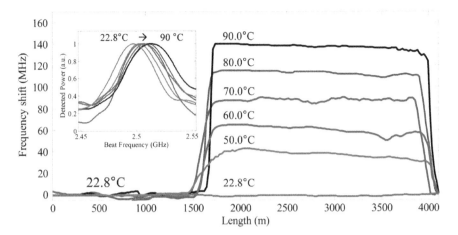

Fig. 17 Distributed sensing measurement at various temperatures ranging from 22.8 to 90.0 °C using the 2nd Stokes order. The inset shows the beat signal between S2, sens, and S2, ref for this temperature range

line) and the 4th Stokes waves (red line) as a function of the temperature of the 2.5 km length of fiber placed in an environmental chamber. The temperature in the chamber is controlled at 70.0 °C while another 1.5 km fiber bundle is kept at room temperature. Top inset in Fig. 16 shows the difference in the beat frequencies for a temperature of 22.8 and 70.0 °C for S4 while the bottom curve shows the change for S2. As can be seen in Fig. 16, sensing with the 4th Stokes wave is more sensitive than using the 2nd Stokes order by a factor of two.

Figure 16 shows this principle as 2.5 km SMF-28 of fiber is placed in an oven at 70.0 °C while the remaining fiber of the sensing oscillator (1.5 km) is kept at 22.8 °C as is the reference oscillator. This temperature corresponds to room temperature. For the same temperature difference ($\Delta T = 47.2$ °C), S4 is shifted from the reference Brillouin frequency shift by 190 MHz which corresponds to a 4.02 MHz/°C variation. S2, however, shifts by 94 MHz which is equivalent to 2 MHz/°C corresponding to half that for S4, but twice the typical temperature sensitivity of BS based sensors of 1.1 MHz/°C using the first Stokes. This means that a smaller variation of temperature could be detected more rapidly, since the frequency shift is larger using the 4th rather than the 2nd order. It should be noted that the 0 °C temperature variation (both reference and sensing oscillator at room temperature) has a beat frequency offset of 2.5 GHz. This is because the reference and probe fibers were judiciously selected to offset the beat frequency away from the DC level, separating the 2nd and 4th Stokes beat note for easier measurement. The REF is a highly nonlinear fiber with small core area from Fibercore with a Brillouin frequency shift of 9.6 GHz while the sensing fiber is a standard SMF (SMF-28 from Corning with a frequency shift of 10.85 GHz. Therefore, the beat frequency for S2 is centered at 2.5 GHz while the beat note for S4 is at 5.0 GHz at room temperature as shown in the bottom and top inset of Fig. 16.

Figure 17 shows a distributed sensing measurement performed with the system for different temperature variation detected with S2. The inset in Fig. 17 shows the displacement of the beat signal between S2, sens, and S2, ref for various temperatures ranging from 22.8 to 90 °C. The different edges (rise and fall) position from these different measurements are within the resolution of the system (~225 m).

3 Conclusion

A SLM-BFL with CR-FP is presented. The BFL is structured by using CR-FP with a longer length fiber (100 m or 50 m) and a shorter length fiber (10 m or 20 m). The autotracking loop constructed by PM-ODL and temperature control is used to stabilize Brillouin frequency shift and FSR shift resulting from environment. The sidemode suppression ratio is improved by 45 and 60 dB. And 6% power fluctuation is observed in 1 h. 0.41 kHz and 3.23 kHz linewidth of BFL which is three and two order of magnitude than that of the pump is measured by self-heterodyne method with 25 km optical fiber and heterodyne beat technique with two independent pump lasers. In addition, the laser is very simple, practical, and cost-effective.

The TR-BFL is structured by TR cavities with 1-km-long SMF ring, 100-m-long SMF ring and SAR with 8 m UP-EDF, respectively. The frequency locked configuration based on PDH is used to steady FSR shift and Brillouin frequency shift caused by the 1 km long cavity and TR setup. Experimental results show that linewidth-reduction ratio and OSNR are 185 and 72 dB, respectively. Meanwhile, 5% power fluctuation in 1 mW laser output power is observed every 1 min in 50 min. Approximately 65 Hz linewidth of TR-BFL is measured by delayed self-heterodyne method. In addition, the laser is practical and has potential application in fiber sensing, long-distance superluminal propagation and coherent optical communication with high resolution.

We have proposed and demonstrated a novel MW SLM-BFL with DR configuration. 7 stable SLM lasing wavelengths including the pump with a 5 dB bandwidth of 0.5 and 0.084 nm spacing are obtained. Because of DR configuration, each Stokes waves work on SLM status. The tunable range is 30 nm from 1535 to 1565 nm. The linewidth of the first Stokes wave is below 4 kHz with 60 dB sidemode suppression value and the power fluctuation of that is 8% in 1 h.

A MW-BEFL sensor with a FUT of 50 m of SMF and a REF of 100 m of SMF for temperature measurement was proposed and demonstrated. The sensitivities of the first-order, second-order, and third-order Stokes wavelengths are obtained with different EDFA powers of 250 mW, 500 mW, and 600 mW, respectively, measured for different FUT temperatures from 20 °C to 50 °C at a step of 5 °C. The temperature elevation sensitivity is approximately 1.036 MHz/°C for the first-order Stokes wavelength, 2.006 MHz/°C for the second-order Stokes wavelength, and 3.104 MHz/°C for the third-order Stokes wavelength. The experimental results are in good agreement with theory. In addition, with the configuration that uses the FFP filter, the SLM status of every Stokes wavelength is guaranteed. Thus, a less than 0.2 °C temperature error, which corresponds to the resolution of the temperature controller, is obtained in the long-term. Moreover, an approximately 10^{-6} °C temperature resolution at the third-order Stokes wavelength is acquired in the short-term, according to the linewidth of the beat frequency. Furthermore, if we consider that, in this scheme, the HNLF, which can obtain more Stokes wavelengths with a uniform intensity distribution, and a temperature chamber with better resolution are adopted, higher temperature sensitivity and resolution can be obtained by detecting a higher order Stokes wavelength.

Iezzi et al. have demonstrated a technique to combine two concepts: use of high-order Brillouin Stokes waves to increase the sensitivity of temperature sensing, and a novel scheme to implement distributed sensing with the high-order SBS Stokes waves. We have demonstrated an increase of $4\times$ over the standard sensitivity achieved in typical Brillouin distributed sensors such as BOTDR and BOTDA systems. The distributed measurement demonstrated had a spatial resolution of approximately 225 m (mostly limited by the rise/fall times and extinction ratio of the AOMs used in the experiments) with more than 4 km sensing reach. Temperature detection from 22.8 to 90.0 °C was shown using a 2nd order Stokes wave which could easily extend to higher temperatures. A complete distributed sensing scheme using 4th order Stokes wave for a discrete variation of temperature from 22.8 to 70.0 °C has also been shown with twice the sensitivity of the 2nd Stokes. The poor extinction ratio of the AOMs and the ASE noise of the amplifiers limited the increase in sensitivity in our current system as generating truly distributed higher order Stokes wavelength was not possible.

References

M.N. Alahbabi, Y.T. Cho, T.P. Newson, J. Opt. Soc. Am. B **22**, 1321–1324 (2005)

G.H. Ames, J. M. Maguire, IEEE Sens. J **7**, 557–561 (2007)

X. Bao, J. Dhliwayo, N. Heron, J. Lightw Technol. **13**, 1340–1348 (1995)

X. Bao, L. Chen, Sensors (Basel) **11**, 4152–4187 (2011)

X. Chen, L. Xian, K. Ogusu, H. Li, Appl. Phys. B **107**, 791–794 (2012)

D. Culverhouse, F. Farahi, C. Pannell, D. Jackson, Electron. Lett. **25**, 913–915 (1989)

A. Debut, S. Randoux, J. Zemmouri, Phys. Rev. A **62**, 023803 (2000)

Y. Dong, H. Zhang, L. Chen, X. Bao, Appl. Opt. **51**, 1229–1235 (2012)

J. Geng, S. Staines, Z. Wang, J. Zong, M. Blake, S. Jiang, IEEE. Photon Technol. Lett. **18**, 1813–1815 (2006)

M.C. Gross, P.T. Callahan, T.R. Clark, D. Novak, R.B. Waterhouse, M.L. Dennis, Opt. Express **18**, 13321–13330 (2010)

K. Hotate, T. Hasegawa, IEICE Trans. Electron **83**, 405–412 (2000)

V.L. Iezzi, S. Loranger, R. Kashyap, Opt. Express **12**, 32591–32601 (2017)

K. Kalli, D. Culverhouse, D. Jackson, Opt. Lett. **16**, 1538–1540 (1991)

P.L. Kelley, in *Nonlinear Optics*, ed. By R.W. Boyd (Academic Press, Rochester, New York, 2008), p. 355

D. Kersey et al., J. Lightw Technol. **15**, 1442–1463 (1997)

A. Kobyakov, M. Sauer, D. Chowdhury, Adv. Ont. Photonics **2**, 1–59 (2009)

T. Kurashima, T. Horiguchi, M. Tateda, Opt. Lett. **15**, 1038–1040 (1990)

S. Liu et al., Opt. Lett. **35**, 835–837 (2010)

Yi Liu, Mingjiang Zhang, J. Lightwave Technol. **35**, 1744–1749 (2017)

Y. Liu, J.L. Yu, W.R. Wang, IEEE Photon. Technol. Lett. **26**, 169–172 (2013)

Y. Liu, J.L. Yu, W.R. Wang, H.G. Pan, E.Z. Yang, IEEE Photon. Technol. Lett. **26**, 169–172 (2014)

Y. Liu, M. Zhang, P. Wang, L. Li, Y. Wang, and X. Bao, IEEE Photon. J. **7**, 1 (2015a)

Y. Liu, J.L. Yu, W.R. Wang, Opt. Rev. **22**, 271–277 (2015b)

W. Li, X. Bao, Y. Li, Opt. Express **16**, 21616–21625 (2008)

J. Li, H. Lee, K.J. Vahala, Nature Commun. **4**, 1–7 (2013)

J. Mandal et al., IEEE Sens. J. **6**, 986–995 (2006)

N. Mohd Nasir, Z. Yusoff, M.H. Al-Mansoori, H.A. Abdul Rashid, P.K. Choudhury, Laser Phys. Lett. **5**, 812–816 (2008)

P.-A. Nicati, K. Toyama, H.J. Shaw, J. Lightw Technol. **13**, 1445–1451 (1995)

M. Nikles, L. Thevenaz, P.A. Robert, J. Lightw Technol. **15**, 1842–1851 (1997)

C.V. Raman, Indian J. Phys. **2**, 387 (1982)

Y.G. Shee, M.H. Al-Mansoori, A. Ismail, S. Hitam, M.A. Mahdi, Opt. Exp. **19**, 1699–1706 (2011)

Y. Shen, X. Zhang, K. Chen, J. Lightw Technol. **23**, 1860–1865 (2005)

K. Shimizu, T. Horiguchi, Y. Koyamada, J. Lightw Technol. **12**, 730–736 (1994)

S. Smith, F. Zarinetchi, S. Ezekiel, Opt. Lett. **16**, 393–395 (1991)

Y.J. Song, L. Zhan, S. Hu, Q.H. Ye, Y.X. Xia, I.E.E.E. Photon, Technol. Lett. **16**, 2015–2017 (2004)

M.A. Soto, G. Bolognini, F.D. Pasquale, Opt. Let. **36**, 232–234 (2011)

V.V. Spirin, C.A. Lopez-Mercado, P. Megret, A.A. Fotiadi, Laser Phys. Lett. **9**, 377 (2012)

L. Stokes, M. Chodorow, H. Shaw, Opt. Lett. **7**, 509–511 (1982)

G. Wang, Opt. Lett. **38**, 19–21 (2013)

J.P.F. Wooler, B. Hodder, R.I. Crickmore, Meas. Sci. Technol. **18**, 884 (2007)

Z. Wu, L. Zhan, Q. Shen, J. Liu, X. Hu, P. Xiao, Opt. Lett. **36**, 3837–3839 (2011)

L. Xing, L. Zhan, S. Luo, Y. Xia, IEEE J. Quantum Electron. **44**, 1133–1138 (2008)

R. Xu, X. Zhang, IEEE Photon. J. **7**, 1–8 (2015)

X. Yang, S. Luo, Z. Chen, J.H. Ng, C. Lu, Opt. Commun. **271**, 203–206 (2007)

X.P. Yang, J.L. Gan, S.H. Xu, Z.M. Yang, Laser Phys. **23**, 045104 (2013)

J. Yong, L. Thévenaz, B. Kim, J. Lightw Technol. **21**, 546–554 (2003)

L.P. Yu, Y.Z. Liu, Z.Y. Dai, Adv. Laser and Optoelectron. **4**, 26–30 (2006)

L. Zhang, L. Zhan, M. Qin, Z. Wang, H. Luo, T. Wang, Opt. Lett. **40**, 4404–4407 (2015a)

L. Zhang, L. Zhan, M. Qin, Z. Zou, Z. Wang, and J. Liu, J. Opt. Soc. Amer. B, Opt. Phys. **32**, 1113–1119 (2015b)

Index

© Springer Nature Singapore Pte Ltd. 2020
L. Wei (ed.), *Advanced Fiber Sensing Technologies*,
Progress in Optical Science and Photonics 9,
https://doi.org/10.1007/978-981-15-5507-7

CPSIA information can be obtained
at www.ICGtesting.com
Printed in the USA
LVHW011600120720
660448LV00002B/194